MATHEMATICAL DISCOVERY

MATHEMATICAL DISCOVERY

*On understanding, learning, and
teaching problem solving*

Combined Edition

George Polya
Professor Emeritus of Mathematics
Stanford University

John Wiley & Sons
New York   Chichester   Brisbane   Toronto

Library of Congress Cataloging in Publication Data:

Polya, George, 1887-
Mathematical discovery.

Bibliography:
Includes index.
1. Mathematics—Problems, exercises, etc.
2. Problem solving.   I. Title.
QA43.P62   1981   510'.76   81-1063
ISBN 0-471-08975-3   AACR2

Printed in the United States of America

10 9 8 7 6 5 4 3 2 1

FOREWORD

You, the readers, have before you a book devoted to explaining how you can improve your understanding of mathematics and how you can better solve mathematical problems. In brief, its intention is that you learn how to do mathematics and how to use mathematics. The examples are designed for those concerned with mathematics at the high school or early undergraduate level.

We claim that this is a wonderful book and that its reissue is timely and necessary. We do not believe that many would dispute this claim but nevertheless we want to take this opportunity to establish its validity. Our proposition then is that *this book is eminently suitable for the needs of all concerned teachers and interested students of mathematics* at the levels indicated above; and we will prove our proposition by contradiction!

If this book were not eminently suitable for your needs, it would have to fail for one of the following reasons, all of which I will easily demonstrate to be absurd! Let me then list eight purely hypothetical, but absolutely false, charges and quickly refute each one!

1. It never was suitable. This assertion would be so ridiculous we scarcely need to rebut it. This book had a tremendous success when it first appeared and was found invaluable by teachers and students alike. It was—and is—a most pleasant and interesting book to read, and it brought not only enlightenment but joy to the reader.

2. Mathematics is no longer relevant. This assertion is equally ridiculous! It is not mathematicians but the leaders of our society who are to be heard most stridently proclaiming the importance of mathematical knowledge and understanding to today's citizens if we are to remain among the leading nations of the world. We mathematicians would also point to the great advantage to individuals in their own lives of being able to reason effectively and solve quantitative problems.

3. The objectives considered have ceased to be crucial. This false view is easily disposed of, but it may be that some could believe the view correct. For it is sometimes thought that, with the availability of the hand-calculator and its big brother, the computer, the use of mathematics is reduced to pushing buttons and feeding in canned programs. This view would be quite false. It is the drudgery that has been eliminated from mathematics by these modern devices and not the need for thought. We assuredly still need to recognize when a problem is suitable—and ripe—for mathematical treatment, and we need to plan a strategy for tackling the problem; these utterly human tasks are the ones with which Polya is concerned.

4. The methods described are now perceived, in many cases, not to be the best or are very seriously disputed. This is a standard reason frequently offered in justification for allowing a once valuable book to go out of print. It does not apply in this case because the author is not concerned with slick solutions but with the whole strategy of problem-solving. That strategy is so closely related to the way human beings think—when they are thinking effectively—that it is not going to undergo any dramatic changes. Polya describes the methods that work for him and other successful problem-solvers. As the old adage goes, "the proof of the pudding is in the eating."

5. The particular mathematical content has been superseded. The mathematical content of this book is drawn from the standard precalculus curriculum, that is, from arithmetic, algebra, and geometry, with some elementary combinatorics. The mathematical material is not, of course, developed systematically, since the author's purpose is to exemplify problem-solving strategies and principles with the aid of interesting and intriguing mathematical questions. It is true that the availability of the hand calculator is not assumed but, on the other hand, the nature of the problems treated is not such as to make its availability essential. Polya shows you how to think about a problem, how to look at special cases, how to generalize in interesting and important directions and how to *solve* a problem. These skills will never be superseded.

6. The content is well known. One wishes this were true, but it is not! The content is not even well known to teachers, let alone to students. The examples Polya gives carry a fresh illumination which must surely inspire any reader. There is fascination in his description of the mathematical ideas, in his elucidation of pedagogical principles, and in his own unique style of exposition. There is nothing in this book that bears the stamp of staleness.

7. The content is now obvious. This assertion differs in an essential way from the previous one. That claimed that the content *has* already been conveyed to the reader from other sources. This claims that it does not need to be conveyed because what the text is saying has now become obvious as a result of a changing climate of thought in mathematical education. But this assertion is just as wrong-headed as its predecessor! The changing climate gives emphasis to the need to teach the art of problem-solving, of mathematical discovery; but we are very far from being a nation of skilled problem-solvers and discoverers! So, far from rendering this book superfluous, our current view of how to teach mathematics renders it absolutely crucial. The world has caught up with Polya; and what he and his lifetime friend, Gabor Szegö, saw so clearly many years ago, long before this book was written, is now the cornerstone of the program of the National Council of Teachers of Mathematics[1] in their campaign to imrove the quality of mathematics education.

8. The book is too expensive for those who would benefit from having it. A mere reissue of the original edition might have led to this charge. The publishers are to be congratulated on bringing the original two volumes together into one volume and in making available a soft-cover edition. The separation of the text material into two volumes was justified at the time of original publication, since it avoided undue delay in the appearance of a much awaited book. There is no case for it today. This book is surely worth its present price to anyone genuinely interested in mathematics.

Thus there can be no possible argument against reissuing this book—and overwhelming arguments in favor of doing so. You, the reader, will surely share my enthusiasm for this wonderfully stimulating, wonderfully contemporary work, written by a man whose youthful approach to new discoveries in mathematics, or any other fields, has never deserted him. Polya succeeds better than almost anybody else in conveying, on the printed page, the excitement he himself feels in the intellectual adventures he so vividly describes. We are privileged and fortunate to be able to share the adventure with him.

The reader's attention should be drawn to two features of this new edition. The bibliography has been updated by Professor Gerald Alexanderson and the index has been expanded by Professor Jean Pedersen. Thus, while the text remains Polya's original, this new edition is, in every sense, a text for our times.

Case Western Reserve University   PETER HILTON
Cleveland, Ohio

Battelle Human Affairs Research Center
Seattle, Washington

---

[1]An Agenda for Action, Reston, VA. (1980).

PREFACE

*A method of solution is perfect if we can foresee
from the start, and even prove, that following
that method we shall attain our aim.*
   LEIBNITZ: *Opuscules,* p. 161.

1. Solving a problem means finding a way out of a difficulty, a way around an obstacle, attaining an aim which was not immediately attainable. Solving problems is the specific achievement of intelligence, and intelligence is the specific gift of mankind: solving problems can be regarded as the most characteristically human activity. The aim of this work is to understand this activity, to propose means to teach it, and, eventually, to improve the problem-solving ability of the reader.

2. This work consists of two parts; let me characterize briefly the role of these two parts.

Solving problems is a practical art, like swimming, or skiing, or playing the piano: you can learn it only by imitation and practice. This book cannot offer you a magic key that opens all the doors and solves all the problems, but it offers you good examples for imitation and many opportunities for practice: if you wish to learn swimming you have to go into the water, and if you wish to become a problem solver you have to solve problems.

If you wish to derive the most profit from your effort, look out for such features of the problem at hand as may be useful in handling the problems to come. A solution that you have obtained by your own effort or one that you have read or heard, but have followed with real interest and insight, may become a *pattern* for you, a model that you can imitate with advantage in solving similar problems. The aim of Part One is to familiarize you with a few useful patterns.

It may be easy to imitate the solution of a problem when solving a closely similar problem; such imitation may be more difficult or scarcely possible if the similarity is not so close. Yet there is a deep-seated human desire for more: for some device, free of limitations, that could solve all problems. This desire may remain obscure in many of us, but it becomes manifest in a few fairy tales and in the writings of a few philosophers. You may remember the tale about the magic word that opens all the doors. Descartes meditated upon a

universal method for solving all problems, and Leibnitz very clearly formulated the idea of a perfect method. Yet the quest for a universal perfect method
has no more succeeded than did the quest for the philosopher's stone which
was supposed to change base metals into gold; there are great dreams that
must remain dreams. Nevertheless, such unattainable ideals may influence
people: nobody has attained the North Star, but many have found the right
way by looking at it. This book cannot offer you (and no book will ever be
able to offer you) a universal perfect method for solving problems, but even a
few small steps toward that unattainable ideal may clarify your mind and
improve your problem-solving ability. Part Two outlines some such steps.

3. I wish to call *heuristics* the study that the present work attempts, the
study of means and methods of problem solving. The term heuristic, which
was used by some philosophers in the past, is half-forgotten and half-
discredited nowadays, but I am not afraid to use it.

In fact, most of the time the present work offers a down-to-earth practical
aspect of heuristic: I am trying, by all the means at my disposal, to entice the
reader to do problems and to think about the means and methods he uses in
doing them.

In most of the following chapters, the greater part of the text is devoted to
the broad presentation of the solution of a few problems. The presentation
may appear too broad to a mathematician who is not interested in methodical
points. In fact, what is presented here are not merely solutions but *case
histories* of solutions. Such a case history describes the sequence of essential
steps by which the solution has been eventually discovered, and tries to
disclose the motives and attitudes prompting these steps. The aim of such a
careful description of a particular case is to suggest some general advice, or
pattern, which may guide the reader in similar situations. The explicit
formulation of such advice or such a pattern is usually reserved for a separate
section, although tentative first formulations may be interspersed between the
incidents of the case history.

Each chapter is followed by examples and comments. The reader who does
the examples has an opportunity to apply, clarify, and amplify the methodical
remarks offered in the text of the chapter. The comments interspersed between
the examples give extensions, more technical or more subtle points, or
incidental remarks.

How far I have succeeded I cannot know, but I have certainly tried hard to
enlist the reader's participation. I have tried to fix on the printed page
whatever modes of oral presentation I found most effective in my classes. By
the case histories, I have tried to familiarize the reader with the atmosphere of
research. By the choice, formulation, and disposition of the proposed
problems (formulation and disposition are much more important and cost me
much more labor than the uninitiated could imagine) I have tried to challenge

the reader, awake his curiosity and initiative, and give him ample opportunity to face a variety of research situations.

4. This book deals most of the time with mathematical problems. Non-mathematical problems are rarely mentioned, but they are always present in the background. In fact, I have carefully taken them into consideration and have tried to treat mathematical problems in a way that sheds light on the treatment of nonmathematical problems whenever possible.

This book deals most of the time with elementary mathematical problems. More advanced mathematical problems, however, although seldom referred to, led me to the conception of the material included. In fact, my main source was my own research, and my treatment of many an elementary problem mirrors my experience with advanced problems which could not be included in this book.

5. This book combines its theoretical aim, the study of heuristics, with a concrete, urgent, practical aim: to improve the preparation of high school mathematics teachers.

I have had excellent opportunity to make observations and form opinions on the preparation of high school mathematics teachers, for all my classes have been devoted to such teachers in the last few years. I hope to be a comparatively unprejudiced observer, and as such I can have but one opinion: *the preparation of high school mathematics teachers is insufficient.* Further-more, I think that all responsible organizations must share the blame, and that especially both the schools of education and the departments of mathematics in the colleges should very carefully revise their offerings to teachers if they wish to improve the present situation.

What courses should the colleges offer to prospective high school teachers? We cannot reasonably answer this question, unless we first answer the related question: *What should the high schools offer to their students?*

Yet this question is of little help, you may think, because it is too controversial; it seems impossible to give an answer that would command sufficient consensus. This is unfortunately so; but there is an aspect of this question about which at least the experts may agree.

Our knowledge about any subject consists of *information* and of *know-how*. If you have genuine *bona fide* experience of mathematical work on any level, elementary or advanced, there will be no doubt in your mind that, in mathematics, know-how is much more important than mere possession of information. Therefore, in the high school, as on any other level, we should impart, along with a certain amount of information, a certain degree of *know-how* to the student.

What is know-how in mathematics? The ability to solve problems—not merely routine problems but problems requiring some degree of independence, judgment, originality, creativity. Therefore, the first and foremost duty of the high school in teaching mathematics is to emphasize *methodical work in problem solving*. This is my conviction; you may not go along with it all the way, but I assume that you agree that problem solving deserves some emphasis—and this will do for the present.

The teacher should know what he is supposed to teach. He should show his students how to solve problems—but if he does not know, how can he show them? The teacher should develop his students' know-how, their ability to reason; he should recognize and encourage creative thinking—but the curriculum he went through paid insufficient attention to his mastery of the subject matter and no attention at all to his know-how, to his ability to reason, to his ability to solve problems, to his creative thinking. Here is, in my opinion, the worst gap in the present preparation of high school mathematics teachers.

To fill this gap, the teachers' curriculum should make room for *creative work on an appropriate level*. I attempted to give opportunity for such work by conducting seminars in problem solving. The present work contains the material I collected for my seminars and directions to use it; see the "Hints to Teachers, and to Teachers of Teachers" at the end of this volume, pp. 209–212. This will, I hope, help to improve the mathematics teacher's preparation; at any rate, this is the practical aim of the present work.

I believe that constant attention to both aims mentioned, the theoretical and the practical, made me write a better book. I believe too that there is no conflict between the interests of the various prospective readers (some concerned with problem solving in general, others with improving their own ability, and still others with improving the ability of their students). What matters to one type of reader has a good chance to be of consequence to the others.

6. The present work is the continuation of two earlier ones, *How to Solve It* and *Mathematics and Plausible Reasoning;* the two volumes of the latter have separate titles: *Induction and Analogy in Mathematics* (vol. 1) and *Patterns of Plausible Inference* (vol. 2). These books complete each other without essential overlapping. A topic considered in one may be reconsidered in another, but then the treatment is different: other examples, other details, or other aspects are offered. And so it does not matter much which one is read first and which one is read later.

For the convenience of the reader, the three works will be compared and corresponding passages listed in a cumulative index at the end of the second volume of this book, *Mathematical Discovery*.

7. To publish the first part of a book when the second part is not yet available entails certain risks. (There is a German proverb: "Don't show a half-built house to a fool.") These risks are not negligible; yet, in the interest of the practical aim of this work, I decided not to delay the publication of this volume; see p. 210.

This first volume contains Part One of the work, Patterns, and two chapters of Part Two, Toward a General Method.

The four chapters of Part One have more extensive collections of problems than the later chapters. In fact, Part One is in many ways similar to a collection of problems in analysis by G. Szegö and the author (see the Bibliography). There are, however, obvious differences: in the present volume the problems proposed are much more elementary, and methodical points are not only suggested but explicitly formulated and discussed.

The second chapter of Part Two is inspired by a recent work of Werner Hartkopf (see the Bibliography). I present here only some points of Hartkopf's work which seem to me the most engaging, and I present them as they best fit my conception of heuristics, with suitable examples and additional remarks.

8. The Committee on the Undergraduate Program in Mathematics supported the preparation of the manuscript of this book by funds granted by the Ford Foundation. I wish to express my thanks, and I wish to thank the Committee also for its moral support. I wish to thank the editor of the *Journal of Education of the Faculty and College of Education, Vancouver and Victoria,* for permission to incorporate parts of an article into the present work. I also wish to thank Professor Gerald Alexanderson, Santa Clara, California, and Professor Alfred Aeppli, Zurich, Switzerland, for their efficient help in correcting the proofs.

Zurich, Switzerland   GEORGE POLYA
December 1961

PREFACE TO THE SECOND VOLUME

The present second volume follows the plan, and attempts to carry into effect the intentions, indicated in the preface to the first volume. The index at the end of this volume refers to both volumes; moreover it contains references to selected parallel passages of my related books, which may be of service, I hope, to interested readers.

The Committee on the Undergraduate Program in Mathematics supported the preparation of the manuscript of this second volume by funds granted by the National Science Foundation. I wish to express my thanks to the Committee for their support and encouragement. I wish to thank also the Publisher for courteous help and the careful printing.

The last chapter of this book was dedicated to Charles Loewner on the occasion of his seventieth birthday; I wish to reiterate here the expression of my high esteem and of my feelings of friendship.

Zurich, Switzerland   GEORGE POLYA
October 1964

PREFACE TO THE CORRECTED PRINTING

The text of the first printing is reprinted here with a few minor changes. An appendix is added which contains 35 problems with solutions. These problems supplement various chapters in both volumes of the work; their numbering indicates where they should be inserted between the problems of the first printing.

Stanford University   GEORGE POLYA
June 1967

PREFACE TO THE COMBINED EDITION

I wish to thank Gerald Alexanderson, Peter Hilton, Dave Logothetti, and Jean Pedersen, who in their kindness to me and in their devotion to the subject matter sacrificed so much of their time to make the printing of this one volume version possible.

Palo Alto, California   GEORGE POLYA
November 1980

A POSTSCRIPT TO HIGH SCHOOL TEACHERS

It was not my intention in writing this book to produce a text that would be used and followed sequentially at the elementary level, as is done with most textbooks. Instead I have endeavored to present sequences of interesting and worthwhile problems. The presentation seeks to emphasize the natural development of the subject and prepares students to ask more questions on their own.

It is my hope that high school teachers will use this book as a reference book in the following way:

1. Look up in the table of contents or index the topic you plan to teach your students.
2. Study the development and problems presented on that topic in these two volumes.
3. Devise a plan that will be appropriate for your students, following as much as possible the Ten Commandments for Teachers on page 116 of Volume II.
4. Carry out your plan.
5. Evaluate the progress of your students. Don't forget to include in this evaluation the enthusiasm your students have for learning more mathematics!

HINTS TO THE READER

Section 5 of chapter 2 is quoted as sect. 2.5, subsection (3) of section 5 of chapter 2 as sect. 2.5(3), example 61 of chapter 3 as ex. 3.61.

HSI and MPR are abbreviations for titles of books by the author which will be frequently quoted; see the Bibliography.

Iff. The abbreviation " iff " stands for the phrase " if and only if."

†. The sign † is prefixed to examples, comments, sections, or shorter passages that require more than elementary mathematical knowledge (see the next paragraph). This sign, however, is not used when such a passage is very short.

Most of the material in this book requires only *elementary mathematical knowledge*, that is, as much geometry, algebra, " graphing " (use of co-ordinates), and (sometimes) trigonometry as is (or ought to be) taught in a good high school.

The problems proposed in this book seldom require knowledge beyond the high school level, but, with respect to difficulty, they are often a little above the high school level. The solution is fully (although concisely) presented for some problems, only a few steps of the solution are indicated for other problems, and sometimes only the result is given.

Hints that may facilitate the solution are added to some problems (in parentheses). The surrounding problems may provide hints. Especial attention should be paid to the introductory lines prefixed to the examples (or to certain groups of examples) in some chapters.

The reader who has spent serious effort on a problem may benefit from the effort even if he does not succeed in solving the problem. For example, he may look at some part of the solution, try to extract some helpful information, and then put the book aside and try to work out the rest of the solution by himself.

The best time to think about methods may be when the reader has finished solving a problem, or reading its solution, or reading a case history. With his task accomplished and his experience still fresh in mind, the reader, in *looking back* at his effort, can profitably explore the nature of the difficulty he has just overcome. He may ask himself many useful questions: " What was the decisive point? What was the main difficulty? What could I have done better? I failed to see this point: which item of knowledge, which attitude of mind should I have had to see it? Is there some trick worth learning, one that I could use the next time in a similar situation? " All these questions are good, and there are many others— but the best question is the one that comes spontaneously to mind.

CONTENTS FOR VOLUME I

CONTENTS FOR VOLUME II

Chapter 12.   The Discipline of the Mind, 77

Chapter 13.   Rules of Discovery?, 89

Chapter 14.   On Learning, Teaching, and Learning Teaching, 99

Chapter 15. Guessing and Scientific Method, 143

Solutions, 169

Appendix, 185

Problems and Solutions, 185

PART ONE

PATTERNS

*Each problem that I solved became a rule
which served afterwards to solve other problems.*
> DESCARTES: *Œuvres*, vol. VI, pp. 20–21; Discours de la Méthode.

*If I found any new truths in the sciences,
I can say that they all follow from, or depend on,
five or six principal problems which I succeeded
in solving and which I regard as so many battles
where the fortune of war was on my side.*
> DESCARTES: *op. cit.*, p. 67.

CHAPTER 1

THE PATTERN OF TWO LOCI

1.1. Geometric constructions

Describing or constructing figures with ruler and compasses has a traditional place in the teaching of plane geometry. The simplest constructions of this kind are used by draftsmen, but otherwise the practical importance of geometric constructions is negligible and their theoretical importance not too great. Still, the place of such constructions in the curriculum is well justified: they are most suitable for familiarizing the beginner with geometric figures, and they are eminently appropriate for acquainting him with the ideas of problem solving. It is for this latter reason that we are going to discuss geometric constructions.

As so many other traditions in the teaching of mathematics, geometric constructions go back to Euclid in whose system they play an important role. The very first problem in Euclid's Elements, Proposition One of Book One, proposes "to describe an equilateral triangle on a given finite straight line." In Euclid's system there is a good reason for restricting the problem to the equilateral triangle but, in fact, the solution is just as easy for the following more general problem: *Describe* (or construct) *a triangle being given its three sides.*

Let us devote a moment to analyzing this problem.

In any problem there must be an *unknown*—if everything is known, there is nothing to seek, nothing to do. In our problem the unknown (the thing desired or required, the *quaesitum*) is a geometric figure, a triangle.

Yet in any problem something must be known or *given* (we call the given things the *data*)—if nothing is given, there is nothing by which we could recognize the required thing: we would not know it if we saw it. In our problem the data are three "finite straight lines" or line segments.

3

Finally, in any problem there must be a *condition* which specifies how the unknown is linked to the data.   In our problem, the condition specifies that the three given segments must be the sides of the required triangle.

The condition is an essential part of the problem.   Compare our problem with the following: "Describe a triangle being given its three altitudes." In both problems the data are the same (three line segments) and the unknown is a geometric figure of the same kind (a triangle).   Yet the connection between the unknown and the data is different, the condition is different, and the problems are very different indeed (our problem is easier).

The reader is, of course, familiar with the solution of our problem. Let a, b, and c stand for the lengths of the three given segments.   We lay down the segment a between the endpoints B and C (draw the figure yourself).   We draw two circles, one with center C and radius b, the other with center B and radius c; let A be one of their two points of intersection. Then ABC is the desired triangle.

1.2. From example to pattern

Let us look back at the foregoing solution, and let us look for promising features which have some chance to be useful in solving similar problems.

By laying down the segment a, we have already located two vertices of the required triangle, B and C; just one more vertex remains to be found. In fact, by laying down that segment we have transformed the proposed problem into another problem equivalent to, but different from, the original problem.   In this new problem

the unknown is a point (the third vertex of the required triangle);
the data are two points (B and C) and two lengths (b and c);
the condition requires that the desired point be at the distance b from
 the given point C and at the distance c from the given point B.

This condition consists of two parts, one concerned with b and C, the other with c and B.   *Keep only one part of the condition, drop the other part; how far is the unknown then determined, how can it vary?*   A point of the plane that has the given distance b from the given point C is neither completely determined nor completely free: it is restricted to a "locus"; it must belong to, but can move along, the periphery of the circle with center C and radius b.   The unknown point must belong to two such loci and is found as their intersection.

We perceive here a pattern (the "pattern of two loci") which we can imitate with some chance of success in solving problems of geometric construction:

First, reduce the problem to the construction of ONE point.

Then, split the condition into TWO parts so that each part yields a locus for the unknown point; each locus must be either a straight line or a circle.

Examples are better than precepts—the mere statement of the pattern cannot do you much good. The pattern will grow in color and interest and value with each example to which you apply it successfully.

1.3. Examples

Almost all the constructions which traditionally belong to the high school curriculum are straightforward applications of the pattern of two loci.

(1) *Circumscribe a circle about a given triangle.* We reduce the problem to the construction of the center of the required circle. In the so reduced problem

the unknown is a point, say X;
the data are three points A, B, and C;
the condition consists in the equality of three distances:

$$XA = XB = XC$$

We split the condition into two parts:

First $\qquad XA = XB$
Second $\qquad XA = XC$

To each part of the condition corresponds a locus. The first locus is the perpendicular bisector of the segment AB, the second that of AC. The desired point X is the intersection of these two straight lines.

We could have split the condition differently: first, $XA = XB$, second, $XB = XC$. This yields a different construction. Yet can the result be different? Why not?

(2) *Inscribe a circle in a given triangle.* We reduce the problem to the construction of the center of the required circle. In the so reduced problem

the unknown is a point, say X;
the data are three (infinite) straight lines a, b, and c;
the condition is that the point X be at the same (perpendicular) distance from all three given lines.

We split the condition into two parts:

First, X is equidistant from a and b.

Second, X is equidistant from a and c.

The locus of the points satisfying the first part of the condition consists of *two* straight lines, perpendicular to each other: the bisectors of the angles included by *a* and *b*. The second locus is analogous. The two loci have four points of intersection: besides the center of the inscribed circle of the triangle we obtain also the centers of the three escribed circles.

Observe that this application calls for a slight modification of our formulation of the pattern at the end of sect. 1.2. What modification?

(3) *Given two parallel lines and a point between them. Draw a circle that is tangent to both given lines and passes through the given point.* If we visualize the required figure (it helps to have it on paper) we may observe that we can easily *solve a part of the problem*: the distance of the two given parallels is obviously the diameter of the required circle and half this distance is the radius.

We reduce the problem to finding the center X of the unknown circle.

Knowing the radius, say r, we split the condition as follows:

First, X is at the distance r from the given point.
Second, X is at the distance r from both given lines.

The first part of the condition yields a circle, the second part a straight line midway between, and parallel to, the two given parallels.

Without knowing the radius of the desired circle, we could have split up the condition as follows:

First, X is at the same distance from the given point and the first given line.
Second, X is at the same distance from the given point and the second given line.

Splitting the condition into these two parts is logically unobjectionable but nevertheless useless: the corresponding loci are *parabolas*; we cannot draw them with ruler and compasses—it is an essential part of the scheme that the loci obtained should be circular or rectilinear.

This example may contribute to a better understanding of the pattern of two loci. This pattern helps in many cases, but not in all, as appropriate examples show.

1.4. Take the problem as solved

Wishful thinking is imagining good things you don't have. A hungry man who had nothing but a little piece of dry bread said to himself: "If I had some ham, I could make some ham-and-eggs if I had some eggs."

People tell you that wishful thinking is bad. Do not believe it, this is just one of those generally accepted errors. Wishful thinking may be bad as too much salt is bad in the soup and even a little garlic is bad in the chocolate pudding. I mean, wishful thinking may be bad if there is too much of it or in the wrong place, but it is good in itself and may be a great help in life and in problem solving. That poor guy may enjoy his dry bread more and digest it better with a little wishful thinking about eggs and ham. And we are going to consider the following problem (see Fig. 1.1).

Given three points A, B, and C. Draw a line intersecting AC in the point X and BC in the point Y so that

$$AX = XY = YB$$

Imagine that we knew the position of one of the two points X and Y (this is wishful thinking). Then we could easily find the other point (by drawing a perpendicular bisector). The trouble is that we know neither of the two—the problem does not look easy.

Let us indulge in a little more wishful thinking and *take the problem as solved.* That is, assume that Fig. 1.1 is drawn according to the condition laid down by our problem, so that the three segments of the broken line $AXYB$ are exactly equal. Doing so we imagine a good thing we have not got yet: we imagine that we have found the required location of the line XY; in fact, we imagine that we *have found the solution.*

Yet it is good to have Fig. 1.1 before us. It shows all the geometric elements we should examine, the elements we have and the elements we want, the data and the unknown, assembled as specified by the condition. With the figure before us, we can speculate as to which useful elements we could construct from the data, and which elements could be used in constructing the unknown. We can start from the data and work forward, or start from the unknown and work backward—even side trips could be instructive.

Could you put together at least a few pieces of the jigsaw puzzle? *Could you solve some part of the problem?* There is a triangle in Fig. 1.1, $\triangle XCY$. Can we construct it? We would need three data but, unfortunately, we have only one (the angle at C).

Use what you have, you cannot use what you have not. *Could you derive something useful from the data?* Well, it is easy to join the given points A and B, and the connecting line has some chance to be useful; let us draw it (Fig. 1.2). Yet it is not so easy to see *how* the line AB can be useful—should we rather drop it?

Figure 1.1 looks so empty. There is little doubt that more lines will be needed in the desired construction—what lines?

The lines AX, XY, and YB are equal (we regard them as equal—wishful thinking!). Yet they are in such an awkward relative position—equal lines can be arranged to form much nicer figures. Perhaps we should add more equal lines—or just one more equal line to begin with.

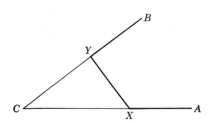

Fig. 1.1. Unknown, data, condition.

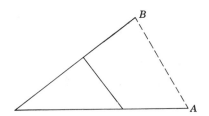

Fig. 1.2. Working forward (from the data).

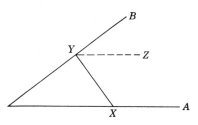

Fig. 1.3. Working backward (from the unknown).

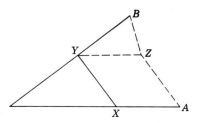

Fig. 1.4. Contacts with previous knowledge.

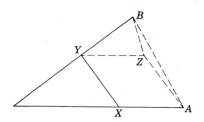

Fig. 1.5. Superposition.

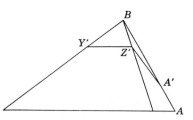

Fig. 1.6. Stepping stone.

Chance or inspiration may prompt us to introduce a line into the picture which, on the face of it, fits quite well into the intended connection: draw *YZ* parallel and equal to *XA*, see Fig. 1.3. (We are starting now from the desired unknown—wishful thinking—and trying to work backward toward the data.)

Introducing the line *YZ* was a trial. Yet the line does not look bad; it brings in familiar shapes. Join *Z* to *A* and *B*, see Fig. 1.4; we obtain the rhombus *XAZY* and the isosceles triangle *BYZ*. *Could you solve some part of the problem?* Can we construct △*BYZ*? We would need two data for an isosceles triangle but, unfortunately, we have only one (the angle at Y is equal to the given angle at C). Still, we have something here. Even if we do not know $\triangle BYZ$ completely, we know its shape; although we do not know its size, we could construct a triangle similar to it.

This may bring us a little nearer to the solution, but we have not got it yet: we must try a few more things. Sooner or later we may remember a former trial, the Fig. 1.2. How about combining it with later remarks? By superposing Figs. 1.2 and 1.4 we obtain Fig. 1.5 in which there is a new triangle, $\triangle BZA$. Can we construct it? We could, if we knew $\triangle BYZ$; in that favorable case, we could muster three data: two sides, ZB and $ZA = ZY$, and the angle at B. Well, we do not know $\triangle BYZ$; at any rate, we do not know it completely, we know only its shape. Yet then, we can...

We can draw the quadrilateral $BY'Z'A'$, see Fig. 1.6, similar to the quadrilateral $BYZA$ in Fig. 1.5, which is an essential part of the desired configuration. This may be a stepping stone!

1.5. The pattern of similar figures

We carry out the construction, the discovery of which is told by the sequence of Figs. 1.1–1.6.

On the given line BC, see Fig. 1.6, we choose a point Y' at random (but not too far from B). We draw the line $Y'Z'$ parallel to CA so that

$$Y'Z' = Y'B$$

Then, we determine a point A' on AB so that

$$A'Z' = Y'Z'$$

Draw a parallel to $A'Z'$ through A and determine its intersection with the prolongation of the line BZ': this intersection is the desired point Z. The rest is easy.

The two quadrilaterals $AZYB$ and $A'Z'Y'B$ are not only similar but also "similarly located" (homothetic). The point B is their center of similarity. That is, any line connecting corresponding points of the two similar figures has to pass through B.

Here is a remark from which we can learn something about problem solving: Of the two similar figures, the one that came to our attention first, $AZYB$, was actually constructed later.[1]

The foregoing example suggests a general pattern: *If you cannot construct the required figure, think of the possibility of constructing a figure* SIMILAR *to the required figure.*

[1] In this "case history" which we have just finished (we started it in sect. 1.4) the most noteworthy step was to "take the problem as solved." For further remarks on this, cf. HSI, Figures 2, pp. 104–105, and Pappus, pp. 141–148, especially pp. 146–147.

There are examples at the end of this chapter which, if you work them through, may convince you of the usefulness of this pattern of "similar figures."

1.6. Examples

The following examples differ from each other in several respects; their differences may show up more clearly the common feature that we wish to disentangle.

(1) *Draw common tangents to two given circles.* Two circles are given in position (plotted on paper). We wish to draw straight lines touching both circles. If the given circles do not overlap they have four common tangents, two exterior and two interior tangents. Let us confine our attention to the exterior common tangents, see Fig. 1.7, which exist unless one of the two given circles lies completely within the other.

If you cannot solve the proposed problem, look around for an appropriate related problem. There is an obvious related problem (of which the reader is supposed to know the solution): to draw tangents to a given circle from an outside point. This problem is, in fact, a limiting case or *extreme case* of the proposed problem: one of the two given circles is shrunken into a point. We arrive at this extreme case in the most natural way by *variation of the data.* Now we can vary the data in many ways: decrease one radius and leave the other unchanged, or decrease one radius and increase the other, or decrease both. And so we may hit upon the idea of letting both radii decrease *at the same rate*, uniformly, so that both are diminished by the same length in the same time. Visualizing this change, we may observe that each common tangent is shifting, but remains parallel to itself while shifting, till ultimately Fig. 1.8 appears—and here is the solution: draw tangents from the center of the smaller given circle to a new circle which is concentric with the larger given circle and the radius of which is

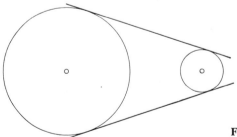

Fig. 1.7. Unknown, data, condition.

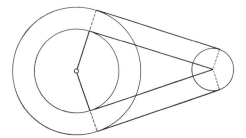

Fig. 1.8. Stepping stone.

the difference of the given radii.   Use the figure so obtained as a stepping stone: the step from it to the desired figure is easy (there are just two rectangles to construct).

(2) *Construct a triangle being given the three medians.* We "take the problem as solved"; that is, we draw the (desired) triangle in which the three (given) medians are duly assembled; see Fig. 1.9.   We should recollect that the three medians meet in one point (the point M in Fig. 1.9, the centroid of the triangle) which divides each median in the proportion $1:2$. To visualize this essential fact, let us mark the midpoint D of the segment AM; the points D and M divide the median AE into three equal parts; see Fig. 1.10.

The desired triangle is divided into six small triangles.   *Could you solve a part of the problem?*   To construct one of those small triangles we need three data; in fact, we know two sides: one side is one third of a given median, another side is two thirds of another given median—but we do not see a third known piece.   Could we introduce some other triangle with three known data?   There is the point D in Fig. 1.10 which is obviously eager for more connections—if we join it to a neighboring point we may notice $\triangle MDG$ each side of which is one third of a median—and so we can construct it, from three known sides—here is a stepping stone!   The rest is easy.

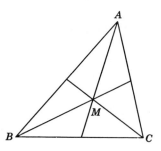

Fig. 1.9. Unknown, data, condition.

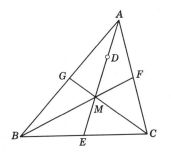

Fig. 1.10. A point eager for more connections.

(3) To each problem concerned with ordinary triangles there corresponds a problem concerned with spherical triangles or trihedral angles. (A trihedral angle is contained between three planes; a sphere described about its vertex as center intersects it in a spherical triangle.) These problems of solid geometry may be reduced to problems of plane geometry. Such reduction of problems about figures in space to drawings in a plane is, in fact, the object of *descriptive geometry*, which is an interesting branch of geometry indispensable to engineers and architects for the accurate drafting of machinery, vessels, buildings, and so on.

The reader needs no knowledge of descriptive geometry, just a little solid geometry and some common sense, to solve the following problem: *Being given the three face angles of a trihedral angle, construct its dihedral angles.*

Let *a*, *b*, and *c* denote the face angles of the trihedral angle (the sides of the corresponding spherical triangle) and α the dihedral angle opposite to the face *a* (α is an angle of the spherical triangle). Being given *a*, *b*, *c*, construct α. (The same method can serve to construct all three dihedral angles, and so we restrict ourselves to one of them, to α.)

To visualize the *data*, we juxtapose the three angles *b*, *a*, and *c* in a plane; see Fig. 1.11. To visualize the *unknown*, we should see the configuration in space. (Reproduce Fig. 1.11 on cardboard, crease the line between *a* and *b* and also that between *a* and *c*, and then fold the cardboard to form the trihedral angle.) In Fig. 1.12, the trihedral angle is seen in perspective; *A* is a point chosen at random on the edge opposite the face *a*; two perpendiculars to this edge starting from *A*, one drawn in the face *b*, the other drawn in the face *c*, include the angle α that we are required to construct.

Look at the unknown! —It is an angle, the angle α in Fig. 1.12.

What can you do to get this kind of unknown? —We often determine an angle from a triangle.

Is there a triangle in the figure? —No, but we can introduce one.

In fact, there is an obvious way to introduce a triangle: the plane that

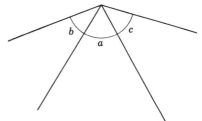

Fig. 1.11. The data.

contains the angle α intersects the trihedral angle in a triangle; see Fig. 1.13. This triangle is a promising auxiliary figure, a likely stepping stone.

In fact, the solution is not far. Return to the figure in the plane, to Fig. 1.11, where the data, the angles a, b, and c, appear in true magnitude. (Unfold the cardboard model we have folded together in passing from Fig. 1.11 to Fig. 1.12.) The point A appears twice, as A_1 and A_2 (by unfolding, we have separated the two faces b and c which are adjacent in space). These points A_1 and A_2 are at the same distance from the vertex V. A perpendicular to $A_1 V$ through A_1 meets the other side of the angle b in C, and B is analogously obtained; see Fig. 1.14. Now we know $A_2 B$, BC, and CA_1, the three sides of the auxiliary triangle introduced in Fig. 1.13, and so we can readily construct it (in dotted lines in Fig. 1.14): it contains the desired angle α.

The problem just discussed is analogous to, and uses the construction of, the simplest problem about ordinary triangles which we discussed in sect. 1.1. We can see herein a sort of justice and a hint about the use of analogy.

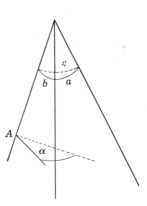

Fig. 1.12. The unknown.

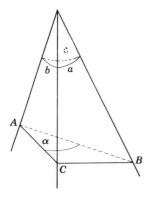

Fig. 1.13. A likely stepping stone.

1.7. The pattern of auxiliary figures

Let us look back at the problems discussed in the foregoing sect. 1.6. They were quite different, and their solutions were quite different too, except that in each case the key to the solution was an *auxiliary figure*: a circle with two tangents from an outside point in (1), a smaller triangle carved out from the desired triangle in (2), another triangle in (3).   In each case we could easily construct the auxiliary figure from the data and, once in possession of the auxiliary figure, we could easily construct the originally required figure by using the auxiliary figure.   And so we attained our goal in two steps; the auxiliary figure served as a kind of stepping stone; its discovery was the decisive performance, the culminating point of our work.   There is a pattern here, the *pattern of auxiliary*

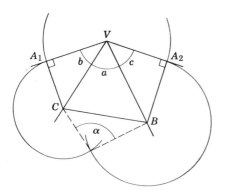

Fig. 1.14. The solution.

figures, which has some promise and which we can describe as follows: *Try to discover some part of the figure or some closely* RELATED FIGURE *which you can construct and which you can use as a stepping stone in constructing the original figure.*

This pattern is very general. In fact, the pattern of similar figures formulated in sect. 1.5 is just a particular case: a figure similar to the required figure is related to it in a particular manner and can serve as a particularly handy auxiliary figure.

Unavoidably, its greater generality renders the pattern of auxiliary figures less concrete, less tangible: it gives no specific advice about what kind of figure we should seek. Experience, of course, can give us some directives (although no hard and fast rules): we should look for figures which are easy to "carve out" from the desired figure, for "simple" figures (as triangles), for "extreme cases," and so on. We may learn procedures, such as the variation of the data, or the use of analogy, which, in certain cases, may indicate an appropriate auxiliary figure.

We have now isolated three different patterns which we may use in dealing with problems of geometric construction. The pattern of *auxiliary figures* leaves us more choice, but offers a less definite target, than the pattern of *similar figures*. The pattern of *two loci* is the simplest—you may try it first just because, in most cases, it is best to try the simplest thing first. Yet do not commit yourself, keep an open mind: take the problem as solved, draw a figure in which the *unknown* and the *data* are appropriately assembled, each element at its right place, all elements connected by the right relations, as required by the *condition*. Study this figure, try to recognize in it some familiar configuration, try to recall any relevant knowledge you may have (related problems, applicable theorems), look out for an opening (a more accessible part of the figure, for instance). You may be lucky: a bright idea may emerge from the figure and suggest an appropriate auxiliary line, the suitable pattern, or some other useful step.

Examples and Comments on Chapter 1

1.1. What is the locus of a variable point that has a given distance from a given point?

1.2. What is the locus of a variable point that has a given distance from a given straight line?

1.3. A variable point remains equidistant from two given points; what is its locus?

1.4. A variable point remains equidistant from two given parallel straight lines; what is its locus?

1.5. A variable point remains equidistant from two given intersecting straight lines; what is its locus?

1.6. Of a triangle, given two vertices, A and B, and the angle γ, opposite to the side AB; the triangle is not determined, its third vertex (that of γ) can vary. What is the locus of this third vertex?

1.7. *Notation.* In dealing with a triangle, it is convenient to use the following notation:

$A,$	$B,$	C	vertices
$a,$	$b,$	c	sides
$\alpha,$	$\beta,$	γ	angles
$h_a,$	$h_b,$	h_c	altitudes ("heights")
$m_a,$	$m_b,$	m_c	medians
$d_\alpha,$	$d_\beta,$	d_γ	bisectors of the angles ("disectors"?!)
	R		radius of circumscribed circle
	r		radius of inscribed circle

It is understood that the side a is opposite the angle α, the vertex of which is the point A which is the common endpoint of the three lines h_a, m_a, and d_α. According to common usage, a stands both for the side (a line segment) and for the length of the side; the reader has to find out from the context which meaning is intended. The same ambiguity is inherent in the symbols $b, c, h_a, \ldots$ d_γ, R, r. We follow this usage although it is objectionable.

The problem "Triangle from a, b, c" means, of course, "construct a triangle being given a, b, and c." Observe that there may be no solution (the figure satisfying the proposed condition may not exist) if the data are adversely chosen; for example, there is no triangle with given sides a, b, and c, if $a > b + c$. Experiment first with data for which the required figure is likely to exist.

1.8. Triangle from a, b, m_a.

1.9. Triangle from a, h_a, m_a.

1.10. Triangle from a, h_a, α.

1.11. Triangle from a, m_a, α.

1.12. Given three (infinite) straight lines. Construct a circle that touches the first two lines and has its center on the third line.

1.13. Given two intersecting infinite straight lines and a line segment of length r. Construct a circle with radius r that touches the two given lines.

1.14. Construct a circle, being given one point on it, one straight line tangent to it, and its radius.

1.15. *Three lighthouses* are visible from a ship; their positions on the map are known, and the angles between the rays of light coming from them have been measured.   Plot the position of the ship on the map.

1.16. Within a given circle, describe three equal circles so that each shall touch the other two and also the given circle.   (This figure can sometimes be seen in Gothic tracery where analogous figures, with four or six inner circles, are more frequent.)

1.17. Inside a given triangle find a point from which all three sides are seen under the same angle.

1.18. Trisect the area of a given triangle.

That is, you should locate a point X inside the given $\triangle ABC$ so that $\triangle XBC$, $\triangle XCA$, and $\triangle XAB$ are equal in area.

[*Keep only a part of the condition, drop the other part:* if only the two triangles $\triangle XCA$ and $\triangle XCB$ are supposed to be equal, what is the locus of X?   The answer to this question may show you a way to the solution, but there are also other approaches.]

1.19. Triangle from a, α, r.

[*Keep only a part of the condition, drop the other part:* disregard r, but keep a and α; what is the locus of the center of the inscribed circle?]

1.20. Triangle from a, h_b, c.

1.21. Triangle from a, h_b, d_γ.

1.22. Triangle from a, h_b, h_c.

1.23. Triangle from h_a, h_b, β.

1.24. Triangle from h_a, β, γ.

1.25. Triangle from h_a, d_α, α.

1.26. Construct a parallelogram, being given one side and two diagonals.

1.27. Construct a trapezoid being given its four sides a, b, c, and d; a and c should be parallel.

1.28. Construct a quadrilateral being given a, b, c, and d, its four sides, and the angle ϵ, included by the opposite sides a and c produced.

1.29. Triangle from a, $b + c$, α.

[Do not fail to introduce *all the data* into the figure.   Where is the "right place" for $b + c$?]

1.30. Triangle from a, $b + c$, $\beta - \gamma$.

1.31. Triangle from $a + b + c$, h_a, α.

[Symmetry: b and c (not given) play interchangeable roles.]

1.32. Given two circles exterior to each other, draw their interior common tangents.   (The two circles are situated in the same halfplane with respect to an exterior common tangent, in different halfplanes with respect to an interior common tangent.)

1.33. Given three equal circles, construct a circle containing, and tangent to, all three given circles.

1.34. Triangle from α, β, d_γ.

1.35. Inscribe a square in a given right triangle. One corner of the square is required to coincide with the right-angle corner of the given triangle, the opposite vertex of the square should lie on the hypotenuse, the two other vertices on the legs of the right triangle, one on each.

1.36. Inscribe a square in a given triangle ABC. Two vertices of the square are required to lie on AB, one on AC, and one on BC.

1.37. Inscribe a square in a given sector of a circle. Two vertices of the square are required to lie on the arc, one vertex on each of the two sides of the central angle of the sector.

1.38. Construct a circle being given two points on it and one straight line tangent to it.

1.39. Construct a circle, being given one point on it and two straight lines tangent to it.

1.40. Construct a pentagon circumscribable about a circle being given its five angles α, β, γ, δ, and ϵ (subject, of course, to the condition $\alpha + \beta + \gamma + \delta + \epsilon = 540°$) and the length of its perimeter l.

1.41. Triangle from h_a, h_b, h_c.

1.42. *A flaw.* It may happen that a problem of geometric construction has no solution: there may be no figure satisfying the proposed condition with the proposed data. For instance, there exists no triangle the sides of which have the given lengths a, b, and c if $a > b + c$. A perfect method of solution will either obtain a figure satisfying the proposed condition or show in failing that there exists no such figure.

There can arise, however, the following situation: the proposed problem itself does possess a solution, yet an auxiliary problem does not—an auxiliary figure, which our scheme would need for the construction of the originally required figure, is impossible to construct. This is, of course, a flaw in our scheme.

Is your method for solving ex. 1.41 perfect in this respect? (The triangle with sides 65, 156, 169 is a right triangle—the sides are proportional to 5, 12, 13—with heights 156, 65, 60.) If the answer is No, can you improve your method?

1.43. Triangle from a, α, R.

1.44. *Looking back* at the solution of ex. 1.43, you may ask some instructive questions and propose some related problems.

(*a*) An analogous problem?

(*b*) A more general problem?

(*c*) Triangle from a, β, R.

(*d*) Triangle from a, r, R.

1.45. *Three listening posts.* The time at which the sound of an enemy gun is heard has been exactly observed at three listening posts A, B, and C. On the basis of these data, plot on the map the position X of that enemy gun.

Regard the velocity of sound as known. Explain the analogy with, and the difference from, the problem of the three lighthouses, ex. 1.15.

1.46. *On the pattern of two loci.* Are the loci with which exs. 1.2, 1.5, and 1.6 are concerned usable in connection with the pattern of two loci? Cf. the statement at the end of sect. 1.2.

1.47. *The pattern of three loci.* A concept of plane geometry may have various analogues in solid geometry. For instance, in sect. 1.6(3) we regarded a spherical triangle or a trihedral angle as analogous to an ordinary plane triangle. Yet we could also regard a tetrahedron as analogous to an ordinary triangle; seen from this viewpoint, the following problem appears as analogous to the problem of sect. 1.3(1).

Circumscribe a sphere about a given tetrahedron.

Let us work out the analogy in some detail. We reduce the problem to obtaining the center of the required sphere. In the so reduced problem

the unknown is a point, say X;

the data are four points (the vertices of the given tetrahedron), say A, B, C, and D;

the condition consists in the equality of four distances

$$XA = XB = XC = XD$$

We may split this condition into three parts:

First	$XA = XB$
Second	$XA = XC$
Third	$XA = XD$

To each part of the condition corresponds a locus. If the point X satisfies the first part of the condition, its locus is (it can vary on) a plane, the perpendicular bisector of the segment AB; to each other part of the condition there corresponds an analogous plane. Finally, the desired center of the sphere is obtained as the intersection of three planes.

Let us assume that we have instruments with which we can determine the points of intersection of three given surfaces when each of these surfaces is either a plane or a sphere. (In fact, we have made this assumption implicitly in the foregoing. By the way, ruler and compasses are such instruments—we can determine with them those points of intersection if we know enough descriptive geometry.) Then we can propose and solve problems of geometric construction in space. The foregoing problem is an example and its solution sets an example: with the help of analogy, we can disentangle from it a pattern for solving problems of construction in space, the *pattern of three loci*.

1.48. In the foregoing ex. 1.47, as in the example of sect. 1.3(1), we could have split the condition differently and so obtain another (although a pretty similar) construction. Yet can the result be different? Why not?

1.49. *On geometric constructions.* There are many problems of geometric construction where the required figure obviously "exists" but cannot be constructed with ruler and compasses (it could be constructed with other—equally idealized—instruments). A famous problem of this kind is the trisection of the angle: a *general* angle cannot be divided into three equal parts by ruler and compasses; see Courant and Robbins, pp. 137–138.

A perfect method for geometric constructions should *either* lead us to a construction of the required figure by ruler and compasses *or* show that such a construction is impossible. Our patterns (two loci, similar figures, auxiliary figures) are not useless (as, I hope, the reader has had opportunity to convince himself) but they yield no perfect method; they frequently suggest a construction but, when they do not suggest one, we are left in the dark about the alternative with which we are most concerned: is the construction impossible in itself, or is it possible and just our effort insufficient?

There is a well known more perfect method for geometric constructions (reduction to algebra—but we need not enter upon details now). Yet for another kind of problem which we may face another day there may be no perfect method known at that time—and still we have to try. And so the patterns considered may contribute to the education of the problem solver just by their inherent imperfection.

1.50. *More problems.* Devise some problems similar to, but different from, the problems proposed in this chapter—especially such problems as you can solve.

1.51. *Sets.* We cannot define the concept of a *set* in terms of more fundamental concepts, because there are no more fundamental concepts. Yet, in fact, everybody is familiar with this concept, even if he does not use the word "set" for it. "Set of elements" means essentially the same as "class of objects" or "collection of things" or "aggregate of individuals." "Those students who will make an A in this course" form a set even if, at this moment, you could not tell all their names. "Those points in space that are equidistant from two given points" form a very clearly defined set of points, a plane. "Those straight lines in a given plane that have a given distance from a given point" form an interesting set consisting of all the tangents of a certain circle. If a, b, and c are any three distinct objects, the set to which just these three objects belong as elements is clearly defined.

Two sets are *equal* if every object that belongs to one of them belongs also to the other. If any element that belongs to the set A belongs also to the set B, we say that A is *contained* in B; there are many ways to say the same thing: B contains A, B includes A, A is a *subset* of B, and so on.

It is often convenient to consider the *empty set*, that is, the set to which no element belongs. For example, the "set of those students who will make A in this course" could well turn out to be the empty set, if no student makes a better grade than B, or if the course should be discontinued without a final examination. The empty set is a useful set as 0 is a useful number. Now, 0 is less than any positive integer; similarly, the empty set is considered as a subset of any set.

The greatest common subset of several sets is termed their *intersection*. That is, the intersection of the sets A, B, C,..., and L consists of those, and only those, elements that belong simultaneously to each of the sets A, B, C,..., and L.

For example, let A and B denote two planes, each considered as a set of points; if they are different and nonparallel, their intersection is a straight line; if they are different but parallel, their intersection is the empty set; if they are identical, their "intersection" is identical with any of them. If A, B, and C are three planes and there is no straight line parallel to all three of them, their intersection is a set containing just one element, a point.

The term "locus" means essentially the same as the term "set": the set (or locus) of those points of a plane that have a given distance from a given point is a circle.

In this example, we define the set (or locus) by stating a *condition* that its elements (points) must satisfy, or a *property* that these elements must possess: the points of a circle satisfy the condition, or have the property, that they are all contained in the same plane and all have the same distance from a given point.

The concepts of "condition" and "property" are indissolubly linked with the concept of a set. In many mathematical examples we can clearly and simply state the condition or property that characterizes the elements of a set. Yet, if a more informative description is lacking we can always say: the elements of the set S have the property of belonging to S, and satisfy the condition that they belong to S.

The consideration of the pattern of three loci (after that of two loci; see ex. 1.47) may have given us already a hint of a wider generalization. The consideration of sets and their intersections intensifies the suggestion. We now leave this suggestion to mature in the mind of the reader and we shall return to it in a later chapter.

(The least extensive set of which each one of several given sets is a subset is called the *union* of those given sets. That is, the union of the sets A, B, ..., and L contains all the elements of A, all the elements of B,..., and all the elements of L, and any element that the union contains, must belong to at least one of the sets A, B,..., and L (it may belong to several of them).

Intersection and union of sets are closely allied concepts (they are "complementary" concepts in a sense which we cannot but hint), and we could not very well discuss one without mentioning the other. In fact, we shall have more opportunity to consider the intersection of given sets than their union. The reader should familiarize himself from some other book with the first notions of the theory of sets which may be introduced into the high schools in the near future.)

CHAPTER 2

THE CARTESIAN PATTERN

2.1. Descartes and the idea of a universal method

René Descartes (1596–1650) was one of the very great. He is regarded by many as the founder of modern philosophy, his work changed the face of mathematics, and he also has a place in the history of physics. We are here mainly concerned with one of his works, the *Rules for the Direction of the Mind* (cf. ex. 2.72).

In his "Rules," Descartes planned to present a universal method for the solution of problems. Here is a rough outline of the scheme that Descartes expected to be applicable to all types of problems:

First, reduce any kind of problem to a mathematical problem.

Second, reduce any kind of mathematical problem to a problem of algebra.

Third, reduce any problem of algebra to the solution of a single equation.

The more you know, the more gaps you can see in this project. Descartes himself must have noticed after a while that there are cases in which his scheme is impracticable; at any rate, he left unfinished his "Rules" and presented only fragments of his project in his later (and better known) work *Discours de la Méthode*.

There seems to be something profoundly right in the intention that underlies the Cartesian scheme. Yet it is more difficult to carry this intention into effect, there are more obstacles and more intricate details than Descartes imagined in his first enthusiasm. Descartes' project failed, but it was a great project and even in its failure it influenced science much more than a thousand and one little projects which happened to succeed.

Although Descartes' scheme does not work in all cases, it does work in an inexhaustible variety of cases, among which there is an inexhaustible variety of *important* cases. When a high school boy solves a "word problem" by "setting up equations," he follows Descartes' scheme and in doing so he prepares himself for serious applications of the underlying idea.

And so it may be worthwhile to have a look at some high school work.

2.2. A little problem

Here is a brain teaser which may amuse intelligent youngsters today as it probably amused others through several centuries.

A farmer has hens and rabbits. These animals have 50 *heads and* 140 *feet. How many hens and how many rabbits has the farmer?*

We consider several approaches.

(1) *Groping.* There are 50 animals altogether. They cannot all be hens, because then they would have only 100 feet. They cannot all be rabbits, because they would then have 200 feet. Yet there should be just 140 feet. If just one half of the animals were hens and the other half rabbits, they would then have.... Let us survey all these cases in a table:

Hens	Rabbits	Feet
50	0	100
0	50	200
25	25	150

If we take a smaller number of hens, we have to take a larger number of rabbits and this leads to more feet. On the contrary, if we take a larger number of hens.... Yes, there must be more than 25 hens—let us try 30:

Hens	Rabbits	Feet
30	20	140

I have got it! Here is the solution!

Yes, indeed, we have got the solution, because the given numbers, 50 and 140, are relatively small and simple. Yet if the problem, proposed with the same wording, had larger or more complicated numbers, we would need more trials or more luck to solve it in this manner, by merely muddling through.

(2) *Bright idea.* Of course, our little problem can be solved less "empirically" and more "deductively"—I mean with fewer trials, less guesswork, and more reasoning. Here is another solution.

The farmer surprises his animals in an extraordinary performance: each hen is standing on one leg and each rabbit is standing on its hind legs. In this remarkable situation just one half of the legs are used, that is, 70 legs. In this number 70 the head of a hen is counted just *once* but the head of a rabbit is counted *twice*. Take away from 70 the number of all heads, which is 50; there remains the number of the rabbit heads—there are

$$70 - 50 = 20$$

rabbits! And, of course, 30 hens.

This solution would work just as well if the numbers in our little problem (50 and 140) were replaced by less simple numbers. This solution (which can be presented less whimsically) is ingenious: it needs a clear intuitive grasp of the situation, a little bit of a bright idea—my congratulations to a youngster of fourteen who discovers it by himself. Yet bright ideas are rare—we need a lot of luck to conceive one.

(3) *By algebra.* We can solve our little problem without relying on chance, with less luck and more system, if we know a little algebra.

Algebra is a language which does not consist of words but of symbols. If we are familiar with it we can translate into it appropriate sentences of everyday language. Well, let us try to translate into it the proposed problem. In doing so, we follow a precept of the Cartesian scheme: "reduce any kind of problem to a problem of algebra." In our case the translation is easy.

State the problem

in English	*in algebraic language*
A farmer has	
a certain number of hens	x
and a certain number of rabbits	y
These animals have fifty heads	$x + y = 50$
and one hundred forty feet	$2x + 4y = 140$

We have translated the proposed question into a system of two equations with two unknowns, x and y. Very little knowledge of algebra is needed to solve this system: we rewrite it in the form

$$x + 2y = 70$$
$$x + \ y = 50$$

and subtracting the second equation from the first we obtain

$$y = 20$$

Using this we find, from the second equation of the system, that

$$x = 30$$

This solution works just as well for large given numbers as for small ones, works for an inexhaustible variety of problems, and needs no rare bright idea, just a little facility in the use of the algebraic language.

(4) *Generalization.* We have repeatedly considered the possibility of substituting other, especially larger, numbers for the given numbers of our problem, and this consideration was instructive. It is even more instructive to substitute *letters for the given numbers.*

Substitute *h* for 50 and *f* for 140 in our problem. That is, let *h* stand for the number of heads, and *f* for the number of feet, of the farmer's animals. By this substitution, our problem acquires a new look; let us consider also the translation into algebraic language.

A farmer has	
a certain number of hens	x
and a certain number of rabbits.	y
These animals have *h* heads	$x + y = h$
and *f* feet.	$2x + 4y = f$

The system of two equations that we have obtained can be rewritten in the form

$$x + 2y = \frac{f}{2}$$

$$x + y = h$$

and yields, by subtraction,

$$y = \frac{f}{2} - h$$

Let us retranslate this formula into ordinary language: the number of rabbits equals one half of the number of feet, less the number of heads: this is the result of the imaginative solution (2).

Yet here we did not need any extraordinary stroke of luck or whimsical imagination; we attained the result by a straightforward routine procedure after a simple initial step which consisted in replacing the given numbers by letters. This step is certainly simple, but it is an important step of generalization.[1]

(5) *Comparison.* It may be instructive to compare different approaches to the same problem. Looking back at the four preceding approaches, we may observe that each of them, even the very first, has some merit, some specific interest.

[1] Cf. HSI, Generalization 3, pp. 109–110; Variation of the problem 4, pp. 210–211; Can you check the result? 2, p. 60.

The first procedure which we have characterized as "groping" and "muddling through" is usually described as a solution by *trial and error*. In fact, it consists of a series of trials, each of which attempts to correct the error committed by the preceding and, on the whole, the errors diminish as we proceed and the successive trials come closer and closer to the desired final result. Looking at this aspect of the procedure, we may wish a better characterization than "trial and error"; we may speak of "successive trials" or "successive corrections" or "successive approximations." The last expression may appear, for various reasons, to be the most suitable. The term *method of successive approximations* naturally applies to a vast variety of procedures on all levels. You use successive approximations when, in looking for a word in the dictionary, you turn the leaves and proceed forward or backward according as a word you notice precedes or follows in alphabetical order the word you are looking for. A mathematician may apply the term successive approximations to a highly sophisticated procedure with which he tries to treat some very advanced problem of great practical importance that he cannot treat otherwise. The term even applies to science as a whole; the scientific theories which succeed each other, each claiming a better explanation of phenomena than the foregoing, may appear as successive approximations to the truth.

Therefore, the teacher should not discourage his students from using "trial and error"—on the contrary, he should encourage the intelligent use of the fundamental method of successive approximations. Yet he should convincingly show that for such simple problems as that of the hens and rabbits, and in many more (and more important) situations, straightforward algebra is more efficient than successive approximations.

2.3. Setting up equations

In the foregoing, cf. sect. 2.2(3), we have translated a proposed problem from the ordinary language of words into the algebraic language of symbols. In our example, the translation was obvious; there are cases, however, where the translation of the problem into a system of equations demands more experience, or more ingenuity, or more work.[2]

What is the nature of this work? Descartes intended to answer this question in the second part of his "Rules" which, however, he left unfinished. I wish to extract from his text and present in contemporary language such parts of his considerations as are the most relevant at this stage of our study. I shall leave aside many things that Descartes did say, and I shall make explicit a few things that he did not quite say, but I still think that I shall not distort his intentions.

---

[2] Cf. HSI, Setting up equations, pp. 174–177.

I wish to follow Descartes' manner of exposition: I shall begin each explanation by a concise "advice" (in fact, it is rather a summary) and then expand that advice (summary) by adding comments.

(1) *First, having well understood the problem, reduce it to the determination of certain unknown quantities* (Rules XIII–XVI).

To spend time on a problem that we do not understand would be foolish. Therefore, our first and most obvious duty is to understand the problem, its meaning, its purpose.

Having understood the problem as a whole, we turn our attention to its principal parts.   We should see very clearly

what kind of thing we have to find (the UNKNOWN or unknowns)

what is given or known (the DATA)

how, by what relations, the unknowns and the data are connected with each other (the CONDITION).

(In the problem of sect. 2.2(4) the unknowns are x and y, and the data h and f, the numbers of hens and rabbits, heads and feet, respectively. The condition is expressed first in words, then in equations.)

Following Descartes, we now confine ourselves to problems in which the unknowns are quantities (that is, numbers but not necessarily integers). Problems of other kinds, such as geometrical or physical problems, may be reduced sometimes to problems of this purely quantitative type, as we shall illustrate later by examples; cf. sects. 2.5 and 2.6.

(2) *Survey the problem in the most natural way, taking it as solved and visualizing in suitable order all the relations that must hold between the unknowns and the data according to the condition* (Rule XVII).

We imagine that the unknown quantities have values fully satisfying the condition of the problem: this is meant essentially by "taking the problem as solved"; cf. sect. 1.4.   Accordingly, we treat unknown and given quantities equally in some respects; we visualize them connected by relations as the condition requires.   We should survey and study these relations in the spirit in which we survey and study the figure when planning a geometric construction; see the end of sect. 1.7.   The aim is to find some indication about our next task.

(3) *Detach a part of the condition according to which you can express the same quantity in two different ways and so obtain an equation between the unknowns.   Eventually you should split the condition into as many parts, and so obtain a system of as many equations, as there are unknowns* (Rule XIX).

The foregoing is a free rendering, or *paraphrase*, of the statement of Descartes' Rule XIX.   After this statement there is a great gap in

Descartes' manuscript: the explanation which should have followed the statement of the Rule is missing (it was probably never written). Therefore, we have to make up our own comments.

The aim is stated clearly enough: we should obtain a system of n equations with n unknowns. It is understood that the computation of these unknowns should solve the proposed problem. Therefore, the system of equations should be equivalent to the proposed condition. If the whole system expresses the whole condition, each single equation of the system should express some part of the condition. Hence, in order to set up the n equations we should split the condition into n parts. But how?

The foregoing considerations under (1) and (2) (which outline very sketchily Descartes' Rules XIII–XVII) give some indications, but no definite instructions. Certainly, we have to understand the problem, we have to see the unknowns, the data, and the condition very, very clearly. We may profit from surveying the various clauses of the condition and from visualizing the relations between the unknowns and the data. All these activities give us a chance to obtain the desired system of equations, but no certainty.

The advice that we are considering (the paraphrase of Rule XIX) stresses an additional point: in order to obtain an equation we have to *express the same quantity in two different ways.* (In the example of sect. 2.2(3) an equation expresses the *number of feet* in two different ways.) This remark, properly digested, often helps to discover an equation between the unknowns—it can always help to explain the equation after it has been discovered.

In short, there are some good suggestions, but there is no foolproof precept for setting up equations. Yet, where no precept helps, practice may help.

(4) *Reduce the system of equations to one equation* (Rule XXI).

The statement of Descartes' Rule XXI which is here paraphrased is not followed by an explanation (in fact, it is the last sentence in Descartes' manuscript). We shall not examine here under which conditions a system of algebraic equations can be reduced to a single equation or how such reduction can be performed; these questions belong to a purely mathematical theory which is more intricate than Descartes' short advice may lead us to suppose, but is pretty well explored nowadays and no concern of ours at this point. Very little algebra will be sufficient to perform the reduction in those simple cases in which we shall need it.

There are other questions which remain unexplored although we should concern ourselves with them. Yet we may take them up more profitably after some examples.

2.4. Classroom examples

The "word problems" of the high school are trivial for mathematicians, but not so trivial for high school boys or girls or teachers. I think, however, that a teacher who makes an earnest effort to bring Descartes' advice, presented in the foregoing, down to classroom level and to put it into practice will avoid many of the usual pitfalls and difficulties.

First of all, the student should not start doing a problem before he has understood it. It can be checked to a certain extent whether the student has really understood the problem: he should be able to repeat the statement of the problem, point out the unknowns and the data, and explain the condition in his own words. If he can do all this reasonably well, he may proceed to the main business.

An equation expresses a *part of the condition*. The student should be able to tell which part of the condition is expressed by an equation that he brings forward—and which part is not yet expressed.

An equation expresses the *same quantity in two different ways*. The student should be able to tell which quantity is so expressed.

Of course, the student should possess the *relevant knowledge* without which he could not understand the problem. Many of the usual high school problems are "rate problems" (see the next three examples). Before he is called upon to do such a problem, the student should acquire in some form the idea of "rate," proportionality, uniform change.

(1) *One pipe can fill a tank in* 15 *minutes, another pipe can fill it in* 20 *minutes, a third pipe in* 30 *minutes. With all three pipes open, how long will it take to fill the empty tank?*

Let us assume that the tank contains g gallons of water when it is full. Then the rate of flow through the first pipe is

$$\frac{g}{15}$$

gallons per minute. Since

$$\text{amount} = \text{rate} \times \text{time}$$

the amount of water flowing through the first pipe in t minutes is

$$\frac{g}{15}\, t$$

If the three pipes together fill the empty tank in t minutes, the *amount of water in the full tank* can be expressed in two ways:

$$\frac{g}{15}\, t + \frac{g}{20}\, t + \frac{g}{30}\, t = g$$

The left-hand side shows the contribution of each pipe separately, the right-hand side the joint result of these three contributions.

Division by g yields the equation for the required time t:

$$\frac{t}{15} + \frac{t}{20} + \frac{t}{30} = 1$$

Of course, the derivation of the equation could be presented differently and the problem itself could be generalized and modified in various ways.

(2) *Tom can do a job in 3 hours, Dick in 4 hours, and Harry in 6 hours. If they do it together* (and do not delay each other), *how long does the job take?*

Tom can do $\frac{1}{3}$ of the whole job in one hour; we can also say that Tom is working at the rate of $\frac{1}{3}$ of the job per hour. Therefore, in t hours Tom does $t/3$ of the job. If the three boys work together and finish the work in t hours (and if they do not delay each other—a very iffy condition), the *full amount of work* can be expressed in two ways:

$$\frac{t}{3} + \frac{t}{4} + \frac{t}{6} = 1$$

in fact, the 1 on the right-hand side stands for "one full job."

This problem is almost identical with the foregoing (1), even numerically since

$$15:20:30 = 3:4:6$$

It is instructive to formulate a common generalization of both (using letters). It is also instructive to compare the solutions and weigh the advantage and disadvantage of introducing the quantity g into the solution (1).

(3) *A patrol plane flies* 220 *miles per hour in still air. It carries fuel for* 4 *hours of safe flying. If it takes off on patrol against a wind of* 20 *miles per hour, how far can it fly and return safely?*

It is understood that the wind is supposed to blow with unchanged intensity during the whole flight, that the plane travels in a straight line, that the time needed for changing direction at the furthest point is negligible, and so on. All word problems contain such unstated simplifying assumptions and demand from the problem solver some preliminary work of *interpretation* and *abstraction*. This is an essential feature of the word problems which is not always trivial and should be brought into the open, at least now and then.

The problem becomes more instructive if for the numbers

$$220 \qquad 20 \qquad 4$$

we substitute general quantities

$$v \qquad\quad w \qquad T$$

which denote the velocity of the plane in still air, the velocity of the wind, and the total flying time, respectively; these three quantities are the *data*. Let x stand for the distance flown in one direction, t_1 for the duration of the outgoing flight, t_2 for the duration of the homecoming flight; these three quantities are *unknowns*. It is useful to display some of these quantities in a neat arrangement:

	Going	Returning
Distance	x	x
Time	t_1	t_2
Velocity	$v - w$	$v + w$

(To fill out the last line we need, in fact, some "unsophisticated" knowledge of kinematics.) Now, as we should know,

$$\text{distance} = \text{velocity} \times \text{time}$$

We express each of the following three quantities in two ways:

$$x = (v - w)t_1$$
$$x = (v + w)t_2$$
$$t_1 + t_2 = T$$

We have here a system of three equations for the three unknowns x, t_1, and t_2. In fact, only x was required by the proposed problem; t_1 and t_2 are *auxiliary unknowns* which we have introduced in order to express neatly the whole condition. Eliminating t_1 and t_2, we find

$$\frac{x}{v - w} + \frac{x}{v + w} = T$$

and hence

$$x = \frac{(v^2 - w^2)T}{2v}$$

There is no difficulty in substituting numerical values for the data v, w, and T. It is more interesting to examine the result, and to check it by the *variation of the data*.

If $w = 0$, then $2x = vT$. This is right, obviously: the whole flight is now supposed to take place in still air.

If $w = v$, then $x = 0$. Again obvious: against a headwind with speed v, the plane cannot start at all.

If w increases from the value $w = 0$ to the value $w = v$, the distance x decreases steadily, according to the formula. And so, again, the formula agrees with what we can foresee without any algebra, just by visualizing the situation.

Working with numerical data instead of general data (letters) we would have missed this instructive discussion of the formula and the valuable checks of our result.   By the way, there are still other interesting checks.

(4) *A dealer has two kinds of nuts; one costs* 90 *cents a pound, the other* 60 *cents a pound.   He wishes to make* 50 *pounds of a mixture that will cost* 72 *cents a pound.   How many pounds of each kind should he use?*

This is a typical, rather simple "mixture problem."   Let us say that the dealer uses x pounds of nuts of the first kind, and y pounds of the second kind; x and y are the unknowns.   We can conveniently survey the unknowns and the data in the array:

	First kind	Second kind	Mixture
Price per pound	90	60	72
Weight	x	y	50

Express in two ways the *total weight of the mixture*:

$$x + y = 50$$

Then express in two ways the *total price of the mixture*:

$$90x + 60y = 72 \cdot 50$$

We have here a system of two equations for the two unknowns x and y. We leave the solution to the reader, who should have no trouble in finding the values

$$x = 20, \qquad y = 30$$

In passing from "numbers" to "letters" the reader obtains a problem which, as it will turn out later, has still other (and more interesting) interpretations.

2.5. Examples from geometry

We shall discuss just two examples.

(1) *A problem of geometric construction.*   It is possible to reduce any problem of geometric construction to a problem of algebra.   We cannot treat here the general theory of such reduction,[3] but here is an example.

A triangular area is enclosed by a straight line AB and two circular arcs, AC and BC.   The center of one circle is A, that of the other is B, and each circle passes through the center of the other.   Inscribe into this triangular area a circle touching all three boundary lines.

The desired configuration, Fig. 2.1, is sometimes seen in Gothic tracery.

Obviously, we can reduce the problem to the construction of one point: the center of the required circle.   One locus for this point is also obvious:

[3] See Courant-Robbins, pp. 117–140.

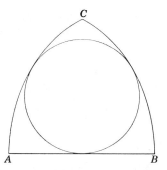

Fig. 2.1. From a Gothic window.

the perpendicular bisector of the segment AB which is a line of symmetry for the given triangular area. And so there remains to find another locus.

Keep only a part of the condition, drop the other part. We consider a (variable) circle touching not three, but only two boundary lines: the straight line AB and the circular arc BC; see Fig. 2.2. In order to find the locus of the center of this variable circle, we use analytic geometry. We let the origin of our rectangular coordinate system coincide with the point A, and let the x axis pass through the point B; see Fig. 2.2. Let x and y denote the coordinates of the center of the variable circle. Join this center to the two essential points of contact, one with the straight line AB, the other with the circular arc BC; see Fig. 2.2. The two radii have the same length which, therefore, can be expressed in two different manners (set $AB = a$):

$$y = a - \sqrt{x^2 + y^2}$$

By getting rid of the square root, we transform this equation into

$$x^2 = a^2 - 2ay$$

And so the locus of the center of the variable circle turns out to be a parabola—a locus of no immediate use in geometric constructions.

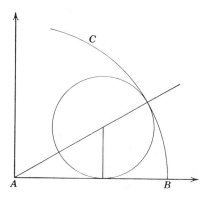

Fig. 2.2. We have dropped a part of the condition.

Yet the obvious locus mentioned at the beginning, the perpendicular bisector of AB, has the equation

$$x = \frac{a}{2}$$

which, combined with the equation of the parabola, yields the ordinate of the desired center of the circle.

$$y = \frac{3a}{8}$$

and this ordinate is easy to construct from the given length $a = AB$.

(2) *The analogue of Pythagoras' theorem in solid geometry.* Analogy is not unambiguous. There are various facts of solid geometry which can be quite properly regarded as analogous to the Pythagorean proposition. We arrive at such a fact if we regard a cube as analogous to a square, and a tetrahedron that we obtain by cutting off a corner of the cube by an oblique plane as analogous to a right triangle (which we obtain by cutting off a corner of a square by an oblique straight line). To the rectangular vertex of the right triangle there corresponds a vertex of the tetrahedron which we shall call a *trirectangular* vertex. In fact, the three edges of the tetrahedron starting from this vertex are perpendicular to each other, forming three right angles.

Pythagoras' theorem solves the following problem: In a triangle that possesses a rectangular vertex O, there are given the lengths a and b of the two sides meeting in O. Find the length c of the side opposite O.

We put the analogous problem: *In a tetrahedron that possesses a trirectangular vertex O, there are given the areas A, B, and C of the three faces meeting in O. Find the area D of the face opposite O.*

We are required to express D in terms of A, B, and C. It is natural to expect a formula analogous to Pythagoras' theorem

$$c^2 = a^2 + b^2$$

which solves the corresponding problem of plane geometry. A high school boy guessed

$$D^3 = A^3 + B^3 + C^3$$

This is a clever guess; the change in the exponent corresponds neatly to the transition from 2 to 3 dimensions.

(3) *What is the unknown?* —The area of a triangle, D.

How can you find such an unknown? *How can you get this kind of thing?* —The area of a triangle can be computed if the three sides are known, by Heron's formula. The area of our triangle is D. Let a, b,

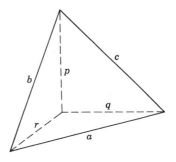

Fig. 2.3. Pythagoras in space.

and c denote the lengths of the sides, and set $s = (a + b + c)/2$; then

$$D^2 = s(s - a)(s - b)(s - c)$$

(This is a form of Heron's formula.)   Let us label the sides of D in the figure; see Fig. 2.3.

Fine!   But are the sides a, b, and c known? —No, but they are in right triangles; if the legs in these right triangles (labeled p, q, r in Fig. 2.3) were known, we could express a, b, and c:

$$a^2 = q^2 + r^2, \qquad b^2 = r^2 + p^2, \qquad c^2 = p^2 + q^2$$

That is good; but are p, q, and r themselves known? —No, but they are connected with the data, the areas A, B, and C:

$$\tfrac{1}{2}qr = A, \qquad \tfrac{1}{2}rp = B, \qquad \tfrac{1}{2}pq = C$$

That is right; but did you achieve anything useful? —I think I did. I now have 7 unknowns

$$D$$
$$a, \quad b, \quad c$$
$$p, \quad q, \quad r$$

but also a system of 7 equations to determine them.

(4) There is nothing wrong with our foregoing reasoning, under (3). We have attained the goal set by Descartes' Rule (freely rendered in sect. 2.3(3)): we have obtained a system with as many equations as there are unknowns.   There is just one thing: the number 7 may seem too high, to solve 7 equations with 7 unknowns may appear as too much trouble. And Heron's formula may not look too inviting.

If we feel so, we may prefer a new start.

What is the unknown? —The area of a triangle, D.

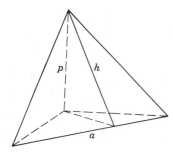

Fig. 2.4. A new departure.

How can you find such an unknown? *How can you get this kind of thing?* —The most familiar way to compute the area of a triangle is

$$D = \frac{ah}{2}$$

where a is the base, and h the altitude, of the triangle with area D; let us introduce h into the figure.   See Fig. 2.4.

Yes, we have seen a before; but what about h? —The height h of the triangle with area D can be computed from a suitable triangle, I hope. In fact, intersect the tetrahedron with a plane passing through h and the trirectangular vertex.   The intersection is a right triangle, its hypotenuse is h, one of its legs is p which we have seen before, and the other leg, say k, is the altitude perpendicular to the side a in the triangle with area A. Therefore,

$$h^2 = k^2 + p^2$$

Very good!   But what about k? —We can get it somehow.   In fact, express the area of the triangle in which, as I have just said, k is an altitude in two different manners:

$$\tfrac{1}{2}ak = A$$

Have you as many equations as you have unknowns? —I also have the former equations, and I have no time to count.   I now see my way, I think. Let me just combine what is before me:

$$
\begin{aligned}
4D^2 &= a^2h^2 \\
&= a^2(k^2 + p^2) \\
&= 4A^2 + a^2p^2 \\
&= 4A^2 + (r^2 + q^2)p^2 \\
&= 4A^2 + (rp)^2 + (pq)^2 \\
&= 4A^2 + 4B^2 + 4C^2
\end{aligned}
$$

Let me bring together the beginning and the end and get rid of that superfluous factor 4. Here it is:

$$D^2 = A^2 + B^2 + C^2$$

This result is, in fact, closely analogous to Pythagoras' theorem. That guess with the exponent 3 was clever—it turned out wrong, but this is not surprising. What is surprising is that the guess came so close to the truth.

It may be quite instructive to compare the two foregoing approaches to the same problem; they differ in various respects.

And could you imagine a different analogue to Pythagoras' theorem?

2.6. An example from physics

We start from the following question.

An iron sphere is floating in mercury. Water is poured over the mercury and covers the sphere. Will the sphere sink, rise, or remain at the same depth?

We have to compare two situations. In both cases the lower part of the iron sphere is immersed in (is under the level of) mercury. The upper part of the sphere is surrounded by air (or vacuum) in the first situation, and by water in the second situation. In which situation is the upper part (the one over the level of the mercury) a greater fraction of the whole volume?

This is a purely qualitative question. Yet we can give it a quantitative twist which renders it more precise (and accessible to algebra): *Compute the fraction of the volume of the sphere that is over the level of the mercury for both situations.*

(1) We can give a plausible answer to the qualitative question by purely intuitive reasoning, just by visualizing a *continuous transition* from one proposed situation to the other. Let us imagine that the fluid poured over the mercury and surrounding the upper part of the iron sphere *changes its density continuously.* To begin with, this imaginary fluid has density zero (we have just vacuum). Then the density increases; it soon attains the density of the air, and after a while the density of water. If you do not see yet how this change affects the floating sphere, let the *density increase still further.* When the density of that imaginary fluid attains the density of iron, the sphere must rise clear out of the mercury. In fact, if the density increased further ever so little, the sphere should pop up and emerge somewhat from that imaginary fluid.

It is natural to suppose that the position of the floating sphere, as the density of the imaginary fluid covering it, changes all the time in the *same direction.* Then we are driven to the conclusion that, in the transition from covering vacuum or air to covering water, the sphere will *rise.*

(2) In order to answer the quantitative question, we need the numerical values of the three specific gravities involved which are

	1.00	13.60	7.84
for	water	mercury	iron

respectively. Yet it is more instructive to substitute letters for these numerical data. Let

$$a \qquad\qquad b \qquad\qquad c$$

denote the specific gravity of the

upper fluid   lower fluid   floating solid

respectively. Let v denote the (given) total volume of the floating solid, x the fraction of v that is over the level separating the two fluids, and y the fraction under that level; see Fig. 2.5. Our data are a, b, c, and v, our unknowns x and y. It is understood that

$$a < c < b$$

We may express the total volume of the floating body in two different ways:

$$x + y = v$$

Now, we cannot proceed beyond this point unless we know the pertinent physical facts. The *relevant knowledge* that we should possess is the law of Archimedes which is usually expressed as follows: the floating body is buoyed up by a vertical force equal in magnitude to the weight of the displaced fluid. The sphere that we are considering displaces fluid in two layers. The weights of the displaced quantities are

	ax	and	by
in the	upper layer	and	lower layer

respectively. These two upward vertical forces must jointly balance the

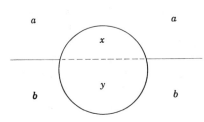

Fig. 2.5. Floating in two fluids.

weight of the floating sphere which we can, therefore, express in two different ways·

$$ax + by = cv$$

Now, we have obtained a system of two equations for our two unknowns x and y. Solving this system, we obtain

$$x = \frac{b - c}{b - a} v, \qquad y = \frac{c - a}{b - a} v$$

(3) Let us return to the original statement of the problem. In the first situation, if there is vacuum over the mercury

$$a = 0, \qquad b = 13.60, \qquad c = 7.84$$

which yield for the fraction of the iron sphere's volume over the level of mercury

$$x = 0.423v$$

In the second situation, when there is water over the mercury,

$$a = 1.00, \qquad b = 13.60, \qquad c = 7.84$$

which yield $$x = 0.457v$$

and the latter fraction is larger, which agrees with the conclusion of our intuitive reasoning.

The general formula (in letters) is, however, more interesting than any particular numerical result that we can derive from it. Especially, it fully substantiates our intuitive reasoning. In fact, keep b, c, and v constant and let a (the density of the upper layer) increase from

$$a = 0 \qquad \text{to} \qquad a = c$$

Then the denominator $b - a$ of x decreases steadily and so x, the fraction of v over the level of the mercury, increases steadily from

$$x = \frac{b - c}{b} v \qquad \text{to} \qquad x = v$$

2.7. An example from a puzzle

How can you make two squares from five? Fig. 2.6 shows a sheet of paper that has the shape of a cross; it is made up of five equal squares. Cut this sheet along a straight line in two pieces, then cut one of the pieces along another straight line again in two, so that the resulting three pieces, suitably fitted together, form two juxtaposed squares.

The cross in Fig. 2.6 is highly symmetric (it has a center of symmetry and four lines of symmetry). The two juxtaposed squares form a rectangle the length of which is twice the width. It is understood that the

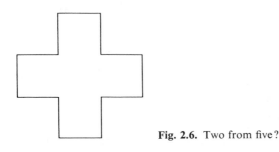

Fig. 2.6. Two from five?

three pieces into which the cross will be divided should fill up this rectangle without overlapping.

Could you solve a part of the problem?   Obviously, the area of the desired rectangle is equal to the area of the given cross, and so it equals $5a^2$ if a denotes the side of one of the five squares forming the cross.   Yet, having obtained its area, we can also find the sides of the rectangle.   Let x denote the length of the rectangle; then its width is $x/2$.   Express the area of the rectangle in two different ways; we obtain

$$x \cdot \frac{x}{2} = 5a^2$$

or
$$x^2 = 10a^2$$

from which we can find both sides of the rectangle.

We now have sufficient information about the rectangle, its shape and size, but the proposed puzzle is not yet solved: we still have to locate the two cuts in the cross.   Yet the expression for x obtained above may yield a hint, especially if we write it in this form:

$$x^2 = 9a^2 + a^2$$

With this indication, I leave the solution to the reader.

We can derive some useful suggestions from the foregoing treatment of the puzzle.

First, it shows that algebra can be useful even when it cannot solve the problem completely: it can solve a part of the problem and the solution of that part can facilitate the remaining work.

Second, the procedure that we have employed may impress us with its peculiar *expanding* pattern.   First, we have obtained only a small part of the solution: the area of the desired rectangle.   We have used, however, this small part to obtain a bigger part: the sides of the rectangle, and hence complete information about the rectangle.   Now, we are trying to use this bigger part to obtain a still bigger part which we may use afterwards, we hope, to obtain the full solution.

2.8. Puzzling examples

The problems that we have considered so far in this chapter are "reasonable." We are inclined to regard a problem as reasonable if its solution is uniquely determined. If we are seriously concerned with our problem, we wish to know (or guess) as early as possible whether it is reasonable or not. And so, from the outset, we may ask ourselves: *Is it possible to satisfy the condition? Is the condition sufficient to determine the unknown? Or is it insufficient? Or redundant? Or inconsistent?*

These questions are important.[4] We postpone a general discussion of their role, but it will be appropriate to look here at a couple of examples.

(1) *A man walked five hours, first along a level road, then up a hill, then he turned round and walked back to his starting point along the same route. He walks 4 miles per hour on the level, 3 uphill, and 6 downhill. Find the distance walked.*[5]

Is this a reasonable problem? Are the data *sufficient to determine the unknown? Or are they insufficient? Or redundant?*

The data seem to be *insufficient*: some information about the extent of the nonlevel part of the route seems to be lacking. If we knew how much time the man spent walking uphill, or downhill, there would be no difficulty. Yet without such information the problem appears indeterminate.

Still, let us try. Let

x stand for the total distance walked,
y for the length of the uphill walk.

The walk had four different phases:

level, uphill, downhill, level.

Now we can easily express the total time spent in walking in two different ways:

$$\frac{x/2 - y}{4} + \frac{y}{3} + \frac{y}{6} + \frac{x/2 - y}{4} = 5$$

Just one equation between two unknowns—it is insufficient. Yet, when we collect the terms, the coefficient of y turns out to be 0, and there remains

$$\frac{x}{4} = 5$$

$$x = 20$$

And so the data are sufficient to determine x, the only unknown required

---

[4] Cf. HSI, p. 122: Is it possible to satisfy the condition?
[5] Cf. "Knot I" of "A Tangled Tale" by Lewis Carroll.

by the statement of the problem. Hence, after all, the problem is not indeterminate: we were wrong.

(2) We were wrong, there is no denying, but we suspect that the author of the problem took pains to mislead us by a tricky choice of those numbers 3, 6, and 4. To get to the bottom of his trick, let us substitute for the numbers

$$3, \qquad\qquad 6, \qquad\qquad 4$$

the letters u, v, w

which stand for the pace of the walk

uphill,   downhill,   on the level,

respectively. We should reread the problem, with the letters just introduced substituted for the original numbers, and then express the total time spent in walking in two different ways, using the appropriate letters:

$$\frac{x/2 - y}{w} + \frac{y}{u} + \frac{y}{v} + \frac{x/2 - y}{w} = 5$$

or

$$\frac{x}{w} + \left(\frac{1}{u} + \frac{1}{v} - \frac{2}{w}\right) y = 5$$

We cannot determine x from this equation, unless the coefficient of y vanishes. And so the problem *is* indeterminate, unless

$$\frac{1}{w} = \frac{1}{2}\left(\frac{1}{u} + \frac{1}{v}\right)$$

If, however, the three rates of walking are chosen at random, they do not satisfy this relation, and so the problem *is* indeterminate. We were put in the wrong by a vicious trick!

(We can express the critical relation also by the formula

$$w = \frac{2uv}{u + v}$$

or by saying that the pace on the level is the harmonic mean of the uphill pace and the downhill pace.)

(3) *Two smaller circles are outside each other, but inside a third, larger circle. Each of these three circles is tangent to the two others and their centers are along the same straight line. Given r, the radius of the larger circle, and t, that piece of the tangent to the two smaller circles in their common point that lies within the larger circle. Find the area that is inside the larger circle but outside the two smaller circles.* See Fig. 2.7.

Is this a reasonable problem? Are the data *sufficient to determine the unknown? Or are they insufficient? Or redundant?*

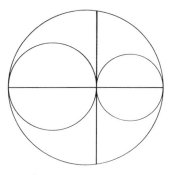

Fig. 2.7. Two data.

The problem seems perfectly reasonable. To determine the configuration of the three circles, it is both necessary and sufficient to know the radii of the two smaller circles, and any two independent data will be just as good. Now, the given r and t are obviously independent: we can vary one without changing the other (except for the inequality $t \leqq 2r$ which we may take for granted). Yes, the two data r and t seem to be just sufficient, neither insufficient nor redundant.

Therefore, let us settle down to work. Let A stand for the required area, x and y for the radii of the two smaller circles. Obviously

$$A = \pi r^2 - \pi x^2 - \pi y^2$$
$$2r = 2x + 2y$$

We have here two equations for our three unknowns, A, x, and y. In order to obtain a third equation, consider the right triangle inscribed in the larger circle, the base of which passes through the three centers and the opposite vertex of which is one of the endpoints of the segment t. The altitude in this triangle, drawn from the vertex of the right angle, is $t/2$; this altitude is a mean proportional (Euclid VI 13):

$$\left(\frac{t}{2}\right)^2 = 2x \cdot 2y$$

Now we have three equations. We rewrite the last two:

$$(x + y)^2 = r^2$$

$$2xy = \frac{t^2}{8}$$

Subtraction yields $x^2 + y^2$, and substitution into the first equation yields

$$A = \frac{\pi t^2}{8}$$

The data turned out to be *redundant*: of the two data, *t* and *r*, only the first is really needed, not the second.   We were wrong again.

The curious relation underlying the example just discussed was observed by Archimedes; see his *Works* edited by T. L. Heath, pp. 304–305.

Examples and Comments on Chapter 2

First Part

2.1. Bob has three dollars and one half in nickels and dimes, fifty coins altogether.   How many nickels has Bob, and how many dimes?   (Have you seen the same problem in a slightly different form?)

2.2. Generalize the problem of sect. 2.4(1) by passing from "numbers" to "letters" and considering several filling and emptying pipes.

2.3. Devise some other interpretation for the equation set up in sect. 2.4(2).

2.4. Find further checks for the solution of the flight problem of sect. 2.4(3).

2.5. In the "mixture problem" of sect. 2.4(4) substitute the letters

$$a \quad b \quad c \quad v$$

for the numerical data

$$90 \quad 60 \quad 72 \quad 50$$

respectively.   Read the problem after this substitution and set up the equations.   Do you recognize them?

2.6. Fig. 2.8 (which is different from, but related to, Fig. 2.1) shows another configuration frequently seen in Gothic tracery.

Find the center of the circle that touches four circular arcs forming a "curvilinear quadrilateral."

Two arcs have the radius *AB*, the center of one is *A*, that of the other *B*.

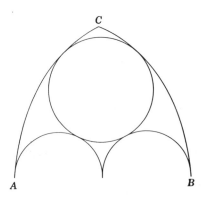

Fig. 2.8. From a Gothic window.

Two semicircles have the radius $AB/4$, the center of each lies on the line AB, one starts from the point A, the other from the point B, both end in the midpoint of the line AB where they are tangent to each other.

2.7. Carry through the plan devised in sect. 2.5(3); it should lead you to the same simple expression for D^2 in terms of A, B, and C that we have obtained in sect. 2.5(4) by other means.

2.8. Compare the approaches of sect. 2.5(3) and 2.5(4).   (Emphasize general viewpoints.)

2.9. Find the volume V of a tetrahedron that has a trirectangular vertex O, being given the areas A, B, and C of the three faces meeting in O.

2.10. *An analogue to Heron's theorem.*   Find the volume V of a tetrahedron that has a trirectangular vertex, being given the lengths a, b, and c of the sides of the face opposite the trirectangular vertex.

(If we introduce the quantity

$$S^2 = \frac{a^2 + b^2 + c^2}{2}$$

into the expression of V in an appropriate, symmetric way, the result assumes a form somewhat similar to Heron's formula.)

2.11. *Another analogue to Pythagoras' theorem.*   Find the length d of the diagonal of a box (a rectangular parallelepiped) being given p, q, and r, the length, the width, and the height of the box.

2.12. *Still another analogue to Pythagoras' theorem.*   Find the length d of the diagonal of a box, being given a, b, and c, the lengths of the diagonals of three faces having a corner in common.

2.13. *Another analogue to Heron's theorem.*   Let V denote the volume of a tetrahedron, a, b, and c the lengths of the three sides of one of its faces, and assume that each edge of the tetrahedron is equal in length to the opposite edge.   Express V in terms of a, b, and c.

2.14. Check the result of ex. 2.10 and that of ex. 2.13 by examining the degenerate case in which V vanishes.

2.15. Solve the puzzle proposed in sect. 2.7.   (The sides x and $x/2$ should result from the cuts—but how can you fit a segment of length x into the cross?)

2.16. Fig. 2.9 shows a sheet of paper of peculiar shape: it is a rectangle with a rectangular hole.   The sides of the outer rectangular boundary measure 9 and 12, those of the inner 1 and 8 units, respectively.   Both rectangular boundaries have the same center and their corresponding sides are parallel.   Cut this sheet along just two lines in two pieces that, fitted together, form a complete square.

(*a*) *Could you solve a part of the problem?*   How long is the side of the desired square?

(*b*) *Take the problem as solved.*   Imagine that the sheet is already cut into

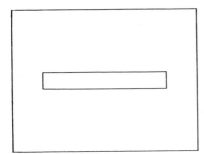

Fig. 2.9. By two cuts a square.

two pieces, the "left piece" and the "right piece." You keep the left piece where it is, and move the right piece into the desired position (where, with the other, it forms a complete square). Knowing the answer to (*a*), what kind of motion do you expect?

(*c*) *Guess a part of the solution.* The given sheet is symmetric with respect to its center and also with respect to two axes perpendicular to each other. Which kind of symmetry do you expect it to retain when cut along the two required lines?

Second Part

Some of the following examples are grouped according to subject matter which is hinted by an indication in front of the first example of the group (*Miscellaneous, Plane Geometry, Solid Geometry,* etc.) Some examples are followed by the name of Newton or Euler in parentheses; these are taken from the following sources, respectively:

Universal Arithmetick: or, a Treatise of Arithmetical Composition and Resolution. Written in Latin by Sir Isaac Newton. Translated by the late Mr. Ralphson. London, 1769. (Examples marked "After Newton" are derived from the same source, but some change is introduced into the formulation or into the numerical data.)

Elements of Algebra. By Leonard Euler. Translated from the French. London, 1797. (In fact, Euler's Algebra was originally written in German.)

Isaac Newton (1643–1727) is regarded by many as the greatest man of science who ever lived. His work encompasses the principles of mechanics, the theory of universal gravitation, the differential and integral calculus, theoretical and experimental optics, and several minor items each of which would be sufficient to secure him a place in the history of science. Leonard Euler (1707–1783) is also one of the very great; he left his traces on almost every branch of mathematics and on several branches of physics; he contributed more than anybody else to the development of the calculus discovered by Newton and Leibnitz. Observe that such great men did not find it beneath their dignity to explain and illustrate at length the application of equations to the solution of "word problems."

2.17. *Miscellaneous.* A mule and an ass were carrying burdens amounting to some hundred weight. The ass complained of his, and said to the mule: "I need only one hundred weight of your load, to make mine twice as heavy as yours." The mule answered: "Yes, but if you gave me a hundred weight of yours, I should be loaded three times as much as you would be."

How many hundred weight did each carry? (Euler)

2.18. When Mr. and Mrs. Smith took the airplane, they had together 94 pounds of baggage. He paid $1.50 and she paid $2 for excess weight. If Mr. Smith made the trip by himself with the combined baggage of both of them, he would have to pay $13.50. How many pounds of baggage can one person take along without charge?

2.19. A father who has three sons leaves them 1600 crowns. The will precises, that the eldest shall have 200 crowns more than the second, and the second shall have 100 crowns more than the youngest. Required the share of each. (Euler)

2.20. A father leaves four sons, who share his property in the following manner:

The first takes the half of the fortune, minus 3000 livres.

The second takes the third, minus 1000 livres.

The third takes exactly the fourth of the property.

The fourth takes 600 livres and the fifth part of the property.

What was the whole fortune, and how much did each son receive? (Euler)

2.21. A father leaves at his death several children, who share his property in the following manner:

The first receives a hundred crowns and the tenth part of what remains.

The second receives two hundred crowns and the tenth part of what remains.

The third takes three hundred crowns and the tenth part of what remains.

The fourth takes four hundred crowns and the tenth part of what remains, and so on.

Now it is found at the end that the property has been divided equally among all the children. Required, how much it was, how many children there were, and how much each received. (Euler)

2.22. Three persons play together; in the first game, the first loses to each of the other two as much money as each of them has. In the next, the second person loses to each of the other two as much money as they have already. Lastly, in the third game, the first and second person gain each from the third as much money as they had before. They then leave off and find that they have all an equal sum, namely, 24 louis each. Required, with how much money each sat down to play. (Euler)

2.23. Three Workmen can do a Piece of Work in certain Times, viz. *A* once in 3 Weeks, *B* thrice in 8 Weeks, and *C* five Times in 12 Weeks. It is desired to know in what Time they can finish it jointly. (Newton)

2.24. The Forces of several Agents being given, to determine the Time wherein they will jointly perform a given Effect. (Newton)

2.25. One bought 40 Bushels of Wheat, 24 Bushels of Barley, and 20 Bushels of Oats together for 15 Pounds 12 Shillings.

Again, he bought of the same Grain 26 Bushels of Wheat, 30 Bushels of Barley, and 50 Bushels of Oats together for 16 Pounds.

And thirdly, he bought the like Kind of Grain, 24 Bushels of Wheat, 120 Bushels of Barley, and 100 Bushels of Oats together for 34 Pounds.

It is demanded at what Rate a Bushel of each of the Grains ought to be valued. (Newton)

2.26. (Continued) Generalize.

2.27. If 12 Oxen eat up $3\frac{1}{3}$ Acres of Pasture in 4 Weeks, and 21 Oxen eat up 10 Acres of like Pasture in 9 Weeks; to find how many Oxen will eat up 24 Acres in 18 Weeks. (Newton)

2.28. *An Egyptian problem.* We take a problem from the Rhind Papyrus which is the principal source of our knowledge of ancient Egyptian mathematics. In the original text, the problem is about hundred loaves of bread which should be divided between five people, but a major part of the condition is not expressed (or not clearly expressed); the solution is attained by "groping": by a guess, and a correction of the first guess.[6]

Here follows the Egyptian problem reduced to abstract form and modern terminology; the reader should go one step further and reduce it to equations: An arithmetic progression has five terms. The sum of all five terms equals 100, the sum of the three largest terms is seven times the sum of the two smallest terms. Find the progression.

2.29. A geometric progression has three terms. The sum of these terms is 19 and the sum of their squares is 133. Find the terms. (After Newton.)

2.30. A geometric progression has four terms. The sum of the two extreme terms is 13, the sum of the two middle terms is 4. Find the terms. (After Newton.)

2.31. Some merchants have a common stock of 8240 crowns; each contributes to it forty times as many crowns as there are partners; they gain with the whole sum as much per cent as there are partners; dividing the profit, it is found that, after each has received ten times as many crowns as there are partners, there remain 224 crowns. Required the number of partners. (Euler)

2.32. *Plane geometry.* Inside a square with side a there are five nonoverlapping circles with the same radius r. One circle is concentric with the square and touches the four other circles each of which touches two sides of the square (is pushed into a corner). Express r in terms of a.

'**2.33.** *Newton on setting up equations in geometric problems.* If the Question be of an Isosceles Triangle inscribed in a Circle, whose Sides and Base are to be compared with the Diameter of the Circle, this may either be proposed of

---

[6] Cf. J. R. Newman, *The World of Mathematics*, vol. 1, pp. 173–174.

the Investigation of the Diameter from the given Sides and Base, or of the Investigation of the Basis from the given Sides and Diameter; or lastly, of the Investigation of the Sides from the given Base and Diameter; but however it be proposed, it will be reduced to an Equation by the same...Analysis. (Newton)

Let d, s, and b stand for the length of the diameter, that of the side, and that of the base, respectively (so that the three sides of the triangle are of length s, s, and b, respectively), and find an equation connecting d, s, and b which solves all three problems: the one in which d, the other in which b, and the third in which s is the unknown. (There are always two data.)

2.34. (Continued) Examine critically the equation obtained as answer to ex. 2.33. (*a*) Are all three problems equally easy? (*b*) The equation obtained yields positive values in the three cases mentioned (for d, b, and s, respectively) only under certain conditions: do these conditions correctly correspond to the geometric situation?

2.35. The four points G, H, V, and U are (in this order) the four corners of a quadrilateral. A surveyor wants to find the length $UV = x$. He knows the length $GH = l$ and measures the four angles

$$\angle GUH = \alpha, \qquad \angle HUV = \beta, \qquad \angle UVG = \gamma, \qquad \angle GVH = \delta.$$

Express x in terms of α, β, γ, δ, and l.

(Remember ex. 2.33 and follow Newton's advice: choose those "Data and Quaesita by which you think it is most easy for you to make out your Equation.")

2.36. The Area and the Perimeter of a right-angled Triangle being given, to find the Hypothenuse. (Newton)

2.37. Having given the Altitude, Base and Sum of the Sides, to find the Triangle. (Newton)

2.38. Having given the Sides of any Parallelogram and one of the Diagonals, to find the other Diagonal. (Newton)

2.39. The triangle with the sides a, a, and b is isosceles. Cut off from it two triangles, symmetric to each other with respect to the altitude perpendicular to the base b, so that the remaining symmetric pentagon is *equilateral*. Express the side x of the pentagon in terms of a and b.

(This problem was discussed by Leonardo of Pisa, called Fibonacci, with the numerical values $a = 10$ and $b = 12$.)

2.40. A hexagon is equilateral, its sides are all of the same length a. Three of its angles are right angles; they alternate with three obtuse angles. (If the hexagon is $ABCDEF$, the angles at the vertices A, C, and E are right angles, those at the vertices B, D, and F obtuse.) Find the area of the hexagon.

2.41. An equilateral triangle is inscribed in a larger equilateral triangle so that corresponding sides of the two triangles are perpendicular to each other.

Thus the whole area of the larger triangle is divided into four pieces each of which is a fraction of the whole area.   Which fraction?

2.42. Divide a given triangle by three straight cuts into seven pieces four of which are triangles (and the remaining three pentagons).   One of the triangular pieces is included by the three cuts, each of the three other triangular pieces is included by a certain side of the given triangle and two cuts.   Choose the three cuts so that the four triangular pieces turn out to be congruent.   Which fraction of the area of the given triangle is the area of a triangular piece in this dissection?

(It may be advantageous to examine first a particular shape of the given triangle for which the solution is particularly easy.)

2.43. The point P is so located in the interior of a rectangle that the distance of P from a corner of the rectangle is 5 yards, from the opposite corner 14 yards, and from a third corner 10 yards.   What is the distance of P from the fourth corner?

2.44. Given the distances a, b, and c of a point in the plane of a square from three vertices of the square; a and c are distances from opposite vertices.

(I) Find the side s of the square.

(II) Test your result in the following four particular cases:

    (1) $a = b = c$
    (2) $b^2 = 2a^2 = 2c^2$
    (3) $a = 0$
    (4) $b = 0$.

2.45. Pennies (equal circles) are arranged in a regular pattern all over a very, very large table (the infinite plane).   We examine two patterns.

In the first pattern, each penny touches four other pennies and the straight lines joining the centers of the pennies in contact dissect the plane into equal squares.

In the second pattern, each penny touches six other pennies and the straight lines joining the centers of the pennies in contact dissect the plane into equal equilateral triangles.

Compute the percentage of the plane covered by pennies (circles) for each pattern.

(For the first pattern see Fig. 3.9, for the second Fig. 3.8.)

2.46. *Solid geometry.*   Inside a cube (a cubical box) with edge a there are nine nonoverlapping spheres (nine balls packed into the box) with the same radius r.   One sphere is concentric with the cube and touches the eight other spheres (the balls are tightly packed), each of which touches three faces of the cube (is pushed into a corner).   Express r in terms of a.

(Or a in terms of r: make a box when you have the balls.   There is an analogous problem in plane geometry, see ex. 2.32: can you use its result or method?)

2.47. Devise a problem of solid geometry analogous to ex. 2.43.

2.48. A pyramid is called "regular" if its base is a regular polygon and the foot of its altitude is the center of its base.

All five faces of a regular pyramid with square base are of the same area. Given the height h of the pyramid, find the area of its surface.

2.49. (Continued) There is some analogy between a regular pyramid and an isosceles triangle; at any rate, if the number of the lateral faces of the pyramid is given, both figures, the solid and the plane, depend on two data.

Devise further problems about regular pyramids.

2.50. Devise a proposition of solid geometry analogous to the result of ex. 2.38. (Ex. 2.12 may serve as a stepping stone to a generalization.)

2.51. Find the area of the surface of the tetrahedron considered in ex. 2.13, being given a, b, and c. (Do you see some analogy?)

2.52. Of twelve congruent equilateral triangles eight are the faces of a regular octahedron and four the faces of a regular tetrahedron. Find the ratio of the volume of the octahedron to the volume of the tetrahedron.

2.53. A triangle rotating first about its side a, then about its side b, and finally about its side c, generates three solids of revolution. Find the ratio of the volumes and that of the surface-areas of these three solids.

2.54. *An inequality.* A rectangle and an isosceles trapezoid are in the relative situation shown in Fig. 2.10: they have a common (vertical) line of symmetry, the same height h and the same area; if $2a$ and $2b$ are the lengths of the lower and upper base of the trapezoid, respectively, the base of the rectangle is of length $a + b$. Rotated about the common line of symmetry, the rectangle describes a cylinder and the trapezoid the frustum of a cone. Which one of these two solids has the greater volume? (Your answer may be suggested by geometry, but should be proved by algebra.)

2.55. *Spherometer.* There are four points A, B, C, and D on the surface of

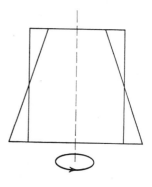

Fig. 2.10. Rotate it!

a sphere. The points A, B, and C form an equilateral triangle with side a. A perpendicular drawn from D to the plane of $\triangle ABC$ has the length h, and its foot is the center of $\triangle ABC$.

Given a and h, compute the radius R of the sphere.

(This geometric situation underlies the construction of the spherometer which is an instrument to determine the curvature of a lens. On the spherometer, A, B, and C are the endpoints of three fixed parallel "legs," whereas the endpoint of a fourth, movable "leg" is screwed into the position D and the distance h is measured by the revolutions of the screw.)

2.56. Graphic time table. In problems about several objects (material points) moving along the same path it is often advantageous to introduce a rectangular coordinate system in which the abscissa represents t, the time, and the ordinate represents s, the distance, measured along the path from a fixed point. To show the use of this device, we reconsider the problem that we have treated in detail in sect. 2.4(3).

We measure the time t and the distance s from the starting time and the starting point of the airplane, respectively. Thus, when the airplane has traveled t hours on its outgoing flight, its distance from its starting point is

$$s = (v - w)t$$

This equation, with fixed v and w and variable s and t, is represented in our coordinate system by a straight line with slope $v - w$ (the velocity) that passes through the origin [the point $(0, 0)$ which represents the start of the airplane]. On the returning flight, the distance s and the time t are connected by the equation

$$s = -(v + w)(t - T)$$

of a straight line with slope $-(v + w)$ passing through the point $(T, 0)$ (which represents the coming back of the airplane to its starting point at the prescribed time T).

The intersection of the two lines represents the point (in space and time) that belongs both to the outgoing, and to the returning flight, where the airplane reverses its direction. At this point, both expressions are valid for s simultaneously, and so

$$(v - w)t = -(v + w)(t - T)$$

This yields

$$t = \frac{(v + w)T}{2v}$$

and therefore (from either expression for s)

$$s = \frac{(v^2 - w^2)T}{2v}$$

as expression for the distance of the farthest point reached by the airplane. This is the result we found in sect. 2.4(3) (with x instead of s).

In Fig. 2.11 (disregard the dotted segments) the flight of the airplane is

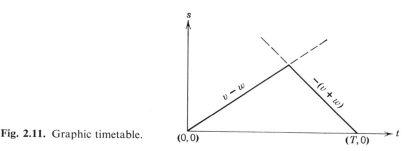

Fig. 2.11. Graphic timetable.

represented by a line consisting of two straight pieces; these pieces form an angle at the point whose ordinate represents the greatest distance reached by the airplane.   The whole line tells the whole story of the flight; it shows where the airplane was at any given time and when it arrived at any given point; this line is appropriately called the *graphic time table* of the flight (of the motion considered).

2.57. Two Post-Boys A and B, at 59 Miles Distance from one another, set out in the Morning in order to meet.   And A rides 7 Miles in two Hours, and B 8 Miles in three Hours, and B begins his Journey one Hour later than A; to find what Number of Miles A will ride before he meets B.   (Newton)

2.58. (Continued) Generalize.

2.59. Al and Bill live at opposite ends of the same street.   Al had to deliver a parcel at Bill's home, Bill one at Al's home.   They started at the same moment; each walked at constant speed and returned home immediately after leaving the parcel at its destination.   They met the first time face to face, coming from opposite directions, at the distance of a yards from Al's home and the second time at the distance of b yards from Bill's home.
(1) How long is the street?
(2) If $a = 300$ and $b = 400$ who walks faster?

2.60. Bob, Peter, and Paul travel together.   Peter and Paul are good hikers; each walks p miles per hour.   Bob has a bad foot and drives a small car in which two people can ride, but not three; the car covers c miles per hour. The three friends adopted the following scheme: they start together, Paul rides in the car with Bob, Peter walks.   After a while, Bob drops Paul who walks on; Bob returns to pick up Peter, and then Bob and Peter ride in the car till they overtake Paul.   At this point, they change: Paul rides and Peter walks just as they started and the whole procedure is repeated as often as necessary.
(1) How much progress (how many miles) does the company make per hour?
(2) Through which fraction of the travel time does the car carry just one man?

2.61. (Continued) Generalize: Bob, with his bad foot and small car, makes a similar arrangement with n friends, A, B, C,..., and L (instead of two) all walking at the pace p miles per hour.

(Draw a graphic time table for $n = 3$. Check the extreme cases $p = 0$, $p = c$, $n = 1$, $n = \infty$.)

2.62. A Stone falling down into a Well, from the Sound of the Stone striking the Bottom, to determine the Depth of the Well. (Newton)

You have to measure the time T between two moments: the first when you let the stone go, and the second when you hear it striking the bottom. You have to know too:

c the velocity of sound and

g the gravitational acceleration.

Being given T, c, and g, find d, the depth of the well.

2.63. To determine the Position of a Comet's Course that moves uniformly in a right Line from three Observations. (Newton)

Let O be the eye of the observer, A, B, and C the place of the comet in the first, the second, and the third observation, respectively. From observations, we know the angles

$$\angle AOB = \omega, \qquad \angle AOC = \omega'$$

and the times, t and t', between the first and the second, and between the first and the third, observations, respectively. From the assumption of uniform motion

$$\frac{AB}{AC} = \frac{t}{t'}$$

Being given ω, ω', t, and t', find $\beta = \angle ABO$.

(Express some trigonometric function of β, for instance, $\cot \beta$, in terms of ω, ω', t, and t'.)

2.64. *As many equations as unknowns.* Find x, y, and z from the system of three equations

$$3x - y - 2z = a$$
$$-2x + 3y - z = b$$
$$-x - 2y + 3z = c$$

a, b, and c are given.

(*Is it possible to satisfy the condition? Is the condition sufficient to determine the unknowns?*)

2.65. *More equations than unknowns.* Find three numbers p, q, and r so that the equation

$$x^4 + 4x^3 - 2x^2 - 12x + 9 = (px^2 + qx + r)^2$$

holds identically for variable x.

(This problem requires the "exact" extraction of a square root of a given polynomial of degree 4, which may be possible in the present case, yet usually it is not. Why not?)

2.66. Show that it is impossible to find (real or complex) numbers a, b, c, A, B, and C such that the equation

$$x^2 + y^2 + z^2 = (ax + by + cz)(Ax + By + Cz)$$

holds identically for independently variable x, y, and z.

2.67. *Fewer equations than unknowns.* A certain person buys hogs, goats, and sheep, to the number of 100, for 100 crowns; the hogs cost him $3\frac{1}{2}$ crowns a piece, the goats $1\frac{1}{3}$ crown, and the sheep $\frac{1}{2}$ a crown; how many had he of each? (Euler)

Euler solves this problem by a procedure which he calls *Regula Caeci* ("Blind Man's Rule") as follows. Let x, y, and z stand for the number of hogs, goats, and sheep bought, respectively; x, y, and z should be, of course, *positive integers*. Expressing first the total number, then the total price, of the animals bought, we obtain

$$x + y + z = 100$$
$$21x + 8y + 3z = 600$$

the second equation has been slightly (and advantageously) transformed. If we eliminate z and solve the resulting equation for y, we obtain

$$y = 60 - \frac{18x}{5}$$

whence we conclude that

$$\frac{x}{5} = t$$

must be a positive integer.

Finish the solution.

2.68. A coiner (we hope that the coins he makes are not counterfeit) has three kinds of silver, the first of 7 ounces, the second of $5\frac{1}{2}$ ounces, the third of $4\frac{1}{2}$ ounces fine per marc (a marc is 8 ounces), and he wishes to form a mixture of the weight of 30 marcs at 6 ounces (fine per marc): how many marcs of each sort must he take? (Euler; additions in parenthesis.)

It is understood that a solution *in integers* is required. A problem is called a *Diophantine problem,* when its condition admits only integral values for the unknowns.

2.69. There is a number (an integer) which yields a square if you add it to 100 and another square if you add it to 168. Which number is it?

2.70. Bob's stamp collection consists of three books. Two tenths of his stamps are in the first book, several sevenths in the second book, and there are 303 stamps in the third book. How many stamps has Bob? (Is the condition sufficient to determine the unknown?)

2.71. A certain make of ball point pen was priced 50 cents in the store opposite the high school but found few buyers. When, however, the store had reduced the price, the whole remaining stock was sold for $31.93. What was the reduced price? (Is the condition sufficient to determine the unknown?)

2.72. *Descartes' Rules.* The work of the great philosopher and mathematician René Descartes, quoted in sect. 2.1, is of particular importance for our study.

The "Rules" were found unfinished in Descartes' manuscripts and published posthumously. Descartes planned 36 sections, but the work consists actually of 18 sections in more or less finished form and of the summaries of three more; the rest was very probably never written. The first twelve sections treat of mental operations useful in solving problems, the next twelve discuss "perfectly understood" problems, and the last twelve sections were intended to deal with "imperfectly understood" problems.[7]

Each section begins with a "Rule," a terse advice to the reader, and the section motivates, explains, exemplifies, or states in greater detail, the idea summarized by the Rule. We shall quote any passage of the section following a certain Rule by giving the number of that Rule.[8]

The words of Descartes will be a valuable guide for us, but the reader would offend the memory of the originator of "Cartesian doubt" if he believed anything that Descartes said merely because Descartes said it. In fact, the reader should not believe what the present author says, or any other author says, or trust too much his own hasty impressions. After having given a fair hearing to the author, the reader should accept only such statements which he can see pretty clearly by himself or of which his well-digested experience can convince him. Doing so, he will act according to the spirit of Descartes' "Rules."

2.73. *Strip the problem.* We quote Descartes: *Strip the question of superfluous notions and reduce it to its simplest form.*[9] This advice is applicable to problems of all kinds and on all levels. Yet let us be specific. Take a usual type of classroom problem, a "rate problem" about motion [such as that discussed in sect. 2.4(3)]. The moving object may be described by the problem as a person, a car, a train, or an airplane. In fact, it makes little difference: in solving the problem on this primitive level, we actually treat the object as a material point moving uniformly along a straight path. Such simplification may be quite reasonable in some cases, ridiculous in other cases. It is certain, however, that we cannot avoid some degree of *simplification* or *abstraction* when we are reducing a problem about real objects to a mathematical problem. For, obviously, a mathematical problem deals with abstractions; it is concerned with real objects only indirectly by virtue of a previous passage from the concrete to the abstract.

Engineers and physicists, in handling their problems, have to devote serious thought to the question, how far should the abstraction and simplification go, which details can be neglected, which small effects disregarded. They have to avoid two opposite dangers: they should not render the mathematical task too formidable, yet they should not oversimplify the physical situation.

---

[7] Rule XII, pp. 429–430. "Perfectly understood problems" can be, "imperfectly understood problems" cannot be, immediately reduced to purely mathematical problems—this seems to be the most relevant difference.

[8] As it was done in footnote 7. We quote the standard edition of the *Œuvres de Descartes* by Charles Adam and Paul Tannery, vol. X, which contains the Latin original of the "Rules," "Regulae ad directionem ingenii" on pp. 359–469. Other quotations from Descartes refer to the same edition, but give also the volume.

[9] Rule XIII, p. 430.

We get a foretaste of their dilemma in dealing with the most primitive word problems. In fact, getting accustomed to a degree of simplification is one of the difficulties that the beginner has to master, and if this difficulty is never brought out into the open, it may become much worse.

There is a related difficulty. The problems proposed in textbooks tacitly assume certain simplifications; real rates may be variable, but all the rates considered in elementary textbooks are constant. The beginner has to become familiar with such tacit assumptions, he has to learn the proper *interpretation* of certain conventionally abbreviated formulations; also this point should be openly discussed, at least now and then.

(There is another related point which is so important that we must mention it, but so far away from our main line of inquiry that we must mention it briefly: We should relate to the extent of simplification and neglect that enters into the formulation of the problem the precision to which we carry the numerical computation of the unknown. We may sin by transgression or omission if we compute more or less decimals than the data warrant. There are few occasions to illustrate this important point on the elementary level, but those few occasions should not be missed.)

For an instructive and not too obvious illustration of some of the points here discussed see the author's paper no. 18 quoted in the Bibliography.

2.74. *Relevant knowledge. Mobilization and organization.* Obviously, we cannot translate a physical problem into equations unless we know (or rather assume as known) the pertinent physical facts. For instance, we could not have solved the problem treated in sect. 2.6 without the knowledge of the law of Archimedes.

In translating a geometrical problem into equations, we use pertinent geometrical facts. For instance, we may apply the theorem of Pythagoras, or the proportionality of sides in similar triangles, or expressions for areas or volumes, and so on.

If we do not possess the relevant knowledge, we cannot translate the problem into equations. Yet even if we have once acquired such knowledge, it may not be present to our mind in the moment we need it; or if, by some chance, it is present we may fail to recognize its utility for the purpose at hand. We can see here clearly a point which should be obvious: It is not enough to possess the needed relevant knowledge in some dormant state; we have to recall it when needed, revive it, *mobilize* it, and make it available for our purpose, adapt it to our problem, *organize* it.

As our work progresses, our conception of the problem continually changes: more and more auxiliary lines appear in the figure, more and more auxiliary unknowns appear in our equations, more and more materials are mobilized and introduced into the structure of the problem till eventually the figure is saturated, we have just as many equations as unknowns, and originally present and successively mobilized materials are merged into an organic whole. (Sect. 2.5(3) yields a good illustration.)

2.75. *Independence and consistence.* Descartes advises us to set up as many equations as there are unknowns.[10]   Let n stand for the number of unknowns, and x_1, x_2, ..., x_n for the unknowns themselves; then we can write the desired system of equations in the form

$$r_1(x_1, x_2, \ldots, x_n) = 0$$
$$r_2(x_1, x_2, \ldots, x_n) = 0$$
$$\cdot \quad \cdot \quad \cdot \quad \cdot \quad \cdot \quad \cdot \quad \cdot$$
$$r_n(x_1, x_2, \ldots, x_n) = 0$$

where $r_1(x_1, x_2, \ldots, x_n)$ indicates a polynomial in x_1, x_2, ..., x_n, and the left-hand sides of the following equations must be similarly interpreted.   Descartes advises us further to reduce this system of equations to one final equation.[11] This is possible "in general" (usually, in the regular case, ...) and "in general" the system has a solution (a system of numerical values, real or complex, for x_1, x_2, ..., x_n satisfying simultaneously the n equations) and only a finite number of solutions (this number depends on the degree of the final equation).

Yet there are exceptional (irregular) cases; we cannot discuss them here in full generality, but let us look at a simple example.

We consider a system of three linear equations with three unknowns:

$$a_1 x + b_1 y + c_1 z + d_1 = 0$$
$$a_2 x + b_2 y + c_2 z + d_2 = 0$$
$$a_3 x + b_3 y + c_3 z + d_3 = 0$$

x, y, z are the unknowns, the twelve symbols a_1, b_1, ..., d_3 represent given real numbers.   We assume that a_1, b_1, and c_1 are not all equal to 0, and similarly for a_2, b_2, c_2 and a_3, b_3, c_3.   Under these assumptions, each of the three equations is represented by a plane if we consider x, y, and z as rectangular coordinates in space; and so the system of three equations is represented by a configuration of three planes.

Concerning the system of our three linear equations, we distinguish various cases.

(1) There exists no solution, that is, no system of three real numbers x, y, z satisfying all three equations simultaneously.   In this case we say that the equations are *incompatible*, and their system is *inconsistent* or *self-contradictory*.

(2) There are infinitely many solutions; then we say that the system is *indeterminate*.   This will be the case if all triplets of numbers x, y, and z that satisfy two of our equations, satisfy also the third, in which case we say that this third equation is *not independent* of the other two.

(3) There is just one solution: the equations are *independent*, the system is *consistent* and *determinate*.

Visualize the corresponding cases for the configuration of the three planes and describe them.

2.76. *Unique solution.   Anticipation.*   If a chess problem or a riddle has more than one solution, we regard it as imperfect.   In general, we seem to

---

[10] Rule XIX, p. 468.
[11] Rule XXI, p. 469.

have a natural preference for problems with a unique solution, they may appear to us as the only "reasonable" or the only "perfect" problems. Also Descartes seems to have shared this preference; he says: "*A perfect question, as we wish to define it, is completely determinate and what it asks can be derived from the data.*"[12]

Is the solution of our problem unique? *Is the condition sufficient to determine the unknown?* We often ask these questions fairly soon (and it is advisable to ask them so) when we are working on a problem. In asking them so early, we do not really need, or expect, a final answer (which will be given by the complete solution), only a preliminary answer, an *anticipation*, a guess (which may deepen our understanding of the problem). This preliminary answer often turns out to be right, but now and then we may fall into a trap as the examples of sect. 2.8 show.[13]

By the way, the unknown may be obtained as a root of an equation of degree n with n different roots where $n > 1$, and the solution may still be unique, if the condition requires a real, or a positive, or an integral value for the unknown and the equation in question has just one root of the required kind.

2.77. *Why word problems?* I hope that I shall shock a few people in asserting that the most important single task of mathematical instruction in the secondary schools is to teach the setting up of equations to solve word problems. Yet there is a strong argument in favor of this opinion.

In solving a word problem by setting up equations, the student *translates* a real situation into mathematical terms: he has an opportunity to experience that mathematical concepts may be related to realities, but such relations must be carefully worked out. Here is the first opportunity afforded by the curriculum for this basic experience. This first opportunity may be also the last for a student who will not use mathematics in his profession. Yet engineers and scientists who will use mathematics professionally, will use it mainly to translate real situations into mathematical concepts. In fact, an engineer makes more money than a mathematician and so he can hire a mathematician to solve his mathematical problems for him; therefore, the future engineer need not study mathematics to *solve* problems. Yet, there is one task for which the engineer cannot fully rely on the mathematician: the engineer must know enough mathematics to *set up* his problems in mathematical form. And so the future engineer, when he learns in the secondary school to set up equations to solve "word problems," has a first taste of, and has an opportunity to acquire the attitude essential to, his principal professional use of mathematics.

2.78. *More problems.* Devise some problems similar to, but different from, the problems proposed in this chapter—especially such problems as you can solve.

[12] Rule XIII, p. 431.

[13] For more examples of this kind see MPR, vol. 1, pp. 190–192, and pp. 200–202, problems 1–12.

CHAPTER 3

RECURSION

3.1. The story of a little discovery

There is a traditional story about the little Gauss who later became the great mathematician Carl Friedrich Gauss. I particularly like the following version which I heard as a boy myself, and I do not care whether it is authentic or not.

"This happened when little Gauss still attended primary school. One day the teacher gave a stiff task: To add up the numbers 1, 2, 3, and so on, up to 20. The teacher expected to have some time for himself while the boys were busy doing that long sum. Therefore, he was disagreeably surprised as the little Gauss stepped forward when the others had scarcely started working, put his slate on the teacher's desk, and said, 'Here it is.' The teacher did not even look at little Gauss's slate, for he felt quite sure that the answer must be wrong, but decided to punish the boy severely for this piece of impudence. He waited till all the other boys had piled their slates on that of little Gauss, and then he pulled it out and looked at it. What was his surprise as he found on the slate just one number and it was the right one! What was the number and how did little Gauss find it?"

Of course, we do not know exactly how little Gauss did it and we shall never be able to know. Yet we are free to imagine something that looks reasonable. Little Gauss was, after all, just a child, although an exceptionally intelligent and precocious child. It came to him probably more naturally than to other children of his age to grasp the purpose of a question, to pay attention to the essential point. He just represented to himself more clearly and distinctly than the other youngsters *what is required*: to find the sum

60

1

2

3

and so on

.

.

.

20

―

He must have seen the problem differently, more completely, than the others, perhaps with some variations as the successive diagrams *A, B, C, D,* and *E* of Fig. 3.1 indicate.   The original statement of the problem emphasizes the beginning of the series of numbers that should be added (*A*). Yet we could also emphasize the end (*B*) or, still better, emphasize the beginning and the end *equally* (*C*).   Our attention may attach itself to the two extreme numbers, the very first and the very last, and we may observe some particular relation between them (*D*).   Then the idea appears (*E*).   Yes, numbers equally removed from the extremes add up all along to the same sum

$$1 + 20 = 2 + 19 = 3 + 18 = \cdots = 10 + 11 = 21$$

and, therefore, the grand total of the whole series is

$$10 \times 21 = 210$$

Did little Gauss really do it this way?   I am far from asserting that. I say only that it would be natural to solve the problem in some such way. How did we solve it?   Eventually we understood the situation (*E*), we "saw the truth clearly and distinctly," as Descartes would say, we saw a convenient, effortless, well-adapted manner of doing the required sum. How did we reach this final stage?   At the outset, we hesitated between two opposite ways of conceiving the problem (*A* and *B*) which we finally

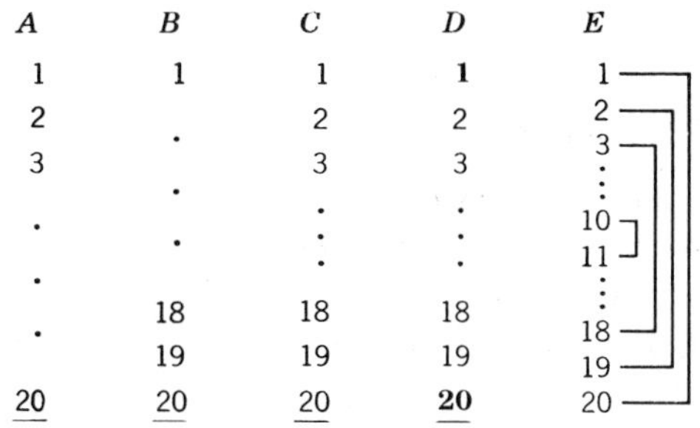

A	*B*	*C*	*D*	*E*
1	1	1	1	1
2	.	2	2	2
3	.	3	3	3
.	.	.	.	.
.	.	.	.	10
.	.	.	.	11
.	18	18	18	.
.	19	19	19	18
20	20	20	20	19
				20

Fig. 3.1. Five phases of a discovery.

succeeded in merging into a better balanced conception (C). The original antagonism resolved into a happy harmony and the transition (D) to the final idea was quite close. Was little Gauss's final idea the same? Did he arrive at it passing through the same stages? Or did he skip some of them? Or did he skip all of them? Did he jump right away at the final conclusion? We cannot answer such questions. Usually a bright idea emerges after a longer or shorter period of hesitation and suspense. This happened in our case, and some such thing may have happened in the mind of little Gauss.

Let us generalize. Starting from the problem just solved and substituting the general positive integer n for the particular value 20, we arrive at the problem: *Find the sum S of the first n positive integers.*

Thus we seek the sum

$$S = 1 + 2 + 3 + \cdots + n$$

The idea developed in the foregoing (which might have been that of little Gauss) was to pair off the terms: a term that is at a certain distance from the beginning is paired with another term at the same distance from the end. If we are somewhat familiar with algebraic manipulations, we are easily led to the following modification of this scheme.

We write the sum twice, the second time reversing the original order:

$$S = 1 + \quad 2 \quad + \quad 3 \quad + \cdots + (n-2) + (n-1) + n$$
$$S = n + (n-1) + (n-2) + \cdots + \quad 3 \quad + \quad 2 \quad + 1$$

The terms paired with each other by the foregoing solution appear here conveniently aligned, one written under the other. Adding the two equations we obtain

$$2S = (n+1) + (n+1) + (n+1) + \cdots + (n+1) + (n+1) + (n+1)$$
$$2S = n(n+1)$$
$$S = \frac{n(n+1)}{2}$$

This is the general formula. For $n = 20$ it yields little Gauss's result, which is as it should be.

3.2. Out of the blue

Here is a problem similar to that solved in the foregoing section: *Find the sum of the first n squares.*

Let S stand for the required sum (we are no longer bound by the notation of the foregoing section) so that now

$$S = 1 + 4 + 9 + 16 + \cdots + n^2$$

The evaluation of this sum is not too obvious. Human nature prompts us to repeat a procedure that has succeeded before in a similar situation; and so, remembering the foregoing section, we may attempt to write the sum twice, reversing the order the second time:

$$S = 1 + \quad 4 \quad + \quad 9 \quad + \cdots + (n-2)^2 + (n-1)^2 + n^2$$
$$S = n^2 + (n-1)^2 + (n-2)^2 + \cdots + \quad 9 \quad + \quad 4 \quad + 1$$

Yet the addition of these two equations, which was so successful in the foregoing case, leads us nowhere in the present case: our attempt fails, we undertook it with more optimism than understanding, our servile imitation of the chosen pattern was, let us confess, silly. (It was an overdose of mental inertia: our mind persevered in the same course, although this course should have been changed by the influence of circumstances.) Yet even such a misconceived trial need not be quite useless; it may lead us to a more adequate appraisal of the proposed problem: yes, it seems to be more difficult than the problem in the foregoing section.

Well, here is a solution. We start from a particular case of a well-known formula:

$$(n+1)^3 = n^3 + 3n^2 + 3n + 1$$

from which follows

$$(n+1)^3 - n^3 = 3n^2 + 3n + 1$$

This is valid for any value of n; write it down successively for $n = 1, 2, 3, \ldots, n$:

$$2^3 \quad - 1^3 = 3 \cdot 1^2 + 3 \cdot 1 + 1$$
$$3^3 \quad - 2^3 = 3 \cdot 2^2 + 3 \cdot 2 + 1$$
$$4^3 \quad - 3^3 = 3 \cdot 3^2 + 3 \cdot 3 + 1$$
$$\cdot \quad \cdot \quad \cdot \quad \cdot \quad \cdot \quad \cdot \quad \cdot \quad \cdot \quad \cdot$$
$$(n+1)^3 - n^3 = 3n^2 \quad + 3n \quad + 1$$

What is the obvious thing to do with these n equations? Add them! Thanks to conspicuous cancellations, the left-hand side of the resulting equation will be very simple. On the right-hand side we have to add three columns. The first column brings in S, the desired sum of the squares— that's good! The last column consists of n units—that is easy. The column in the middle brings in the sum of the first n numbers—but we know this sum from the foregoing section. We obtain

$$(n+1)^3 - 1 = 3S + 3\frac{n(n+1)}{2} + n$$

and in this equation everything is known (that is, expressed in terms of n)

except S, and so we can determine S from the equation. In fact, we find by straightforward algebra

$$2(n^3 + 3n^2 + 3n) = 6S + 3(n^2 + n) + 2n$$

$$S = \frac{2n^3 + 3n^2 + n}{6}$$

or finally

$$S = \frac{n(n + 1)(2n + 1)}{6}$$

How do you like this solution?

I shall be highly pleased with the reader who is displeased with the foregoing solution provided that he gives the right reason for his displeasure. What is wrong with the solution?

The solution is certainly correct. Moreover, it is efficient, clear, and short. Remember that the problem appeared difficult—we cannot reasonably expect a much clearer or shorter solution. There is, as far as I can see, just one valid objection: the solution appears *out of the blue*, pops up from nowhere. It is like a rabbit pulled out of a hat. Compare the present solution with that in the foregoing section. There we could visualize to some extent how the solution was discovered, we could learn a little about the ways of discovery, we could even gather some hope that some day we shall succeed in finding a similar solution by ourselves. Yet the presentation of the present section gives no hint about the sources of discovery, we are just hit on the head with the initial equation from which everything follows, and there is no indication how we could find this equation by ourselves. This is disappointing; we came here to learn problem solving—how could we learn it from the solution just presented?[1]

3.3. We cannot leave this unapplied

Yes, we *can* learn something important about problem solving from the foregoing solution. It is true, the presentation was not enlightening: the source of the invention remained hidden and so the solution appeared as a trick, a cunning device. Do you wish to know what is behind the trick? Try to *apply that trick* yourself and then you may find out. The device was so successful that we really cannot afford to leave it unapplied.

Let us start by generalizing. We bring both problems considered in the foregoing (in sections 3.1 and 3.2) under the same viewpoint by considering the sum of the kth powers of the first n natural numbers

$$S_k = 1^k + 2^k + 3^k + \cdots + n^k$$

[1] Cf. MPR, vol. 2, pp. 146–148, the sections on *"deus ex machina"* and "heuristic justification."

We found in the foregoing section

$$S_2 = \frac{n(n + 1)(2n + 1)}{6}$$

and before that

$$S_1 = \frac{n(n + 1)}{2}$$

to which we may add the obvious, but perhaps not useless, extreme case

$$S_0 = n$$

Starting from the particular cases $k = 0$, 1, and 2 we may raise the general problem: express S_k similarly. Surveying those particular cases, we may even conjecture that S_k can be expressed as a polynomial of degree $k + 1$ in n.

It is natural to try on the general case the trick that served us so well in the case $k = 2$. Yet let us first examine the next particular case $k = 3$. We have to imitate what we have seen in sect. 3.2 on the next higher level— this cannot be very difficult.

In fact, we start by applying the binomial formula with the next higher exponent 4:

$$(n + 1)^4 = n^4 + 4n^3 + 6n^2 + 4n + 1$$

from which follows

$$(n + 1)^4 - n^4 = 4n^3 + 6n^2 + 4n + 1$$

This is valid for any value of n; write it down successively for $n = 1, 2, 3, \ldots, n$:

$$
\begin{aligned}
2^4 \quad &- 1^4 = 4 \cdot 1^3 + 6 \cdot 1^2 + 4 \cdot 1 + 1 \\
3^4 \quad &- 2^4 = 4 \cdot 2^3 + 6 \cdot 2^2 + 4 \cdot 2 + 1 \\
4^4 \quad &- 3^4 = 4 \cdot 3^3 + 6 \cdot 3^2 + 4 \cdot 3 + 1 \\
\cdots \quad & \\
(n + 1)^4 - n^4 &= 4n^3 \quad + 6n^2 \quad + 4n \quad + 1
\end{aligned}
$$

As before, we add these n equations. There are conspicuous cancellations on the left-hand side. On the right-hand side, there are four columns to add, and each column involves a sum of like powers of the first n integers; in fact, each column introduces another particular case of S_k:

$$(n + 1)^4 - 1 = 4S_3 + 6S_2 + 4S_1 + S_0$$

Yet we can already express S_2, S_1, and S_0 in terms of n, see above. Using those expressions, we transform our equation into

$$(n + 1)^4 - 1 = 4S_3 + 6 \frac{n(n + 1)(2n + 1)}{6} + 4 \frac{n(n + 1)}{2} + n$$

and in this equation everything is expressed in terms of n except S_3. What is needed now to determine S_3 is merely a little straightforward algebra:

$$4S_3 = (n + 1)^4 - (n + 1) - 2n(n + 1) - n(n + 1)(2n + 1)$$
$$= (n + 1)[n^3 + 3n^2 + 3n - 2n - n(2n + 1)]$$
$$S_3 = \left[\frac{n(n + 1)}{2}\right]^2$$

We have arrived at the desired result, and even the route seems instructive: having used the trick a second time, we may foresee a general outline. Remember that dictum of a famous pedagogue: "A method is a device which you use twice."[2]

3.4. Recursion

What was the salient feature of our work in the preceding sect. 3.3? In order to obtain S_3, we went back to the previously determined S_2, S_1, and S_0. This illuminates the "trick" of sect. 3.2 where we obtained S_2 by recurring to the previously determined S_1 and S_0.

In fact, we could use the same scheme to derive S_1 which we obtained in sect. 3.1 by a quite different method. By a most familiar formula

$$(n + 1)^2 = n^2 + 2n + 1$$
$$(n + 1)^2 - n^2 = 2n + 1$$

We list particular cases:

$$2^2 \quad - 1^2 = 2 \cdot 1 + 1$$
$$3^2 \quad - 2^2 = 2 \cdot 2 + 1$$
$$4^2 \quad - 3^2 = 2 \cdot 3 + 1$$
$$\cdot \quad \cdot \quad \cdot \quad \cdot \quad \cdot \quad \cdot \quad \cdot \quad \cdot$$
$$(n + 1)^2 - n^2 = 2n \quad + 1$$

By adding we obtain

$$(n + 1)^2 - 1 = 2S_1 + S_0$$

Of course, $S_0 = n$ and so

$$S_1 = \frac{(n + 1)^2 - 1 - n}{2} = \frac{n(n + 1)}{2}$$

which is the final result of sect. 3.1.

After having worked the scheme in the particular cases $k = 1, 2$, and 3,

---

[2] HSI, The traditional mathematics professor, p. 208.

we apply it without hesitation to the general sum S_k. We now need the binomial formula with the exponent $k + 1$:

$$(n + 1)^{k+1} = n^{k+1} + \binom{k+1}{1}n^k + \binom{k+1}{2}n^{k-1} + \cdots + 1$$

$$(n + 1)^{k+1} - n^{k+1} = (k + 1)n^k + \binom{k+1}{2}n^{k-1} + \cdots + 1$$

We list particular cases:

$$2^{k+1} \quad - 1^{k+1} = (k + 1)1^k + \binom{k+1}{2}1^{k-1} + \cdots + 1$$

$$3^{k+1} \quad - 2^{k+1} = (k + 1)2^k + \binom{k+1}{2}2^{k-1} + \cdots + 1$$

$$4^{k+1} \quad - 3^{k+1} = (k + 1)3^k + \binom{k+1}{2}3^{k-1} + \cdots + 1$$

$$\cdot \quad \cdot \quad \cdot \quad \cdot \quad \cdot \quad \cdot \quad \cdot \quad \cdot \quad \cdot \quad \cdot \quad \cdot \quad \cdot \quad \cdot \quad \cdot \quad \cdot \quad \cdot$$

$$(n + 1)^{k+1} - n^{k+1} = (k + 1)n^k + \binom{k+1}{2}n^{k-1} + \cdots + 1$$

By adding we obtain

$$(n + 1)^{k+1} - 1 = (k + 1)S_k + \binom{k+1}{2}S_{k-1} + \cdots + S_0$$

From this equation we can determine (express in terms of n) S_k provided that we have previously determined S_{k-1}, $S_{k-2}, \ldots$, S_1 and S_0. For example, as we have obtained in the foregoing expressions for S_0, S_1, S_2, and S_3, we could derive an expression for S_4 by straightforward algebra. Having obtained S_4, we could proceed to S_5, and so on.[3]

Thus, by following up the "trick" of sect. 3.2, which appeared "out of the blue," we have arrived at a pattern which deserves to be formulated and remembered with a view to further applications. When we are facing a well-ordered sequence (such as S_0, S_1, S_2, $S_3, \ldots$, $S_k, \ldots$) there is a chance to evaluate the terms of the sequence one at a time. We need two things.

First, the initial term of the sequence should be known somehow (the evaluation of S_0 was obvious).

Second, there should be some relation linking the general term of the sequence to the foregoing terms (S_k is linked to S_0, $S_1, \ldots$, S_{k-1} by the final equation of the present section, foreshadowed by the "trick" of sect. 3.2).

Then we can find the terms one after the other, successively, *recursively*,

[3] This method is due to Pascal; see *Œuvres de Blaise Pascal*, edited by L. Brunschvicg and P. Boutroux, vol. 3, pp. 341–367.

by going back or recurring each time to the foregoing terms. This is the important pattern of *recursion*.

3.5. Abracadabra

The word "abracadabra" means something like "complicated nonsense." We use the word contemptuously today, but there was a time when it was a magic word, engraved on amulets in mysterious forms (like Fig. 3.2 in some respect), and people believed that such an amulet would protect the wearer from disease and bad luck.

In how many ways can you read the word "abracadabra" in Fig. 3.2? It is understood that we begin with the uppermost *A* (the north corner) and read down, passing each time to the next letter (southeast or southwest) till we reach the last *A* (the south corner).

The question is curious. Yet your interest may be really aroused if you notice that there is something familiar behind it. It may remind you of walking or driving in a city. Think of a city that consists of perfectly square blocks, where one-half of the streets run from northwest to southeast and the other streets (or avenues) crossing the former run from northeast to southwest. Reading the magic word of Fig. 3.2 corresponds to a zigzag path in the network of such streets. When you walk along the zigzag path emphasized in Fig. 3.3, you walk ten blocks from the initial *A* to the final *A*. There are several other paths which are ten blocks long between these two endpoints in this network of streets, but there is no path that would be shorter. *Find the number of the different shortest paths in the network between the given endpoints*—this is the general, really

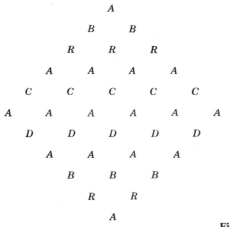

Fig. 3.2. A magic word.

Fig. 3.3. The zigzag path is the shortest.

interesting, problem behind the curious particular problem about the magic word of Fig. 3.2.

A general formulation may have various advantages. It sometimes suggests an approach to the solution, and this happens in our case. *If you cannot solve the proposed problem* about Fig. 3.2 (probably you cannot), *try first to solve some simpler related problem.* At this point the general formulation may help: it suggests trying simpler cases that fall under it. In fact, if the two given corners are close enough to each other in the network (closer than the extreme *A*'s in Fig. 3.3) it is easy to count the different zigzag paths between the two: you can draw each one after the other and survey all of them. Listen to this suggestion and pursue it systematically. Start from the point *A* and go downward. Consider first the points that you can reach by walking one block, then those to which you have to walk two blocks, then those which are three or four or more blocks away. Survey and count for each point the shortest zigzag paths

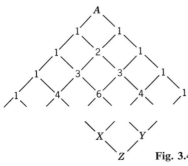

Fig. 3.4. Count the number of shortest zigzag paths.

that connect it with A.   In Fig. 3.4 a few numbers so obtained are marked (but you should have obtained these numbers and a few more by yourself—check them at least).   Observe these numbers—do you notice something?

If you have enough previous knowledge you may notice many things. Yet even if you have never before seen this array of numbers displayed by Fig. 3.4 you may notice an interesting relation: any number in Fig. 3.4 that is different from 1 is the sum of two other numbers in the array, of its northwest and northeast neighbors.   For instance,

$$4 = 1 + 3, \qquad 6 = 3 + 3$$

You may discover this law by observation as a naturalist discovers the laws of his science by observation.   Yet, after having discovered it, you should ask yourself: Why is that so?   What is the reason?

The reason is simple enough.   Consider three corners in your network, the points X, Y, and Z, the relative position of which is shown by Fig. 3.4: X is the northwest neighbor and Y the northeast neighbor of Z.   If we wish to reach Z coming from A along a shortest path in the network, we must pass either through X or through Y.   Once we have reached X, we can proceed hence to Z in just one way, and the same is true for proceeding from Y to Z.   Therefore, the *total number of shortest paths from A to Z* is a sum of two terms: it *equals the number of shortest paths from A to X added to the number of those from A to Y.*   This explains fully our observation and proves the general law.

Having clarified this basic point, we can extend the array of numbers in Fig. 3.4 by simple additions till we obtain the larger array in Fig. 3.5, the south corner of which yields the desired answer: we can read the magic word in Fig. 3.2 in exactly 252 different ways.

3.6. The Pascal triangle

By now the reader has probably recognized the numbers and their peculiar configuration which we have examined in the foregoing section. The numbers in Fig. 3.4 are *binomial coefficients* and their triangular arrangement is usually called the *Pascal triangle*. (Pascal himself called it the "arithmetical triangle.") Further lines can be added to the triangle of Fig. 3.4 and, in fact, it can be extended indefinitely. The array in Fig. 3.5 is a square piece cut out of a larger triangle.

Some of the binomial coefficients and their triangular arrangement can be found in the writings of other authors before Pascal's *Traité du triangle arithmétique*. Still, the merits of Pascal in this matter are quite sufficient to justify the use of his name.

(1) We have to introduce a suitable *notation* for the numbers contained in the Pascal triangle; this is a step of major importance. For us each number attached to a point of this triangle has a geometric meaning: it indicates the number of different shortest zigzag paths from the apex of the triangle to that point. Each of these paths passes along the same number of blocks, let us say *n* blocks. Moreover, all these paths agree in the number of blocks described in the southwesterly direction and in the number of those in the southeasterly direction. Let *l* and *r* stand for these numbers, respectively (*l* to the left and *r* to the right—of course, downward in both cases). Obviously

$$n = l + r$$

If we give any two of the three numbers *n*, *l*, and *r*, the third is fully determined and so is the point to which they refer. (In fact, *l* and *r* are the rectangular coordinates of the point with respect to a system the origin of which is the apex of the Pascal triangle; one of the axes points southwest,

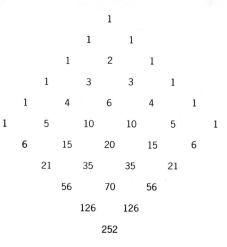

Fig. 3.5. A square from a triangle.

the other southeast.)   For instance, for the last A of the path shown in Fig. 3.3

$$l = 5, \qquad r = 5, \qquad n = 10$$

and for the second B of the same path

$$l = 5, \qquad r = 3, \qquad n = 8$$

We shall denote by $\binom{n}{r}$ (this notation is due to Euler) the number of shortest zigzag paths from the apex of the Pascal triangle to the point specified by n (total number of blocks) and r (blocks to the right downward).   For instance, see Fig. 3.5,

$$\binom{8}{3} = 56, \qquad \binom{10}{5} = 252$$

The symbols for the numbers contained in Fig. 3.4 are assembled in Fig. 3.6.   The symbols with the same number upstairs (the same n) are horizontally aligned (along the nth "base"—the base of a right triangle). The symbols with the same number downstairs (the same r) are obliquely aligned (along the rth "avenue").   The fifth avenue forms one of the sides of the square in Fig. 3.5—the opposite side is formed by the 0th avenue (but you may call it the borderline, or Riverside Drive, if you prefer to do so).   The fourth base is emphasized in Fig. 3.4.

(2) Besides the geometric aspect, the Pascal triangle also has a computational aspect.   All the numbers along the boundary (0th street, 0th avenue, and their common starting point) are equal to 1 (it is obvious that

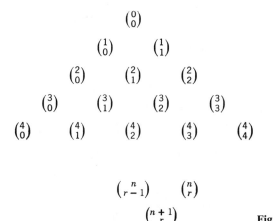

Fig. 3.6. Symbolic Pascal triangle.

there is just one shortest path to these street corners from the starting point). Therefore,

$$\binom{n}{0} = \binom{n}{n} = 1$$

It is appropriate to call this relation the *boundary condition* of the Pascal triangle.

Any number inside the Pascal triangle is situated along a certain horizontal row, or base. We compute a number of the $(n + 1)$th base by going back, or recurring, to two neighboring numbers of the nth base:

$$\binom{n + 1}{r} = \binom{n}{r} + \binom{n}{r - 1}$$

see Fig. 3.6. It is appropriate to call this equation the *recursion formula* of the Pascal triangle.

From the computer's standpoint the numbers $\binom{n}{r}$ are determined (or defined, if you wish) by the recursion formula and the boundary condition of the Pascal triangle.

3.7. Mathematical induction

When we compute a number in the Pascal triangle by using the recursion formula, we have to rely on the previous knowledge of two numbers of the foregoing base. It would be desirable to have a scheme of computation independent of such previous knowledge. There is a well-known formula, which we shall call the *explicit formula* for binomial coefficients, that yields such an independent computation:

$$\binom{n}{r} = \frac{n(n - 1)(n - 2)\cdots(n - r + 1)}{1 \cdot 2 \cdot 3 \cdots r}$$

Pascal's treatise contains the explicit formula (stated in words, not in our modern notation). Pascal does not say how he has discovered it and we shall not speculate too much how he might have discovered it. (Perhaps he just guessed it first—we often find such things by observation and tentative generalization of the observed; see the remark in the solution of ex. 3.39.) Yet Pascal gives a remarkable proof for the explicit formula and we wish to devote our full attention to his method of proof.[4]

We need a preliminary remark. The explicit formula does not apply,

[4] Cf. Pascal's *Œuvres l.c.* footnote 3, pp. 455–464, especially pp. 456–457. The following presentation takes advantage of modern notation and modifies less essential details.

as it stands, to the case $r = 0$. Yet we lay down the rule that, if $r = 0$, it should be interpreted as

$$\binom{n}{0} = 1$$

The explicit formula does apply to the case $r = n$ and yields

$$\binom{n}{n} = \frac{n(n-1)\cdots\ 2\cdot 1}{1\cdot 2\ \ \ \cdots(n-1)n} = 1$$

which is the correct result. Therefore, we have to prove the explicit formula only for $0 < r < n$, that is, in the interior of the Pascal triangle where we can use the recursion formula. Now, we quote Pascal, with unessential modifications some of which will be included in square brackets [ ].

Although this proposition [the explicit formula] contains infinitely many cases I shall give for it a very short proof, supposing two lemmas.

The first lemma asserts that the proposition holds for the first base, which is obvious. [The explicit formula is valid for $n = 1$, because, in this case, all possible values of r, $r = 0$ and $r = 1$, fall under the preliminary remark.]

The second lemma asserts this: if the proposition happens to be valid for any base [for any value n] it is necessarily valid for the next base [for $n + 1$].

We see hence that the proposition holds necessarily for all values of n. For it is valid for $n = 1$ by virtue of the first lemma; therefore, for $n = 2$ by virtue of the second lemma; therefore, for $n = 3$ by virtue of the same, and so on *ad infinitum*.

And so nothing remains but to prove the second lemma.

In accordance with the statement of the second lemma, we assume that the explicit formula is valid for the nth base, that is, for a certain value of n and all compatible values of r (for $r = 0, 1, 2, \ldots, n$). In particular, along with

$$\binom{n}{r} = \frac{n(n-1)\cdots(n-r+2)(n-r+1)}{1\cdot 2\ \ \ \cdots\ \ (r-1)\ \ \cdot\ \ \ \ r}$$

we also have (if $r \geq 1$)

$$\binom{n}{r-1} = \frac{n(n-1)\cdots(n-r+2)}{1\cdot 2\ \ \ \cdots\ \ (r-1)}$$

Adding these two equations and using the recursion formula, we derive as a necessary consequence

$$\binom{n+1}{r} = \binom{n}{r} + \binom{n}{r-1} = \frac{n(n-1)\cdots(n-r+2)}{1\cdot 2\ \ \ \cdots\ \ (r-1)}\left[\frac{n-r+1}{r} + 1\right]$$

$$= \frac{n(n-1)\cdots(n-r+2)}{1\cdot 2\ \ \ \cdots\ \ (r-1)}\cdot\frac{n+1}{r}$$

$$= \frac{(n+1)n(n-1)\cdots(n-r+2)}{1\cdot 2\cdot 3\ \ \ \cdots\ \ \ \ r}$$

That is, the validity of the explicit formula for a certain value of n involves its validity for $n + 1$. This is precisely what the second lemma asserts— we have proved it.

The words of Pascal which we have quoted are of historic importance because his proof is the first example of a fundamental pattern of reasoning which is usually called *mathematical induction*.

This pattern of reasoning deserves further study.[5] If carelessly introduced, reasoning by mathematical induction may puzzle the beginner; in fact, it may appear as a devilish trick.

You know, of course, that the devil is dangerous: if you give him the little finger, he takes the whole hand. Yet Pascal's second lemma does exactly this: by admitting the first lemma you give just one finger, the case $n = 1$. Yet then the second lemma also takes your second finger (the case $n = 2$), then the third finger ($n = 3$), then the fourth, and so on, and finally takes all your fingers even if you happen to have infinitely many.

3.8. Discoveries ahead

After the work in the three foregoing sections, we now have three different approaches to the numbers in the Pascal triangle, the binomial coefficients.

(1) *Geometrical approach.* A binomial coefficient is the number of the different shortest zigzag paths between two given corners in a network of streets.

(2) *Computational approach.* The binomial coefficients can be defined by their recursion formula and their boundary condition.

(3) *Explicit formula.* We have proved it, by Pascal's method, in sect. 3.7.

The name of the numbers considered reminds us of another approach.

(4) *Binomial theorem.* For indeterminate (or variable) x and any non-negative integer n we have the identity

$$(1 + x)^n = \binom{n}{0} + \binom{n}{1}x + \binom{n}{2}x^2 + \cdots + \binom{n}{n}x^n$$

For a proof, see ex. 3.1.

There are still other approaches to the numbers in the Pascal triangle which play, in fact, a role in a great many interesting questions and possess a great many interesting properties. "This table of numbers has eminent

---

[5] HSI, Induction and mathematical induction, pp. 114–121; MPR, vol. 1, pp. 108–120.

and admirable properties" wrote Jacob Bernoulli in his *Ars Conjectandi* (Basle 1713; see Second Part, Chapter III, p. 88). "We have just shown that the essence of combinations is concealed in it [see ex. 3.22–3.27] but those who are more intimately acquainted with Geometry know also that capital secrets of all Mathematics are hidden in it." Times have changed and many things hidden in Bernoulli's time are clearly seen today. Still, the reader who wants instructive, and perhaps fascinating, exercise has an excellent opportunity: he has an excellent chance to discover something by observing the numbers in the Pascal triangle and combining his observations with one or the other or several approaches. There are so many possibilities—some of them should be favorable.

By the way, we have broached another subject in the first four sections of the chapter (sum of like powers of the first *n* integers). Moreover, we have encountered two important general patterns (recursion and mathematical induction) which we still should apply to more examples if we wish to understand them thoroughly. And so there are still more prospects ahead.

3.9. Observe, generalize, prove, and prove again

Let us return to our starting point and have another look at it.

(1) We started from the magic word of Fig. 3.2 and Fig. 3.3, or rather from a problem concerning that word. What was the unknown? The number of shortest zigzag paths in that network of streets from the first A to the last A, that is, from the north corner of the square to its south corner. Such a zigzag path must cross somewhere the horizontal diagonal of the square. There are six possible crossing points (street corners, A's) along the horizontal diagonal. There are, therefore, six different kinds of zigzag paths in our problem—how many paths are there of each kind? We have here a *new problem*.

Let us be specific. Take a definite crossing point on that horizontal diagonal, for instance the third point from the left ($l = 3$, $r = 2$, $n = 5$ in the notation of sect. 3.6). A zigzag path crossing this chosen point consists of two sections: the upper section starts from the north corner of the square and ends in the chosen point, the lower section starts from the chosen point and ends in the south corner; see Fig. 3.3. We have found before (see Fig. 3.5) the number of the different upper sections; it is

$$\binom{5}{2} = 10$$

The number of the different lower sections is the same. Now any upper section can be combined with any lower section to form a full path [as suggested by Fig. 3.7(III)]. Therefore, the number of such paths is

$$\binom{5}{2}^2 = 100$$

Of course, the number of zigzag paths crossing the horizontal diagonal at any other given point can be similarly computed. Hence we find a new solution of our original problem: we can read the magic word of Fig. 3.2 in exactly

$$1 + 25 + 100 + 100 + 25 + 1$$

different ways. This sum must agree with the result found at the end of sect. 3.5; in fact, it equals 252.

(2) *Generalization.* One side of the square considered in Fig. 3.3 consists of five blocks. In generalizing (passing from 5 to n) we find that

$$\binom{n}{0}^2 + \binom{n}{1}^2 + \binom{n}{2}^2 + \cdots + \binom{n}{n}^2 = \binom{2n}{n}$$

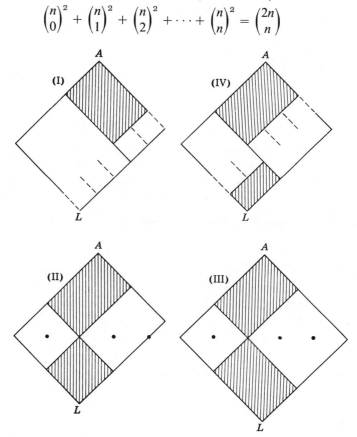

Fig. 3.7. Suggestions.

"The sum of the squares of the numbers in the nth base of the Pascal triangle is equal to the number in the middle of the $2n$th base." Our reasoning under (1) essentially proves this general statement. It is true, we have considered the special case $n = 5$ (we have even considered a special point of the fifth base) but there is no particular virtue (and no misleading peculiarity) in the special case considered. And so our reasoning is generally valid. Yet it may be a useful exercise for the reader to repeat the reasoning with special attention to its generality—he has to say n instead of 5.[6]

(3) *Another approach.* Still, the result is surprising. We would understand it better if we could attain it from another side.

Surveying the various approaches listed in sect. 3.8, we may try to link our result to the binomial formula. There is, in fact, a connection:

$$(1 + x)^{2n} = \cdots + \binom{2n}{n}x^n + \cdots$$

$$= (1 + x)^n(1 + x)^n$$

$$= \left[\binom{n}{0} + \binom{n}{1}x + \binom{n}{2}x^2 + \cdots + \binom{n}{n}x^n\right] \cdot$$

$$\left[\binom{n}{n} + \cdots + \binom{n}{2}x^{n-2} + \binom{n}{1}x^{n-1} + \binom{n}{0}x^n\right]$$

Let us focus the coefficient of x^n. On the right-hand side of the first line the coefficient of x^n is the right-hand side of the general equation given under (2) for which we are seeking a second proof. Now let us turn to the product of the two factors which are displayed on the last two lines; in writing them we made use of the symmetry of the binomial coefficients:

$$\binom{n}{r} = \binom{n}{n-r}$$

Now, in this product, the coefficient of x^n is obviously the left-hand side of the equation under (2) which we are about to prove. And here is the proof: the coefficient of x^n must be the same in both cases since we have here an identity in x.

Examples and Comments on Chapter 3

First Part

The examples and comments of this first part are connected with the first four sections.

[6] We have here a *representative* special case; see MPR, vol. 1, p. 25, ex. 10.

3.1. Prove the binomial theorem stated in sect. 3.8(4) (and used in sect. 3.4). (Use mathematical induction. Which one of the first three approaches mentioned in sect. 3.8 appears the most appropriate for the present purpose?)

3.2. *A particular case equivalent to the general case.* The identity asserted in sect. 3.8(4) and proved in ex. 3.1 follows as a particular case ($a = 1, b = x$) from the more general identity

$$(a + b)^n = \binom{n}{0}a^n + \binom{n}{1}a^{n-1}b + \binom{n}{2}a^{n-2}b^2 + \cdots + \binom{n}{n}b^n$$

Show that, conversely, the general identity also follows from that particular case.[7]

3.3. In the first three sections of this chapter we have computed S_k (defined in sect. 3.3) for $k = 1, 2, 3$; the case $k = 0$ is obvious. Comparing these expressions, we may be led to the general theorem: S_k *is a polynomial in n of degree $k + 1$ and the coefficient of its highest term is $1/(k + 1)$.*

This theorem which asserts that

$$S_k = \frac{n^{k+1}}{k+1} + \cdots$$

(where the dots indicate terms of lower degree in n) played an important role in the history of the integral calculus.

Prove the theorem; use mathematical induction.

3.4. We can guess an expression for S_4 by computing numerically the ratio S_4/S_2 for a few small values of n. In fact, for

$$n = 1, \quad 2, \quad 3, \quad 4, \quad 5$$

$$\frac{S_4}{S_2} = 1, \quad \frac{17}{5}, \quad 7, \quad \frac{59}{5}, \quad \frac{89}{5}$$

For the sake of uniformity we write rather

$$\frac{5}{5}, \quad \frac{17}{5}, \quad \frac{35}{5}, \quad \frac{59}{5}, \quad \frac{89}{5}$$

The numerators are close to multiples of 6; in fact, they are

$$6 \cdot 1 - 1, \quad 6 \cdot 3 - 1, \quad 6 \cdot 6 - 1, \quad 6 \cdot 10 - 1, \quad 6 \cdot 15 - 1$$

You should recognize the numbers

$$1, \quad 3, \quad 6, \quad 10, \quad 15$$

If you succeed in constructing an expression for S_4, prove it, independently of sect. 3.4, by mathematical induction.[8]

3.5. Compute S_4, independently of ex. 3.4, by the method indicated in sect. 3.4.

---

[7] Such equivalence of the particular and the general may seem bewildering to the philosopher or to the beginner, but is, in fact, quite usual in mathematics; see MPR, vol. 1, p. 23, ex. 3 and ex. 4.

[8] For broader discussion of a very similar simpler case see MPR, vol. 1, pp. 108–110.

3.6. Show that

$$n = S_0$$
$$n^2 = 2S_1 - S_0$$
$$n^3 = 3S_2 - 3S_1 + S_0$$
$$n^4 = 4S_3 - 6S_2 + 4S_1 - S_0$$

and generally

$$n^k = \binom{k}{1}S_{k-1} - \binom{k}{2}S_{k-2} + \binom{k}{3}S_{k-3} - \cdots + (-1)^{k-1}\binom{k}{k}S_0$$

(This is similar to, but different from, the principal formula of sect. 3.4.)

3.7. Show that

$$S_1 = S_1$$
$$2S_1{}^2 = 2S_3$$
$$4S_1{}^3 = 3S_5 + S_3$$
$$8S_1{}^4 = 4S_7 + 4S_5$$

and generally, for $k = 1, 2, 3, \ldots$,

$$2^{k-1}S_1{}^k = \binom{k}{1}S_{2k-1} + \binom{k}{3}S_{2k-3} + \binom{k}{5}S_{2k-5} + \cdots$$

The last term on the right-hand side is S_k or kS_{k+1} according as k is odd or even.

(This is similar to ex. 3.6 where, in fact, we could substitute $S_0{}^k$ for n^k.)

3.8. Show that

$$3S_2 = 3S_2$$
$$6S_2S_1 = 5S_4 + S_2$$
$$12S_2S_1{}^2 = 7S_6 + 5S_4$$
$$24S_2S_1{}^3 = 9S_8 + 14S_6 + S_4$$

and generally, for $k = 1, 2, 3, \ldots$,

$$3 \cdot 2^{k-1}S_2S_1{}^{k-1} = \left[\binom{k}{0} + 2\binom{k}{1}\right]S_{2k} + \left[\binom{k}{2} + 2\binom{k}{3}\right]S_{2k-2} + \cdots$$

the last term on the right-hand side is $(k + 2)S_{k+1}$ or S_k according as k is odd or even.

3.9. Show that

$$S_3 = S_1{}^2$$
$$S_5 = S_1{}^2(4S_1 - 1)/3$$
$$S_7 = S_1{}^2(6S_1{}^2 - 4S_1 + 1)/3$$

and generally that S_{2k-1} is a polynomial in $S_1 = n(n + 1)/2$, of degree k, divisible by $S_1{}^2$ provided that $2k - 1 \geq 3$. (This generalizes the result of sect. 3.3.)

3.10. Show that

$$S_4 = S_2(6S_1 - 1)/5$$
$$S_6 = S_2(12S_1{}^2 - 6S_1 + 1)/7$$
$$S_8 = S_2(40S_1{}^3 - 40S_1{}^2 + 18S_1 - 3)/15$$

and generally that S_{2k}/S_2 is a polynomial in S_1 of degree $k - 1$. (This generalizes a result encountered in the solution of ex. 3.4.)

3.11. We introduce the notation

$$1^k + 2^k + 3^k + \cdots + n^k = S_k(n)$$

which is more explicit (or specific) than the one introduced in sect. 3.3; k stands for a non-negative integer and n for a positive integer.

We now extend the range of n (but not the range of k): we let $S_k(x)$ denote the polynomial in x of degree $k + 1$ that coincides with $S_k(n)$ for $x = 1, 2, 3, \ldots$; for example,

$$S_3(x) = \frac{x^2(x + 1)^2}{4}$$

Prove that for $k \geq 1$ (not for $k = 0$)

$$S_k(-x - 1) = (-1)^{k-1} S_k(x)$$

3.12. Find $1 + 3 + 5 + \cdots + (2n - 1)$, the sum of the first n odd numbers. (List as many different approaches as you can.)

3.13. Find $1 + 9 + 25 + \cdots + (2n - 1)^2$.

3.14. Find $1 + 27 + 125 + \cdots + (2n - 1)^3$.

3.15. (Continued) Generalize.

3.16. Find $2^2 + 5^2 + 8^2 + \cdots + (3n - 1)^2$

3.17. (Continued) Generalize.

3.18. Find a simple expression for

$$1 \cdot 2 + (1 + 2)3 + (1 + 2 + 3)4 + \cdots + [1 + 2 + \cdots + (n - 1)]n.$$

(Of course, you should try to use suitable points from the foregoing work. What has better prospects to be usable: the results or the method?)

3.19. Consider the $\dfrac{n(n - 1)}{2}$ differences

$$2 - 1,$$
$$3 - 1, \quad 3 - 2$$
$$4 - 1, \quad 4 - 2, \quad 4 - 3$$
$$\cdot \quad \cdot \quad \cdot \quad \cdot \quad \cdot \quad \cdot$$
$$n - 1, \quad n - 2, \quad n - 3 \quad , \ldots, \quad n - (n - 1)$$

and compute (*a*) their sum, (*b*) their product, and (*c*) the sum of their squares.

3.20. Define $E_1, E_2, E_3 \ldots$ by the identity

$$x^n - E_1 x^{n-1} + E_2 x^{n-2} - \cdots + (-1)^n E_n$$
$$= (x - 1)(x - 2)(x - 3) \cdots (x - n)$$

Show that

$$E_1 = \frac{n(n + 1)}{2}$$

$$E_2 = \frac{(n - 1)n(n + 1)(3n + 2)}{24}$$

$$E_3 = \frac{(n - 2)(n - 1)n^2(n + 1)^2}{48}$$

$$E_4 = \frac{(n - 3)(n - 2)(n - 1)n(n + 1)(15n^3 + 15n^2 - 10n - 8)}{5760}$$

and show in general that E_k [which should rather be denoted by $E_k(n)$ since it depends on n] is a polynomial of degree $2k$ in n.

[The knowledge of a certain proposition of algebra may be a great help; E_k is the so-called kth elementary symmetric function of the first n integers the sum of the kth powers of which is $S_k = S_k(n)$.   Check $E_k(k) = k!$]

3.21. *Two forms of mathematical induction.*   A typical proposition A that is accessible to proof by mathematical induction has an infinity of cases $A_1, A_2, A_3, \ldots, A_n, \ldots$; in fact, A is equivalent to the simultaneous assertion of $A_1, A_2, A_3, \ldots$.   For instance, if A is the binomial theorem, A_n asserts the validity of the identity.

$$(1 + x)^n = \binom{n}{0} + \binom{n}{1}x + \binom{n}{2}x^2 + \cdots + \binom{n}{n}x^n$$

see ex. 3.1; the binomial theorem asserts, in fact, that this identity holds for $n = 1, 2, 3, 4, \ldots$.

Let us consider three statements about the sequence of propositions $A_1, A_2, A_3, \ldots$:

(I) A_1 is true.

(IIa) A_n implies A_{n+1}.

(IIb) $A_1, A_2, A_3, \ldots A_{n-1}$ and A_n jointly imply A_{n+1}.

Now we can distinguish two procedures.

(a) We can conclude from (I) and (IIa) that A_n is true generally, for $n = 1, 2, 3, \ldots$; we drew this conclusion, with Pascal, in sect. 3.7.

(b) We can conclude from (I) and (IIb) that A_n is true generally, for $n = 1, 2, 3, \ldots$; we proceeded so in the solution of ex. 3.3.

You may feel that the difference between the procedures (a) and (b) is more in the form than in the essence.   Could you clarify this feeling and propose a clear argument?

Second Part

3.22. Ten boys went camping together, Bernie, Ricky, Abe, Charlie, Al, Dick, Alex, Bill, Roy, and Artie.   In the evening they divided into two teams of five boys each: one team put up the tent, the other team cooked the supper.

In how many different ways is such a division into two teams possible? (Can a magic word help you?)

3.23. Show that a set of n individuals has $\binom{n}{r}$ different subsets of r individuals. [In more traditional language: the number of *combinations* of n objects taken r at a time is $\binom{n}{r}$.]

3.24. Given n points in the plane in "general position" so that no three points lie on the same straight line. How many straight lines can you draw by joining two given points? How many triangles can you form with vertices chosen among the given points?

3.25. (Continued) Formulate and solve an analogous problem in space.

3.26. Find the number of the diagonals of a convex polygon with n sides.

3.27. Find the number of intersections of the diagonals of a convex polygon of n sides. Consider only points of intersection inside the polygon, and assume that the polygon is "general" so that no three diagonals have a common point.

3.28. A polyhedron has six faces. (We may consider the polyhedron as irregular so that no two of its faces are congruent.) The faces should be painted, one red, two blue, and three brown. In how many different ways can this be done?

3.29. A polyhedron has n faces (no two of which are congruent.) Of these faces, r should be painted red, s sapphire, and t tan; we suppose that $r + s + t = n$. In how many different ways can this be done?

3.30. (Continued) Generalize.

Third Part

In solving some of the following problems, the reader may consider, and choose between, several approaches. (See sect. 3.8; the combinatorial interpretation of the binomial coefficients, cf. ex. 3.23, provides one more access.) The importance of approaching the same problem from several sides was emphasized by Leibnitz. Here is a free translation of one of his remarks: "In comparing two different expressions of the same quantity, you may find an unknown; in comparing two different derivations of the same result, you may find a new method."

3.31. Show in as many ways as you can that

$$\binom{n}{r} = \binom{n}{n-r}$$

3.32. Consider the sum of the numbers along a base of the Pascal triangle:

$$
\begin{array}{ll}
1 & = 1 \\
1 + 1 & = 2 \\
1 + 2 + 1 & = 4 \\
1 + 3 + 3 + 1 & = 8
\end{array}
$$

These facts seem to suggest a general theorem. Can you guess it? Having guessed it, can you prove it? Having proved it, can you devise another proof?

3.33. Observe

$$
\begin{array}{ll}
1 - 1 & = 0 \\
1 - 2 + 1 & = 0 \\
1 - 3 + 3 - 1 & = 0 \\
1 - 4 + 6 - 4 + 1 & = 0
\end{array}
$$

generalize, prove, and prove again.

3.34. Consider the sum of the first six numbers along the third avenue of the Pascal triangle:

$$1 + 4 + 10 + 20 + 35 + 56 = 126$$

Locate this sum in the Pascal triangle, try to observe analogous facts, generalize, prove, and prove again.

3.35. Add the thirty-six numbers displayed in Fig. 3.5, try to locate their sum in the Pascal triangle, formulate a general theorem, and prove it. (Adding so many numbers is a boring task—in doing it cleverly, you may easily catch the essential idea.)

3.36. Try to recognize and locate in the Pascal triangle the numbers involved in the following relation:

$$1 \cdot 1 + 5 \cdot 4 + 10 \cdot 6 + 10 \cdot 4 + 5 \cdot 1 = 126$$

Observe (or remember) analogous cases, generalize, prove, prove again.

3.37. Try to recognize and locate in the Pascal triangle the numbers involved in the following relation:

$$6 \cdot 1 + 5 \cdot 3 + 4 \cdot 6 + 3 \cdot 10 + 2 \cdot 15 + 1 \cdot 21 = 126$$

Observe (or remember) analogous cases, generalize, prove, prove again.

3.38. Fig. 3.8 shows the first four from an infinite sequence of figures each of which is an assemblage of equal circles into an equilateral triangular shape. Any circle that is not on the rim of the assemblage touches six surrounding circles. In the nth figure there are n circles aligned along each side of the triangular assemblage and the total number of circles in this nth figure is termed the nth *triangular number*. Express the nth triangular number in terms of n and locate it in the Pascal triangle.

3.39. Replace in Fig. 3.8 each circle by a sphere (a marble) of which the

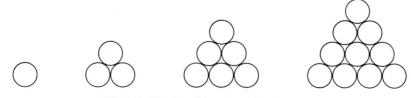

Fig. 3.8. The first four triangular numbers.

circle forms the equator. Fix 10 marbles arranged as in Fig. 3.8 on a horizontal plane, place 6 marbles on top (they fit neatly into the interstices) as a second layer, add 3 marbles on top of these as a third layer and place finally 1 marble on the very top. This configuration of

$$1 + 3 + 6 + 10 = 20$$

marbles is so related to a regular tetrahedron as each of the assemblages of circles shown by Fig. 3.8 is related to a certain equilateral triangle: 20 is the fourth *pyramidal number*. Express the nth pyramidal number in terms of n and locate it in the Pascal triangle.

3.40. You can build a pyramidal pile of marbles in another manner: begin with a layer of n^2 marbles, arranged in a square as in Fig. 3.9, place on top of it a second layer of $(n - 1)^2$ marbles, then $(n - 2)^2$ marbles, and so on, and finally just one marble on the very top. How many marbles does the pile contain?

3.41. Interpret the product

$$\binom{n_1}{r_1}\binom{n_2}{r_2}\binom{n_3}{r_3}\cdots\binom{n_h}{r_h}$$

as the number of a certain set of zigzag paths in a network of streets.

3.42. All the shortest zigzag paths from the apex of the Pascal triangle to the point specified by n (the total number of blocks) and r (blocks to the right downward) have a point in common with the line of symmetry of the Pascal triangle (from the first A to the last A in Fig. 3.3) namely their common initial

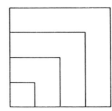
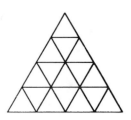

Fig. 3.9. The fourth square number.

point, the apex.   In this set of paths, consider the subset of such paths as have no further point in common with the line of symmetry and find their number N.

In order to realize the meaning of our problem, consider easy particular cases: for

$$r = 0, \quad n, \quad n/2 \ (n \text{ even})$$
$$N = 1, \quad 1, \quad 0$$

Solution.   It will suffice to consider the case $r > n/2$; that is, the common lower endpoint of our zigzag paths lies in the right-hand half of the plane bisected by the line of symmetry.   There are $\binom{n}{r}$ paths in the full set which we divide into three nonoverlapping subsets.

(1) The subset defined above of which we have to find the number of members, N.   A path of the set that does *not* belong to this subset has, besides A, another point on the line of symmetry.

(2) Paths beginning with a block to the left downward; such a path must cross the line of symmetry somewhere since its endpoint lies in the other half plane.   The number of paths in this subset is obviously $\binom{n-1}{r}$.

(3) Such paths as belong neither to (1) nor to (2); they begin with a block to the right downward but subsequently attain somewhere the line of symmetry.

Show that there are just as many paths in subset (2) as in subset (3) (Fig. 3.10 hints the decisive idea of a one-one correspondence between these subsets) and

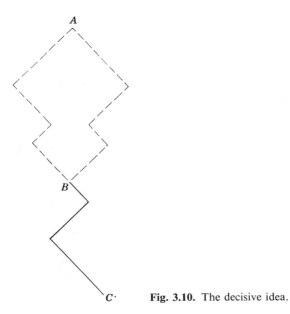

Fig. 3.10. The decisive idea.

Fig. 3.11. A modification of the decisive idea.

derive hence that

$$N = \frac{|2r - n|}{n} \binom{n}{r}$$

3.43. (Continued) The number of all shortest zigzag paths from the apex to the nth base, that have only the initial point in common with the line of symmetry, is $\binom{2m}{m}$ if $n = 2m$ is even and $2 \binom{2m}{m}$ if $n = 2m + 1$ is odd.

3.44. *Trinomial coefficients.* Fig. 3.12 shows a fragment of an infinite triangular array of numbers defined by two conditions.

(1) *Boundary condition.* Each horizontal line or "base" (this term has been similarly used in sect. 3.6) begins with 0, 1 and ends with 1, 0. (The nth base consists of $2n + 3$ numbers and so the boundary condition leaves undefined $2n - 1$ numbers of the nth base, for $n = 1, 2, 3, \ldots .$)

					0	1	0					
				0	1	1	1	0				
			0	1	2	3	2	1	0			
		0	1	3	6	7	6	3	1	0		
	0	1	4	10	16	19	16	10	4	1	0	
0	1	5	15	30	45	51	45	30	15	5	1	0

Fig. 3.12. Trinomial coefficients.

(2) *Recursion formula.* Any number of the $(n + 1)$th base left undefined by (1) is computed as the sum of three numbers of the nth base: of its north-western, northern, and northeastern neighbors. (For instance, $45 = 10 + 16 + 19$.)

Compute the numbers of the seventh base. (They are, with three exceptions, divisible by 7.)

3.45. (Continued) Show that the numbers of the nth base, beginning and ending with 1, are the coefficients in the expansion of $(1 + x + x^2)^n$ in powers of x. (Hence the name "trinomial coefficient.")

3.46. (Continued) Explain the symmetry of Fig. 3.12 with respect to its middle vertical line.

3.47. (Continued) Observe that

$$
\begin{array}{ll}
1 + 1 + 1 & = 3 \\
1 + 2 + 3 + 2 + 1 & = 9 \\
1 + 3 + 6 + 7 + 6 + 3 + 1 & = 27
\end{array}
$$

generalize and prove.

3.48. (Continued) Observe that

$$
\begin{array}{ll}
1 - 1 + 1 & = 1 \\
1 - 2 + 3 - 2 + 1 & = 1 \\
1 - 3 + 6 - 7 + 6 - 3 + 1 & = 1
\end{array}
$$

generalize and prove.

3.49. (Continued) Observe that the value of the sum

$$ 1^2 + 2^2 + 3^2 + 2^2 + 1^2 = 19 $$

is a trinomial coefficient, generalize, and prove.

3.50. (Continued) Find lines in Fig. 3.12 agreeing with lines in the Pascal triangle.

3.51. *Leibnitz's Harmonic Triangle.* Fig. 3.13 shows a section of this little known but remarkable arrangement of numbers. It has properties which are so to say "analogous by contrast" to those of the Pascal triangle. That triangle contains integers, this one (as far as visible) the reciprocals of integers. In Pascal's triangle, each number is the sum of its northwestern and north-eastern neighbors. In Leibnitz's triangle, each number is the sum of its southwestern and southeastern neighbors; for instance

$$ \frac{1}{2} = \frac{1}{3} + \frac{1}{6}, \qquad \frac{1}{3} = \frac{1}{4} + \frac{1}{12}, \qquad \frac{1}{6} = \frac{1}{12} + \frac{1}{12} $$

This is the *recursion formula* of the Leibnitz triangle. This triangle has also a *boundary condition*: the numbers along the northwest borderline (the "0th avenue") are the reciprocals of the successive integers, $1/1, 1/2, 1/3, \ldots$. (The boundary condition of the Pascal triangle is of a different nature: values are prescribed along the whole boundary, 0th avenue *and* 0th street.)

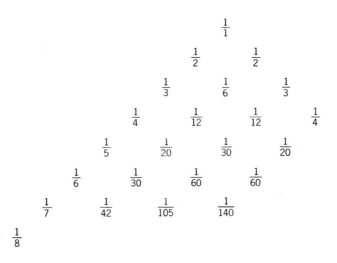

Fig. 3.13. A fragment of Leibnitz's Harmonic Triangle.

Starting from the given boundary values, we obtain the others by addition in the case of the Pascal triangle, but by subtraction in the present case; in Fig. 3.13 there are some gaps that can be filled immediately with the help of the recursion formula, for instance

$$\frac{1}{4} - \frac{1}{20} = \frac{1}{5} \quad \text{and} \quad \frac{1}{7} - \frac{1}{8} = \frac{1}{56}$$

Using the boundary condition and the recursion formula, extend the table of Fig. 3.13 to the eighth base inclusively.

3.52. *Pascal and Leibnitz.* Try to recognize a connection between corresponding numbers of the two triangles and, having recognized it, prove it.

†**3.53.** Prove[9]

$$\frac{1}{1} = \frac{1}{2} + \frac{1}{6} + \frac{1}{12} + \frac{1}{20} + \frac{1}{30} + \cdots$$

$$\frac{1}{2} = \frac{1}{3} + \frac{1}{12} + \frac{1}{30} + \frac{1}{60} + \frac{1}{105} + \cdots$$

$$\frac{1}{3} = \frac{1}{4} + \frac{1}{20} + \frac{1}{60} + \frac{1}{140} + \frac{1}{280} + \cdots$$

. . . .

(Locate these numbers in the harmonic triangle.)

[9] To accommodate readers who have not had the opportunity to acquire precise ideas about infinite series (limits, convergence,...) the solutions to this and similar problems given on the next pages will not insist on precise details. Yet such details (they are easy in most cases) should be supplied by readers with more complete knowledge.

†3.54. Find the sum

$$\frac{1}{12} + \frac{1}{30} + \frac{1}{60} + \frac{1}{105} + \cdots$$

and generalize. (Do you know an analogous problem?)

†3.55. Find the sum of the series

$$\frac{1}{1\cdot 2} + \frac{1}{2\cdot 3} + \frac{1}{3\cdot 4} + \frac{1}{4\cdot 5} + \cdots$$

$$\frac{1}{1\cdot 2\cdot 3} + \frac{1}{2\cdot 3\cdot 4} + \frac{1}{3\cdot 4\cdot 5} + \frac{1}{4\cdot 5\cdot 6} + \cdots$$

. . . .

$$\frac{1}{1\cdot 2 \ldots (r - 1)r} + \frac{1}{2\cdot 3 \ldots r(r + 1)} + \frac{1}{3\cdot 4 \ldots (r + 1)(r + 2)} + \cdots$$

Fourth Part

Several problems in this part are connected with ex. 3.61, and some others with ex. 3.70.

†3.56. *Power series.* The decimal fraction 3.14159...of the number π is, in fact, an "infinite series"

$$3 + 1\left(\frac{1}{10}\right) + 4\left(\frac{1}{10}\right)^2 + 1\left(\frac{1}{10}\right)^3 + 5\left(\frac{1}{10}\right)^4 + 9\left(\frac{1}{10}\right)^5 + \cdots$$

By substituting for $\frac{1}{10}$ the *variable x,* and for the successive digits

$$3, \quad 1, \quad 4, \quad 1, \quad 5, \quad 9, \quad \cdots$$

the *constant coefficients*

$$a_0, \quad a_1, \quad a_2, \quad a_3, \quad a_4, \quad a_5, \quad \cdots$$

respectively, we obtain a *power series*

(1) $$a_0 + a_1 x + a_2 x^2 + a_3 x^3 + \cdots$$

We are not prepared to consider here the convergence of power series and related important matters, only formal operations with such series (see the footnote appended to ex. 3.53). Multiplication of the proposed power series by a constant *c* yields

$$ca_0 + ca_1 x + ca_2 x^2 + ca_3 x^3 + \cdots$$

Addition of the proposed power series (1) to another

(2) $$b_0 + b_1 x + b_2 x^2 + b_3 x^3 + \cdots$$

yields

$$(a_0 + b_0) + (a_1 + b_1)x + (a_2 + b_2)x^2 + (a_3 + b_3)x^3 + \cdots$$

and the product of our two power series (1) and (2) is

$$a_0 b_0 + (a_0 b_1 + a_1 b_0)x + (a_2 b_0 + a_1 b_1 + a_0 b_2)x^2 + \cdots$$

Our two power series (1) and (2) are equal if, and only if,

$$a_0 = b_0, \quad a_1 = b_1, \quad a_2 = b_2, \quad \ldots, \quad a_n = b_n, \quad \ldots$$

We conceive a polynomial as a power series in which infinitely many coefficients, in fact, all coefficients beyond a certain one, vanish. For instance, the polynomial $3x - x^3$ has to be considered as the particular case of our power series (1) in which

$$a_0 = 0, \quad a_1 = 3, \quad a_2 = 0, \quad a_3 = -1 \quad \text{and} \quad a_n = 0 \quad \text{for} \quad n \geq 4$$

Convince yourself that the foregoing rules are valid for polynomials.

†**3.57.** Compute the product

$$(1 - x)(1 + x + x^2 + \cdots + x^n + \cdots)$$

†**3.58.** Find the coefficient of x^n in the product

$$(a_0 + a_1x + a_2x^2 + \cdots + a_nx^n + \cdots)(1 + x + x^2 + \cdots + x^n + \cdots)$$

†**3.59.** A gap in the solution of ex. 3.57 may suggest the consideration of the series

$$
\begin{array}{lllll}
1 + & x + & x^2 + & x^3 + \cdots \\
1 + & 2x + & 3x^2 + & 4x^3 + \cdots \\
1 + & 3x + & 6x^2 + & 10x^3 + \cdots \\
1 + & 4x + & 10x^2 + & 20x^3 + \cdots
\end{array}
$$

Do you know the sum of one of these series? Could you find the sum of the others?

†**3.60.** Give another proof for the result of ex. 3.37.

†**3.61.** *The binomial theorem for fractional and negative exponents.* In a letter of October 24, 1676, addressed to the Secretary of the Royal Society, Newton described how he discovered the (general) binomial theorem; he wrote this letter to answer an inquiry of Leibnitz about his (Newton's) method of discovery.[10] Newton considered the areas under certain curves; he was strongly influenced by ideas of Wallis about interpolation; and eventually he arrived at a conjecture: *The expansion*

$$(1 + x)^a = 1 + \frac{a}{1}x + \frac{a(a-1)}{1 \cdot 2}x^2 + \frac{a(a-1)(a-2)}{1 \cdot 2 \cdot 3}x^3 + \cdots$$

is valid not only for positive integral values of the exponent a, but also for fractional and negative values, in fact, for all numerical values of a.[11]

Newton did not produce a formal proof for his conjecture; rather he relied on examples and analogy. He investigated the question, we may say, as a physicist, "experimentally" or "inductively." In order to understand his viewpoint, we shall try to retrace some of the steps that convinced him of the

[10] Cf. J. R. Newman, *The World of Mathematics*, vol. 1, pp. 519–524.

[11] We know today that some restriction concerning x is necessary but we disregard it here. Such neglect agrees with Newton's standpoint in whose time the convergence of a series was not explicitly defined, and it agrees also with the standpoint of footnote 9.

soundness of his conjecture, which we shall call, for the sake of brevity, the "conjecture N."

If a is a non-negative integer, the coefficient of x^{a+1} vanishes on the right-hand side of the proposed series, and all the following coefficients vanish (thanks to the presence of a factor 0 in the numerator): the series terminates. If, however, a has a value not contained in the sequence $0, 1, 2, 3, \ldots$, the series does not terminate but goes to infinity.   For example, for $a = \frac{1}{2}$ the expansion under scrutiny turns out to be

$$(1 + x)^{\frac{1}{2}} = 1 + \frac{\frac{1}{2}}{1}x + \frac{\frac{1}{2}(-\frac{1}{2})}{1 \cdot 2} x^2 + \frac{\frac{1}{2}(-\frac{1}{2})(-\frac{3}{2})}{1 \cdot 2 \cdot 3} x^3 + \cdots$$

$$= 1 + \frac{x}{2} - \frac{x^2}{8} + \frac{x^3}{16} - \frac{5x^4}{128} + \cdots$$

Newton did not appear to be disturbed by an infinity of nonvanishing terms. He knew very well the analogy, which he mentions elsewhere, between power series and decimal fractions; see ex. 3.56.   Now, some decimal fractions terminate (as that for $\frac{1}{2}$ or $\frac{3}{5}$) and others do not (as that for $\frac{1}{3}$ or $\frac{7}{11}$, for instance).

Is the above series for $(1 + x)^{\frac{1}{2}}$ valid?   To examine this question, Newton multiplied the series by itself; the result *should* be

$$(1 + x)^{\frac{1}{2}}(1 + x)^{\frac{1}{2}} = 1 + x$$

To check this, compute the coefficients of x, x^2, x^3, and x^4 in the product series (ex. 3.56).

†**3.62.** Compute the coefficients of x, x^2, x^3, and x^4 in the square of the series

$$1 + \frac{x}{3} - \frac{x^2}{9} + \frac{5x^3}{81} - \frac{10x^4}{243} + \cdots$$

which is the expansion of $(1 + x)^{\frac{1}{3}}$ according to conjecture N.   The result should be the expansion, according to the same conjecture, of $(1 + x)^{\frac{2}{3}}$. Check it!

†**3.63.** (Continued) Compute the coefficients of x, x^2, x^3, and x^4 in the cube of the given series.   Predict the result and check your prediction.

†**3.64.** Expand $(1 + x)^{-1}$ according to conjecture N.   Any comment?

3.65. *Extending the range.*   In sect. 3.6, we have defined the symbol $\binom{n}{r}$

for non-negative integers n and r subject to the inequality $r \leqq n$.   We now extend the range of n (but not that of r, cf. ex. 3.11): we set

$$\binom{x}{0} = 1, \qquad \binom{x}{r} = \frac{x(x - 1)(x - 2)\cdots(x - r + 1)}{1 \cdot \ 2 \ \cdot \ 3 \ \cdots \ r}$$

for $r = 1, 2, 3, \cdots$ and an arbitrary number x.   This definition implies:

(I) $\binom{x}{r}$ is a polynomial in x, of degree r, for $r = 0, 1, 2, 3, \cdots$.

(II) $\binom{x}{r} = (-1)^r \binom{r - 1 - x}{r}$

(III) If n and r are non-negative integers and $r > n$, $\binom{n}{r} = 0$.

(IV) Conjecture N can be written in the form

$$(1 + x)^a = \binom{a}{0} + \binom{a}{1}x + \binom{a}{2}x^2 + \cdots + \binom{a}{n}x^n + \cdots$$

(I), (III), and (IV) are obvious; prove (II).

†**3.66.** Generalize ex. 3.64: examine whether the full result of ex. 3.59 agrees with the conjecture N.

†**3.67.** Apply a device which we have already used three times (sect. 3.9, ex. 3.36, and ex. 3.60) once more: Assuming the conjecture N, compute in two different ways the coefficient of x^r in the expansion of

$$(1 + x)^a(1 + x)^b = (1 + x)^{a+b}$$

†**3.68.** Try to assess the result of ex. 3.67: is it proved? Is some part of it proved? Are there other means to prove it? If we took it for granted, could we prove conjecture N? Or could we prove some part of conjecture N?

†**3.69.** Try to recognize the coefficients of the expansion

$$(1 - 4x)^{-\frac{1}{2}} = 1 + 2x + 6x^2 + 20x^3 + \cdots$$

and express the general term in a familiar form (which should render obvious the fact that the coefficients are integers).

†**3.70.** *The method of undetermined coefficients.* Expand the ratio of two given power series in a power series.

We have to expand in powers of the variable x the ratio

$$\frac{b_0 + b_1x + b_2x^2 + \cdots + b_nx^n + \cdots}{a_0 + a_1x + a_2x^2 + \cdots + a_nx^n + \cdots}$$

where the coefficients $a_0, a_1, a_2, \ldots, b_0, b_1, b_2, \ldots$ are given numbers; we assume that $a_0 \neq 0$. (This assumption, which we have not mentioned in the first short statement of the problem, is essential.)

We are required to exhibit the given ratio in the form

$$\frac{b_0 + b_1x + b_2x^2 + \cdots}{a_0 + a_1x + a_2x^2 + \cdots} = u_0 + u_1x + u_2x^2 + \cdots$$

The coefficients $u_0, u_1, u_2, \ldots, u_n, \ldots$ are not yet determined at this moment when we are just introducing them (hence the name of the method which we are about to apply), yet we hope to determine them eventually; in fact, to determine them is precisely the task imposed upon us by the problem; the coefficients $u_0, u_1, u_2, \ldots$ are the unknowns of our problem (which has, as we now see, an infinity of unknowns).

We rewrite the relation between the three power series (two are given, one we are required to find) in the form

$$(a_0 + a_1x + a_2x^2 + \cdots)(u_0 + u_1x + u_2x^2 + \cdots) = b_0 + b_1x + b_2x^2 + \cdots$$

and now we can see the situation in a more familiar light (ex. 3.56): by equating the coefficients of like powers of x on opposite sides, we obtain a system of equations

$$
\begin{aligned}
a_0 u_0 &= b_0 \\
a_1 u_0 + a_0 u_1 &= b_1 \\
a_2 u_0 + a_1 u_1 + a_0 u_2 &= b_2 \\
a_3 u_0 + a_2 u_1 + a_1 u_2 + a_0 u_3 &= b_3
\end{aligned}
$$

$$\cdot \quad \cdot \quad \cdot \quad \cdot \quad \cdot \quad \cdot \quad \cdot \quad \cdot \quad \cdot \quad \cdot$$

This system of equations presents a familiar pattern: it is recursive, that is, it can be solved by recursion. We obtain the initial unknown u_0 from the initial equation and, having obtained $u_0, u_1, \ldots, u_{n-2}$ and u_{n-1}, we find the next unknown u_n from the next equation not formerly used.

Express u_0, u_1, u_2, and u_3 in terms of a_0, a_1, a_2, a_3, b_0, b_1, b_2, and b_3.

(The foregoing solution can advantageously serve as a pattern. Note the typical steps:

we introduce the unknowns as the coefficients of a power series;
we derive a system of equations by comparing coefficients of like powers on opposite sides of a relation between power series;
we compute the unknowns recursively.

These steps characterize the pattern, or method, of "undetermined coefficients" which yields some of the most remarkable and most useful systems of equations that can be solved by recursion.)

†**3.71.** We consider the product of powers

$$a_i{}^{\alpha_i} a_j{}^{\alpha_j} a_k{}^{\alpha_k} b_l{}^{\beta_l} b_m{}^{\beta_m}$$

of which, by definition,

$\alpha_i + \alpha_j + \alpha_k + \beta_l + \beta_m$ is the *degree,*
$\alpha_i + \alpha_j + \alpha_k$ the *degree in the a's.*
$\beta_l + \beta_m$ the *degree in the b's*
$i\alpha_i + j\alpha_j + k\alpha_k + l\beta_l + m\beta_m$ the *weight.*

Of course, the terms just defined are applicable if any number of a's and b's are involved, and not just three of the one kind and two of the other.

Observe the expressions for u_0, u_1, u_2, and u_3 you found in answering ex. 3.70 and explain the regularities observed.

†**3.72.** Expand the ratio

$$\frac{b_0 + b_1 x + b_2 x^2 + \cdots + b_n x^n + \cdots}{1 + x + x^2 + \cdots + x^n + \cdots}$$

(The result is simple—can you use it?)

†**3.73.** Expand the ratio

$$\frac{b_0 + b_1 x + b_2 x^2 + \cdots + b_n x^n + \cdots}{1 - x}$$

(The result is simple—can you use it?)

†3.74. Expand the ratio

$$\frac{1 + \dfrac{x}{3} + \dfrac{x^2}{15} + \dfrac{x^3}{105} + \cdots + \dfrac{x^n}{3 \cdot 5 \cdot 7 \cdots (2n+1)} + \cdots}{1 + \dfrac{x}{2} + \dfrac{x^2}{8} + \dfrac{x^3}{48} + \cdots + \dfrac{x^n}{2 \cdot 4 \cdot 6 \cdots 2n} + \cdots}$$

(Compute a few terms and try to guess the general term.)

†3.75. *Inversion of a power series.* Being given the power series of a function, find the power series of the inverse function.

In other words: being given the expansion of x in powers of y, expand y in powers of x.

More precisely: being given

$$x = a_1 y + a_2 y^2 + \cdots + a_n y^n + \cdots$$

assume that $a_1 \neq 0$ and find the expansion

$$y = u_1 x + u_2 x^2 + \cdots + u_n x^n + \cdots$$

We follow the pattern of ex. 3.70. In the given expansion of x in powers of y, we substitute for y its (desired) power series:

$$\begin{aligned}
x = \; & a_1(u_1 x + u_2 x^2 + u_3 x^3 + \cdots) \\
& + a_2(u_1{}^2 x^2 + 2u_1 u_2 x^3 + \cdots) \\
& + a_3(u_1{}^3 x^3 + \cdots) \\
& + \quad . \quad . \quad .
\end{aligned}$$

In equating the coefficients of like powers of x on opposite sides of this relation, we obtain a system of equations for $u_1, u_2, u_3, \ldots$:

$$\begin{aligned}
1 &= a_1 u_1 \\
0 &= a_1 u_2 + a_2 u_1{}^2 \\
0 &= a_1 u_3 + 2a_2 u_1 u_2 + a_3 u_1{}^3 \\
& \cdot \quad \cdot \quad \cdot \quad \cdot \quad \cdot \quad \cdot \quad \cdot \quad \cdot \quad \cdot
\end{aligned}$$

and the system so obtained is recursive (although not linear.)

Express u_1, u_2, u_3, u_4, and u_5 in terms of a_1, a_2, a_3, a_4, and a_5.

†3.76. Examine the degree and weight of the expressions you found in answering ex. 3.75.

†3.77. Being given that

$$x = y + y^2 + y^3 + \cdots + y^n + \cdots$$

expand y in powers of x.

(The result is simple—can you use it?)

†3.78. Being given that

$$4x = 2y - 3y^2 + 4y^3 - 5y^4 + \cdots$$

expand y in powers of x. (Try to guess the form of the general term and, having guessed it, try to explain it.)

†3.79. Being given that

$$x = y + ay^2$$

expand y in powers of x. (The result can be used to clarify a detail of the general situation considered in ex. 3.75.)

†3.80. Being given that

$$x = y + \frac{y^2}{2} + \frac{y^3}{6} + \frac{y^4}{24} + \cdots + \frac{y^n}{n!} + \cdots$$

expand y in powers of x.

†3.81. *Differential equations.* Expand in powers of x the function y that satisfies the *differential equation*

$$\frac{dy}{dx} = x^2 + y^2$$

and the *initial condition*

$$y = 1 \quad \text{for} \quad x = 0.$$

Following the pattern of ex. 3.70, we set

$$y = u_0 + u_1 x + u_2 x^2 + u_3 x^3 + \cdots$$

with coefficients $u_0, u_1, u_2, \ldots$ which we have still to determine. The differential equation requires

$$u_1 + 2u_2 x + 3u_3 x^2 + 4u_4 x^3 + \cdots$$
$$= u_0^2 + 2u_0 u_1 x + (2u_0 u_2 + u_1^2 + 1)x^2 + \cdots$$

Comparing coefficients of like powers on opposite sides of this relation, we obtain the system of equations.

$$\begin{aligned}
u_1 &= u_0^2 \\
2u_2 &= 2u_0 u_1 \\
3u_3 &= 2u_0 u_2 + u_1^2 + 1 \\
4u_4 &= 2u_0 u_3 + 2u_1 u_2
\end{aligned}$$

$$\cdot \quad \cdot \quad \cdot \quad \cdot \quad \cdot \quad \cdot \quad \cdot$$

From this system we can find $u_1, u_2, u_3, \ldots$, recursively, since the initial condition yields

$$u_0 = 1$$

Compute numerically u_1, u_2, u_3, and u_4.

(The solution of differential equations by the method of undetermined coefficients, which is exemplified by our problem, is of great importance both in theory and in practice.)

†3.82. (Continued) Show that $u_n > 1$ for $n \geq 3$.

†3.83. Expand in powers of x the function y that satisfies the differential equation

$$\frac{d^2 y}{dx^2} = -y$$

and the initial conditions

$$y = 1, \quad \frac{dy}{dx} = 0 \quad \text{for} \quad x = 0$$

†**3.84.** Find the coefficient of x^{100} in the power series expansion of the function

$$(1 - x)^{-1}(1 - x^5)^{-1}(1 - x^{10})^{-1}(1 - x^{25})^{-1}(1 - x^{50})^{-1}$$

There is little doubt that, in order to solve the proposed problem, we have to generalize it and look for ways and means to compute the general coefficient (that of x^n) in the expansion under consideration. It is also advisable to examine the easier analogous problems implied by the proposed problem. Some meditation on these lines may eventually suggest a plan: introduce *several* power series with "undetermined" coefficients. We set

$$(1 - x)^{-1} = A_0 + A_1 x + A_2 x^2 + A_3 x^3 + A_4 x^4 + \cdots$$
$$(1 - x)^{-1}(1 - x^5)^{-1} = B_0 + B_1 x + B_2 x^2 + B_3 x^3 + \cdots$$
$$(1 - x)^{-1}(1 - x^5)^{-1}(1 - x^{10})^{-1} = C_0 + C_1 x + C_2 x^2 + \cdots$$
$$(1 - x)^{-1}(1 - x^5)^{-1}(1 - x^{10})^{-1}(1 - x^{25})^{-1} = D_0 + D_1 x + \cdots$$

and finally

$$(1 - x)^{-1}(1 - x^5)^{-1}(1 - x^{10})^{-1}(1 - x^{25})^{-1}(1 - x^{50})^{-1}$$
$$= E_0 + E_1 x + E_2 x^2 + \cdots + E_n x^n + \cdots$$

In this notation, the proposed problem requires to find E_{100}. Instead of our single original unknown E_{100}, we have introduced infinitely many new unknowns: we should find A_n, B_n, C_n, D_n, and E_n for $n = 0, 1, 2, 3, \ldots$. Yet, for some of these, the result is well known or obvious.

$$A_0 = A_1 = A_2 = \cdots = A_n = \cdots = 1$$
$$B_0 = C_0 = D_0 = E_0 = 1$$

Moreover, the unknowns introduced are not unrelated:

$$A_0 + A_1 x + A_2 x^2 + \cdots = (B_0 + B_1 x + B_2 x^2 + \cdots)(1 - x^5)$$

from which we conclude, by looking at the coefficient of x^n, that

$$A_n = B_n - B_{n-5}$$

Find analogous relations and find the intermediaries through which the unknown E_{100} is connected with the values already known. Eventually, you should obtain a numerical value for E_{100}.

†**3.85.** Find the nth derivative $y^{(n)}$ of the function $y = x^{-1} \log x$.

We obtain by straightforward differentiation and algebraic transformation

$$y' = - x^{-2} \log x + x^{-2}$$
$$y'' = 2x^{-3} \log x - 3x^{-3}$$
$$y''' = -6x^{-4} \log x + 11x^{-4}$$

and from these (or more) cases we may guess that the desired nth derivative is of the form

$$y^{(n)} = (-1)^n n! x^{-n-1} \log x + (-1)^{n-1} c_n x^{-n-1}$$

where c_n is an integer depending on n (but independent of x). Prove this and express c_n in terms of n.

3.86. Find a short expression for
$$1 + 2x + 3x^2 + \cdots + nx^{n-1}$$
(Do you know a related problem? Could you use its result—or its method?)

3.87. Find a short expression for
$$1 + 4x + 9x^2 + \cdots + n^2 x^{n-1}$$
(Do you know a related problem? Could you use its result—or its method?)

3.88. (Continued) Generalize.

3.89. Being given that
$$a_{n+1} = a_n \frac{n + \alpha}{n + 1 + \beta}$$
for $n = 1, 2, 3, \ldots$ and $\alpha \neq \beta$, show that
$$a_1 + a_2 + a_3 + \cdots + a_n = \frac{a_n(n + \alpha) - a_1(1 + \beta)}{\alpha - \beta}$$

3.90. Find
$$\frac{p}{q} + \frac{p}{q}\frac{p+1}{q+1} + \frac{p}{q}\frac{p+1}{q+1}\frac{p+2}{q+2} + \cdots + \frac{p}{q}\frac{p+1}{q+1}\frac{p+2}{q+2}\cdots\frac{p+n-1}{q+n-1}$$

3.91. *On the number π.* We consider the unit circle (its radius $=1$); we circumscribe about it, and inscribe in it, a regular polygon with n sides; let C_n (circumscribed) and I_n (inscribed) stand for the perimeters of these two polygons, respectively.

Introduce the abbreviations
$$\frac{a+b}{2} = A(a, b), \qquad \sqrt{ab} = G(a, b), \qquad \frac{2ab}{a+b} = H(a, b)$$
(arithmetic, geometric, and harmonic mean, respectively).

(1) Find C_4, I_4, C_6, I_6.
(2) Show that
$$C_{2n} = H(C_n, I_n), \qquad I_{2n} = G(I_n, C_{2n})$$
(Thus, starting from C_6, I_6 we can compute the sequence of numbers
$$C_6, I_6; \qquad C_{12}, I_{12}; \qquad C_{24}, I_{24}; \qquad C_{48}, I_{48}; \quad \ldots$$
by recursion as far as we wish, and so we can enclose π between two numerical bounds whose difference is arbitrarily small. Archimedes, in computing the first ten numbers of the sequence, that is, proceeding to regular polygons with 96 sides, found that
$$3\tfrac{10}{71} < \pi < 3\tfrac{1}{7}$$
See his *Works*, edited by T. L. Heath (Dover), pp. 91–98.)

3.92. *More problems.* Devise some problems similar to, but different from, the problems proposed in this chapter—especially such problems as you can solve.

CHAPTER 4

SUPERPOSITION

4.1. Interpolation

We need several steps to arrive at the final formulation of our next problem.

(1) We are given n different abscissas

$$x_1, \qquad x_2, \qquad x_3, \qquad \cdots, \qquad x_n$$

and n corresponding ordinates

$$y_1, \qquad y_2, \qquad y_3, \qquad \cdots, \qquad y_n$$

and so we are given n different points

$$(x_1, y_1), \qquad (x_2, y_2), \qquad (x_3, y_3), \qquad \cdots, \qquad (x_n, y_n)$$

We are required to find a function $f(x)$ the values of which at the given abscissas are the corresponding ordinates:

$$f(x_1) = y_1, \quad f(x_2) = y_2, \quad f(x_3) = y_3, \quad \cdots, \quad f(x_n) = y_n$$

In other words, we are required to find a curve, with the equation $y = f(x)$, that passes through the n given points; see Fig. 4.1. This is the problem of *interpolation*. Let us explore the background of this problem; such exploration may increase our interest in it and so our chances to solve it.

(2) The problem of interpolation may arise whenever we consider a quantity y depending on another quantity x. Let us take a more concrete case: let x be the temperature and y the length of a homogeneous rod, kept under constant pressure. To each temperature x there corresponds a certain length y of the rod; this is what we express by saying that y *depends*

99

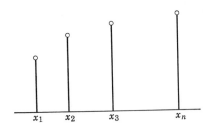

Fig. 4.1. Interpolation.

on x, or y is a *function* of x, or by writing $y = f(x)$. A physicist, in investigating experimentally the dependence of y on x, subjects the rod to different temperatures

$$x_1, \quad x_2, \quad x_3, \quad \ldots, \quad x_n$$

and, by measuring the length of the rod at each of these temperatures, he finds the values

$$y_1, \quad y_2, \quad y_3, \quad \ldots, \quad y_n$$

respectively. The physicist, of course, would like to know the length y also at some such temperature x as he has not yet had the opportunity to observe. That is, the physicist wants to know, on the basis of his n observations, the function $y = f(x)$ in its full extent, for the whole range of the independent variable x—and so he poses the problem of interpolation.

(3) Let us remark parenthetically that the physicist's problem is, in fact, more complicated. His values $x_1, y_1, x_2, y_2, \ldots, x_n, y_n$ are not the "true values" of the quantities measured but are affected by unavoidable errors of measurement. Therefore, his curve need not pass *through*, it should only pass *close* to, the given points.

Moreover, it is usual to distinguish two cases: the hitherto unobserved abscissa x, to which the physicist wants to find the corresponding ordinate y, may lie in the interval *between* the extreme observed values (x_1 and x_n in Fig. 4.1) or it may lie *outside* this interval: in the first case it is customary to speak of *interpolation* and in the latter of *extrapolation*. (It is usual to regard interpolation as more reliable than extrapolation.)

Yet let us disregard this distinction and the other remarks of this subsection, for the time being; let us close the parenthesis and return to the standpoint of the subsections (1) and (2).

(4) The problem posed in subsection (1) is utterly indeterminate: there is an inexhaustible variety of curves passing through the n given points. His n observations, by themselves, do not entitle the physicist to prefer one of those curves to the others. If the physicist decides to draw a curve, he

must have some reason for his choice *outside* his n observations—what reason?

Thus the problem of interpolation raises (and this adds a good deal to its interest) a general question: What suggests, or what justifies, the transition to a mathematical formulation from given observations and a given mental background? This is a major philosophical question—yet, as it is rather unlikely that major philosophical questions can be satisfactorily answered, we turn to another aspect of the problem of interpolation.

(5) It would be natural to modify the problem stated in subsection (1) by asking for the *simplest* curve passing through the n given points. This modification, however, leaves the problem indeterminate, even vague, since "simplicity" is hardly an objective quality: we may judge simplicity according to our personal taste, standpoint, background, or mental habits.

Yet, in the case of our problem, we may give an interpretation to the term "simple" that looks acceptable and leads to a determinate and useful formulation. First, let us regard addition, subtraction, and multiplication as the simplest computational operations. Then, let us regard such functions as the simplest the values of which can be computed by the simplest computational operations. Accepting both points, we have to regard the polynomials as the simplest functions; a polynomial is of the form

$$a_0 + a_1 x + a_2 x^2 + \cdots + a_n x^n$$

Its value can be computed by the three simplest computational operations from the numerically given coefficients $a_0, a_1, \ldots, a_n$ and the value of the independent variable x. If we assume that $a_n \neq 0$, the degree of the polynomial is n.

Finally, being given two polynomials of different degree, let us regard the one with the lower degree as the simpler. If we accept this point too, the problem of passing the simplest possible curve through n points becomes a determinate problem, the problem of *polynomial interpolation*, which we formulate as follows:

Being given n different numbers $x_1, x_2, \ldots, x_n$ and n further numbers $y_1, y_2, \ldots, y_n$, find the polynomial $f(x)$ of the lowest possible degree satisfying the n conditions

$$f(x_1) = y_1, \quad f(x_2) = y_2, \quad \ldots, \quad f(x_n) = y_n$$

4.2. A special situation

If we see no other approach to the proposed problem, we may try to *vary the data*. For instance, we may keep one given ordinate fixed and let the others decrease; and so we may hit upon a special situation that looks more accessible. We need not touch the given abscissas, we accept any n different numbers

$$x_1, \quad x_2, \quad x_3, \quad \ldots, \quad x_n$$

but we choose a particularly simple system of ordinates:

$$0, \quad 1, \quad 0, \quad \ldots, \quad 0$$

respectively. (All given ordinates vanish, except the one corresponding to the abscissa x_2; see Fig. 4.2.)

There is an interesting piece of information: the polynomial assuming these values vanishes at $n - 1$ given points, has the $n - 1$ different roots $x_1, x_3, x_4, \ldots, x_n$, and, therefore, it must be divisible by each of the following $n - 1$ factors:

$$x - x_1, \quad x - x_3, \quad x - x_4, \quad \ldots, \quad x - x_n$$

Therefore, it must be divisible by the product of these $n - 1$ factors, and so it is at least of degree $n - 1$. If the polynomial attains this lowest possible degree $n - 1$, it must be of the form

$$f(x) = C(x - x_1)(x - x_3)(x - x_4)\ldots(x - x_n)$$

where C is some constant.

Have we used all the data? There remains the ordinate corresponding to the abscissa x_2 to be taken into account:

$$f(x_2) = C(x_2 - x_1)(x_2 - x_3)(x_2 - x_4)\ldots(x_2 - x_n) = 1$$

We compute C from this equation, substitute the value computed for C in the expression of $f(x)$, and find so

$$f(x) = \frac{(x - x_1)\,(x - x_3)\,(x - x_4)\ldots(x - x_n)}{(x_2 - x_1)(x_2 - x_3)(x_2 - x_4)\ldots(x_2 - x_n)}$$

Obviously, this polynomial $f(x)$ takes the required values for all given abscissas. We have succeeded in solving the problem of polynomial interpolation in a particular case, in a special situation.

4.3. Combining particular cases to solve the general case

We were lucky to spot such an especially accessible particular case. To

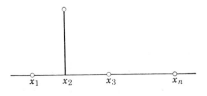

Fig. 4.2. A special situation.

deserve our luck, we should try now to make good use of the solution obtained.

By modifying a little the solution obtained, we can handle a slightly more extended particular case: to the given abscissas

$$x_1, \quad x_2, \quad x_3, \quad \ldots, \quad x_n$$

we let correspond the ordinates

$$0, \quad y_2, \quad 0, \quad \ldots, \quad 0$$

respectively. We obtain the polynomial that assumes these values by multiplying the expression obtained in sect. 4.2 by an obvious factor:

$$y_2 \, \frac{(x - x_1)(x - x_3)(x - x_4)\ldots(x - x_n)}{(x_2 - x_1)(x_2 - x_3)(x_2 - x_4)\ldots(x_2 - x_n)}$$

In this expression, the abscissa x_2 plays a special role, distinct from the common role that falls to the other abscissas. Yet there is no peculiar virtue in the abscissa x_2: we can let any other given abscissa play that special role. And so, if to the abscissas

$$x_1, \quad x_2, \quad x_3, \quad \ldots, \quad x_n$$

we let correspond the values displayed on any one of the n following lines

$$
\begin{array}{ccccc}
y_1, & 0, & 0, & \ldots, & 0 \\
0, & y_2, & 0, & \ldots, & 0 \\
0, & 0, & y_3, & \ldots, & 0 \\
\cdot & \cdot & \cdot & \cdot & \cdot \\
0, & 0, & 0, & \ldots, & y_n
\end{array}
$$

we can write down an expression for the polynomial of degree $n - 1$ that assumes the values on that line at the corresponding abscissas.

We have here outlined the solution in n different particular cases of our problem. Can you combine them so as to obtain the solution of the general case from the combination? Of course, you can, by adding the n expressions outlined:

$$f(x) = y_1 \, \frac{(x - x_2)(x - x_3)(x - x_4)\ldots(x - x_n)}{(x_1 - x_2)(x_1 - x_3)(x_1 - x_4)\ldots(x_1 - x_n)}$$

$$+ \, y_2 \, \frac{(x - x_1)(x - x_3)(x - x_4)\ldots(x - x_n)}{(x_2 - x_1)(x_2 - x_3)(x_2 - x_4)\ldots(x_2 - x_n)}$$

$$+ \, y_3 \, \frac{(x - x_1)(x - x_2)(x - x_4)\ldots(x - x_n)}{(x_3 - x_1)(x_3 - x_2)(x_3 - x_4)\ldots(x_3 - x_n)}$$

$$\cdot \quad \cdot \quad \cdot \quad \cdot \quad \cdot \quad \cdot \quad \cdot \quad \cdot$$

$$+ \, y_n \, \frac{(x - x_1)(x - x_2)(x - x_3)\ldots(x - x_{n-1})}{(x_n - x_1)(x_n - x_2)(x_n - x_3)\ldots(x_n - x_{n-1})}$$

is a polynomial, of degree not exceeding $n - 1$, that satisfies the condition

$$f(x_i) = y_i \quad \text{for} \quad i = 1, 2, 3, \ldots, n$$

as we can see at a glance if we realize the structure of its expression.

(Are there any questions?)

4.4. The pattern

The foregoing solution of the interpolation problem, which is due to Lagrange, has a highly suggestive general plan. *Have you seen it before?*

(1) The reader is probably familiar with, and by the foregoing may be reminded of, the usual proof of a well-known theorem of plane geometry: "The angle at the center of a circle is double the angle at the circumference on the same base, that is, on the same arc." (The arc is emphasized by a double line in Figs. 4.3 and 4.4.) The proof is based on two remarks, and proceeds in two steps; cf. Euclid III 20.

(2) There is a more accessible *special situation*: If one of the sides of the angle at the circumference is a diameter, see Fig. 4.3, the angle at the center α is obviously the sum of two angles of an isosceles triangle; these two angles are equal to each other, and one of them is the angle at the circumference, β. This proves the desired equation

$$\alpha = 2\beta$$

for the special situation of Fig. 4.3.

(3) Now, we have no more the special situation of Fig. 4.3 before us. We can, however, draw a diameter (dotted line in Fig. 4.4) through the vertex of the angle at the circumference, and then the special situation arises twice in the figure. Let the equations

$$\alpha' = 2\beta', \qquad \alpha'' = 2\beta''$$

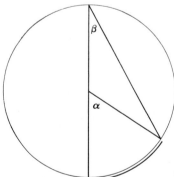

Fig. 4.3. A special situation.

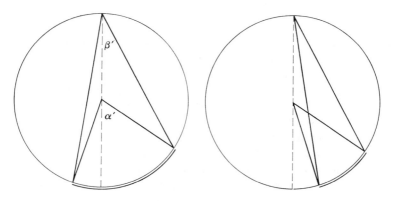

Fig. 4.4. The general case.

refer to these special situations (see Fig. 4.3). These equations are firmly established by the considerations of subsection (2). The angles α and β, at the center and at the circumference, respectively, with which the desired theorem deals, can be exhibited as sum or as difference, according as we have one or the other case represented by Fig. 4.4 before us:

$$\alpha = \alpha' + \alpha'', \quad \beta = \beta' + \beta'' \quad \text{or} \quad \alpha = \alpha' - \alpha'', \quad \beta = \beta' - \beta''$$

Now, by adding and subtracting our two already established equations, we obtain

$$\alpha' + \alpha'' = 2(\beta' + \beta''), \quad \alpha' - \alpha'' = 2(\beta' - \beta'')$$

respectively, and this proves the desired theorem

$$\alpha = 2\beta$$

in full generality.

(4) Now, let us compare the two problems discussed in this chapter: the problem to find, from algebra, treated in sects. 4.1, 4.2, and 4.3; and the problem to prove, from plane geometry, treated in subsections (1), (2), and (3) of the present section. Although these problems differ in several respects, their solutions show the same *pattern*. In both examples, the result was attained in two steps.

First, we were lucky enough to spot a particularly accessible case, a *special situation*, and gave a solution well adapted, but restricted, to this special situation; see sect. 4.2 and subsection (2), Fig. 4.2 and Fig. 4.3.

Then, by *combining particular cases* to which the restricted solution is applicable, we obtained the full, unrestricted solution, applicable to the *general case*; see sect. 4.3 and subsection (3).

Let us introduce two terms which underline certain features of this pattern.

The first step deals with a particular case which is not only especially accessible, but also especially useful; we can appropriately call it a *leading particular case*: it leads the way to the general solution.[1]

The second step combines particular cases by a specific algebraic operation. In sect. 4.3, n particular solutions, after being multiplied by given constants, are added to form the general solution. In subsection (3), we add and subtract equations dealing with the special situation to obtain the general proof. Let us call the algebraic operation employed in sect. 4.3 [there is more generality there than in subsection (3)] *linear combination* or *superposition*. (More about this concept in ex. 4.11.)

We may use the terms introduced to outline our pattern: *Starting from a leading special situation we attain the general solution by superposition of particular cases.*

Other comments and more examples may enable the reader to fill in this outline. He may even burst this outline and enlarge the scope of the pattern.

Examples and Comments on Chapter 4

First Part

4.1. In proving the expression $Bh/3$ for the volume of a pyramid (B is the base, h the height) we may regard the case of the tetrahedron (which is a pyramid with triangular base) as leading particular case and use superposition. How?

4.2. If $f(x)$ is a polynomial of degree k, there exists a polynomial $F(x)$ of degree $k + 1$ such that, for $n = 1, 2, 3, \ldots,$

$$f(1) + f(2) + f(3) + \cdots + f(n) = F(n)$$

In proving this theorem, we may regard the result of ex. 3.3 as leading particular case and use superposition. How?

4.3. (Continued) There is, however, another way: we may regard the result of ex. 3.34 as leading particular case and give so, by using superposition, a different proof. How?

4.4. Being given the coefficients $a_0, a_1, a_2, \ldots, a_k$, find numbers $b_0, b_1, b_2, \ldots, b_k$ such that the equation

$$a_0 x^k + a_1 x^{k-1} + \cdots + a_k = b_0 \binom{x}{k} + b_1 \binom{x}{k-1} + \cdots + b_k \binom{x}{0}$$

holds identically in x (notation of ex. 3.65).

Show that this problem has just one solution.

[1] MPR, vol. 1, p. 24, ex. 7, 8, 9.

4.5. By applying the method of ex. 4.3, give a new derivation for the expression of S_3 obtained in sect. 3.3.

4.6. By applying the result of ex. 4.3 (the theorem stated in ex. 4.2) give a new derivation for the expression of S_3 obtained in sect. 3.3.

4.7. What does ex. 4.3 yield for the problem of ex. 3.3?

4.8. A question on sect. 4.1: What about the particular case $n = 2$?   When just two points are given, it would be natural to say that the simplest line passing through them is the straight line, which is uniquely determined.   Is this in agreement with the standpoint at which we eventually arrived in sect. 4.1(5)?

4.9. A question on sect. 4.2: What about the particular case in which

$$y_i = 0 \qquad \text{for} \quad i = 1, 2, \ldots \, n$$

that is, all the given ordinates vanish?

4.10. A question on sect. 4.3: Does the polynomial $f(x)$ obtained satisfy all clauses of the condition?   Is its degree the lowest possible?

4.11. *Linear combination or superposition.*   Let

$$V_1, \qquad V_2, \qquad V_3, \qquad \cdots, \qquad V_n$$

be n mathematical objects of some clearly stated nature (belonging to some clearly defined set) such that their *linear combination*

$$c_1 V_1 + c_2 V_2 + c_3 V_3 + \cdots + c_n V_n$$

formed with any n numbers

$$c_1, \qquad c_2, \qquad c_3, \qquad \cdots, \qquad c_n$$

is of the same nature (belongs to the same set).

Here are two examples.

(*a*) V_1, V_2, $V_3, \ldots$, V_n are polynomials of degree not exceeding a certain given number d; their linear combination is again a polynomial of degree not exceeding d.

(*b*) V_1, V_2, $V_3, \ldots$, V_n are vectors parallel to a given plane; their linear combination (addition means here vector-addition) is again a vector parallel to the given plane.

Example (*a*) plays a role in sect. 4.3.   With regard to sect. 4.4(3) let us observe that addition and subtraction are special cases of linear combination ($n = 2$; $c_1 = c_2 = 1$ and $c_1 = -c_2 = 1$, respectively).

Example (*b*) is suggestive; such objects as can be linearly combined subject to the "usual" laws of algebra are called "vectors," and their set is called a "vector-space," in abstract algebra.

Linear combinations (vector-spaces) play a role in several advanced branches of mathematics.   We can consider here only a few not too advanced examples (ex. 4.12, 4.13, 4.14, and 4.15).

We use here the two terms "linear combination" and "superposition" in the same meaning, but we use the latter term more often.   The term "super-position" is often employed in physics (especially in wave theory).   We take here just one example from physics, ex. 4.16, which is simple enough for us and important in several respects.

†**4.12.** *Homogeneous linear differential equations with constant coefficients.* Such an equation is of the form

$$y^{(n)} + a_1 y^{(n-1)} + \cdots + a_{n-1} y' + a_n y = 0$$

$a_1, a_2, \ldots, a_n$ are given numbers, called the *coefficients* of the equation; n is the *order* of the equation; y is a function of the independent variable x; $y', y'', \ldots,$ $y^{(n)}$ denote, as usual, the successive derivatives of y.   A function y satisfying the equation is called a *solution*, or an "integral."

(*a*) Show that a linear combination of solutions is a solution.

(*b*) Show that there is a particular solution of the special form

$$y = e^{rx}$$

where r is an appropriately chosen number.

(*c*) Combine particular solutions of such special form to obtain a solution as general as possible.

†**4.13.** Find a function y satisfying the differential equation

$$y'' = -y$$

and the initial conditions

$$y = 1, \quad y' = 0 \quad \text{for} \quad x = 0$$

4.14. *Homogeneous linear difference equations with constant coefficients.* Such an equation is of the form

$$y_{k+n} + a_1 y_{k+n-1} + \cdots + a_{n-1} y_{k+1} + a_n y_k = 0$$

$a_1, a_2, \ldots, a_n$ are given numbers, called the *coefficients* of the equation; n is the *order* of the equation; an infinite sequence of numbers

$$y_0, y_1, y_2, \ldots, y_k, \ldots$$

which satisfies the equation for $k = 0, 1, 2, \ldots$ is called a *solution*.

(We may regard y_x as a function of the independent variable x defined for non-negative integral values of x.   On the other hand, we may regard the proposed equation as a recursion formula, that is, a uniform rule by virtue of which we can compute any term y_{k+n} of the sequence from n foregoing terms $y_{k+n-1}, y_{k+n-2}, \ldots, y_k$—or y_k from $y_{k-1}, y_{k-2}, \ldots, y_{k-n}$.)

(*a*) Show that a linear combination of solutions is a solution.

(*b*) Show that there is a particular solution of the special form

$$y_k = r^k$$

where r is an appropriately chosen number.

(*c*) Combine particular solutions of such special form to obtain a solution as general as possible.

4.15. The sequence of *Fibonacci numbers*

$$0, 1, 1, 2, 3, 5, 8, 13, \ldots$$

is defined by the difference equation (recursion formula)

$$y_k = y_{k-1} + y_{k-2}$$

valid for $k = 2, 3, 4, \ldots$ and the initial conditions

$$y_0 = 0, \quad y_1 = 1$$

Express y_k in terms of k.

4.16. *Superposition of motions.* Galileo, having found the law of falling bodies and the law of inertia, combined these laws to discover the trajectory of (the curve described by) a projectile. The reader who realizes how much he is helped by modern notation may relive, in a fashion, this discovery of Galileo.

Let x and y denote rectangular coordinates in a vertical plane; the x axis is horizontal, the y axis points upward. A projectile (a material point devoid of friction) moves in this plane starting from the origin at the instant $t = 0$; t is the time. The initial velocity of the projectile is v, its initial direction includes the angle α with the positive x axis. With the actual motion of the projectile, we may associate three virtual motions, starting from the same point at the same time.

(*a*) A heavy material point, starting from rest and falling freely, has, at the time t, the coordinates

$$x_1 = 0, \qquad y_1 = -\tfrac{1}{2} g t^2$$

(*b*) A material point free from gravity, which has received the vertical component $v \sin \alpha$ of the initial velocity, has, at the time t, by virtue of the law of inertia, the coordinates

$$x_2 = 0, \qquad y_2 = t v \sin \alpha$$

(*c*) A material point, free from gravity, which has received the horizontal component of the initial velocity, has, at the time t, by virtue of the law of inertia, the coordinates

$$x_3 = t v \cos \alpha, \qquad y_3 = 0$$

If the actual motion is compounded from these three virtual motions according to the "simplest" assumption, what is its trajectory?

Second Part

Opportunity is offered to the reader to participate in an investigation, important phases of which are indicated by ex. 4.17 and ex. 4.24.

4.17. *The multiplicity of approaches.* In a tetrahedron two opposite edges have the same length a, they are perpendicular to each other, and each is perpendicular to the line of length *b* that joins their midpoints. Find the volume of the tetrahedron.

There are several different approaches to this problem. The reader who needs help may look at some, or all, of the following ex. 4.18–23. If he wishes to visualize the spatial relations involved, he may look for a simple orthogonal projection, or for a simple cross-section.

4.18. *What is the unknown?* The unknown in ex. 4.17 is the volume of a tetrahedron.

How can you find this kind of unknown? The volume of a tetrahedron can be computed if its base and height are given—but neither of these two quantities is given in ex. 4.17.

Well, *what is the unknown?*

4.19. (Continued) You need the area of a triangle—*how can you find this kind of unknown?* The area of a triangle can be computed if its base and height are given—but only one of these two quantities is given for the triangle that forms the base of the tetrahedron of ex. 4.17.

You need the length of a line—*how can you find this kind of unknown?* The usual thing is to compute the length of a line from some triangle—but, in the figure, there is no triangle in which the height of the tetrahedron of ex. 4.17 would be contained.

In fact, there is no such triangle; but *could you introduce one?* At any rate, *introduce suitable notation* and collect whatever you have.

4.20. *Here is a problem related to yours and solved before*: "The volume of a tetrahedron can be computed if its base and height are given." You cannot apply this to ex. 4.17 immediately, because the base and height of that tetrahedron are not given. There may be, however, other, more accessible tetrahedra *around*.

4.21. (Continued) And there may be more accessible tetrahedra *within*.

4.22. *More knowledge* may help. Ex. 4.17 is easy for you, if you know the *prismoidal formula*.

A *prismoid* is a polyhedron. Two faces of the prismoid, called the *lower base* and the *upper base*, are parallel to each other; all the other faces are *lateral faces*. The prismoid has three kinds of edges: edges surrounding the lower base, edges surrounding the upper base, and *lateral edges*. Any lateral edge of the prismoid (this is an important part of its definition) joins a vertex of the lower base to a vertex of the upper base. A prism is a special prismoid.

The distance between the two bases is the *height* of the prismoid. A plane that is parallel to, and equidistant from, the two bases, the lower and the upper, intersects the prismoid in a polygon called the *midsection*.

Let *V* stand for the volume of the prismoid, *h* for the height,

$$L, \quad M, \quad \text{and} \quad N$$

for the areas of its lower base, its midsection, and its upper base, respectively. Then (this expression for V is called the prismoidal formula)

$$V = \frac{(L + 4M + N)h}{6}$$

Apply it to ex. 4.17.

4.23. Perhaps you have abandoned the path to the solution of ex. 4.17 that starts from ex. 4.18 and leads through ex. 4.19, but attained the result following some other route. If so, look at the result, return to the abandoned path, and follow it to the end.

4.24. *The prismoidal formula.* Study all sides of the question, consider it under various aspects, turn it over and over in your mind—we did so, look at Fig. 4.5. Having found four different derivations for the same result, we should be able to profit from their comparison.[2]

Out of our four derivations, three do not use the prismoidal formula, but one does (ex. 4.22). Hence, for that particular case of the prismoidal formula that intervenes in the problem treated, we have in fact, implicitly at least, three different proofs. Could we render one of these proofs explicit and extend it so that it proves that formula not only in a particular case, but generally?

On the face of it, which one of the three derivations in question (ex. 4.20, ex. 4.21, and ex. 4.18, 4.19, 4.23) appears to have the best chances?

4.25. Verify the prismoidal formula for the prism (which is a very special prismoid).

4.26. Verify the prismoidal formula for the pyramid (which, in an appropriate position, can be regarded as a prismoid—a degenerate, or limiting, case of a prismoid if you prefer, with an upper base shrunken into a point).

4.27. In generalizing the situation that underlies the solution of ex. 4.20, we consider a prismoid P split into n prismoids $P_1, P_2, \ldots, P_n$ which are nonoverlapping and fill P completely, their lower bases fill the lower base of P, their upper bases the upper base of P. (In the case of ex. 4.20, Fig. 4.5b, P is a prism with a square base, $n = 5$, $P_1, P_2, P_3,$ and P_4 are congruent tetrahedra,

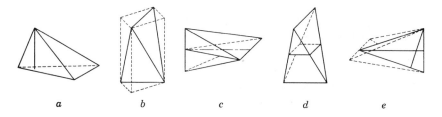

a b c d e

Fig. 4.5. Turn it over and over, consider it under various aspects, study all sides.

[2] In this, we follow Leibnitz's opinion; see the quotation preceding ex. 3.31.

P_5 another tetrahedron.)   Show: If the prismoidal formula is valid for n prismoids out of the $n + 1$ considered, it is necessarily valid also for the remaining one.

4.28. In generalizing the situation that underlies the solution of ex. 4.22, Fig. 4.5d, we let l and n denote opposite edges of a tetrahedron (l lower, n upper).   Pass a plane through l parallel to n, and another plane through n parallel to l; let h denote the distance of these two (parallel) planes.   The tetrahedron can be regarded as a prismoid (a degenerate prismoid, if you prefer) of which the edges l and n are the bases, the lower and the upper, respectively, and h the height.   (The midsection is a parallelogram.)

Verify the prismoidal formula for this kind of prismoid.

4.29. Prove the prismoidal formula generally (by superposition of particular cases treated previously).

4.30. *No chain is stronger than its weakest link.*   Reexamine the solution of ex. 4.28.

4.31. Reexamine the solution of ex. 4.29.

†4.32. *Simpson's rule.*   Let $f(x)$ be a function defined (and continuous) in the interval

$$a \leqq x \leqq a + h$$

put

$$\int_a^{a+h} f(x)dx = I$$

$$f(a) = L, \qquad f\left(a + \frac{h}{2}\right) = M, \qquad f(a + h) = N$$

Then, under certain conditions which we intend to explore,

$$I = \frac{L + 4M + N}{6} h$$

this expression for I is called *Simpson's rule.*

Let n denote a non-negative integer, take

$$f(x) = x^n, \qquad a = -1, \qquad h = 2$$

and determine those values of n for which the expression of the integral I by Simpson's rule is valid.

(Even when Simpson's rule is not exactly valid, it may be "approximately valid," that is, the difference between the two sides may be relatively small. This is frequently the case, and hence Simpson's rule is important for the approximate evaluation of integrals.)

†4.33. Prove that Simpson's rule is valid for any polynomial of degree not exceeding 3, provided that $a = -1$ and $h = 2$.

†4.34. Prove that Simpson's rule is valid for any polynomial of degree not exceeding 3 and unrestricted a and h.

†**4.35.** Derive the prismoidal formula from ex. 4.34, using solid analytic geometry and integral calculus. ("To appreciate the easy way do it first the hard way," said the traditional mathematics professor.)

4.36. *Widening the scope.* In solving some of the foregoing problems we actually went beyond the outline of the pattern of superposition formulated in sect. 4.4(4). We have, in fact, attained the general solution by superposing more accessible particular cases, but these particular cases were not all of the same type, they did not all belong to the same special situation. (In the solution of ex. 4.29, some of the superposed solids are pyramids, treated in ex. 4.26, others are tetrahedra in a special position, treated in ex. 4.28. Also in the solution of ex. 4.33, we superpose cases of different nature.) Essentially, we deviated from the formulation of sect. 4.4(4) in just one point: we did not start from one leading special situation, but from several such situations. Let us, therefore, enlarge the outline of our pattern: *Starting from a leading special situation, or from several such situations, we attain the general solution by superposition of particular cases.*

The pattern of superposition points out a path from a leading special case (or from a few such cases) to the general case. There is a very different connecting path between the same endpoints with which the ambitious problem-solver should be equally acquainted: it is often possible to reduce the general case to a leading special case by an appropriate *transformation*. (The general case of ex. 4.34 is reduced to the special case of ex. 4.33 by a transformation of the variable of integration.) For a suggestive discussion of this topic see J. Hadamard, *Leçons de géométrie élémentaire. Géométrie plane*, 1898; Méthodes de transformation, pp. 272–278.

PART TWO

TOWARD A GENERAL METHOD

Human wisdom remains always one and the same although applied to the most diverse objects and it is no more changed by their diversity than the sunshine is changed by the variety of objects which it illuminates.

DESCARTES: Rule I, *Œuvres*, vol. X, p. 360.

CHAPTER 5

PROBLEMS

The solution of problems is the most characteristic and
peculiar sort of voluntary thinking.
WILLIAM JAMES

5.1. What is a problem?

In what follows, the word "problem" will be taken in a very comprehensive meaning. Our first task is to outline this meaning.

Getting food is usually no problem in modern life. If I get hungry
at home, I grab something in the refrigerator, and I go to a coffeeshop or
some other shop if I am in town. It is a different matter, however, when
the refrigerator is empty or I happen to be in town without money; in such
a case, getting food becomes a problem. In general, a desire may or may
not lead to a problem. If the desire brings to my mind immediately,
without any difficulty, some obvious action that is likely to attain the
desired object, there is no problem. If, however, no such action occurs
to me, there is a problem. Thus, to have a problem means: *to search
consciously for some action appropriate to attain a clearly conceived, but not
immediately attainable, aim.* To solve a problem means to find such action.

A problem is a "great" problem if it is very difficult, it is just a "little"
problem if it is just a little difficult. Yet some degree of difficulty belongs
to the very notion of a problem: where there is no difficulty, there is no
problem.

A typical problem is to find the way to a preassigned spot in some little
known region. We can easily imagine how serious this problem was for
our primitive ancestors who dwelt in a primeval forest. This may or may
not be the reason that the solution of any problem appears to us somehow
as finding a way: a way out of a difficulty, a way around an obstacle.

The greater part of our conscious thinking is concerned with problems.
When we do not indulge in mere musing or daydreaming, our thoughts
are directed towards some end, we seek ways and means to that end, we
try to think of some course following which we could attain our aim.

117

Solving problems is the specific achievement of intelligence, and intelligence is the specific gift of man. The ability to go round an obstacle, to undertake an indirect course where no direct course presents itself, raises the clever animal above the dull one, raises man far above the most clever animals, and men of talent above their fellow men.

Nothing is more interesting for us humans than human activity. The most characteristically human activity is solving problems, thinking for a purpose, devising means to some desired end. Our aim is to understand this activity—it seems to me that this aim deserves a good deal of interest.

In the foregoing, we have studied elementary mathematical problems, in grouping together problems accessible to the same method of solution. We have acquired so a certain experimental basis and, from this basis, we shall now try to ascend to higher generality, attempting to embrace, as far as possible, also nonmathematical problems. To aim at a general method applicable to all sorts of problems may seem too ambitious, but it is quite natural: although the variety of problems we may face is infinite, each of us has just one head to solve them, and so we naturally desire just one method to solve them.

5.2. Classification of problems

A student is taking a written examination in mathematics; he is just an average student, but he did some work to prepare himself for this examination. After having read a proposed problem, he may ask himself: "What kind of problem is this?" In fact, it could be to his advantage to ask this question: if he can classify his problem, recognize its type, place it in such and such a chapter of his textbook, he has made some progress: he may now recall the method he has learned for solving this type of problem.

The same holds, in a sense, for all levels of problem solving. The question "What kind of problem is this?" leads to the next question "What can be done about this kind of problem?" and these questions may be asked with profit even in quite advanced research.

And so it may be useful to classify problems, to distinguish problems of various types. A good classification should introduce such types that *the type of problem may suggest the type of solution.*

We shall not enter now upon a detailed, or attempt a perfect, classification. In interpreting with some latitude a tradition which goes back to Euclid and his commentators, we just wish to characterize two very general types of problems.

Euclid's Elements contain axioms, definitions, and "propositions."

His commentators and some of his translators distinguish two kinds of "propositions": the aim of the first kind (the Latin name is "problema") is to construct a figure, the aim of the second kind (the Latin name is "theorema") is to prove a theorem.   In extending this distinction, we shall consider two kinds of problems, problems "to find" and problems "to prove."   The aim of a *problem to find* is to find (construct, produce, obtain, identify,...) a certain object, the unknown of the problem.   The aim of a *problem to prove* is to decide whether a certain assertion is true or false, to prove it or disprove it.

For instance, when you ask "What did he say?" you pose a problem to find.   Yet, when you ask "Did he say that?" you pose a problem to prove.

For more details about these two kinds of problems see the next two sections.

5.3. Problems to find

The aim of a "problem to find" is to find a certain object, the *unknown* of the problem, satisfying the *condition* of the problem which relates the unknown to the *data* of the problem.   Let us consider two examples.

"Being given two line-segments a and b, and the angle γ, construct the parallelogram of which the given line-segments are adjacent sides including the angle γ."

"Being given two line-segments a and b, and the angle γ, construct the parallelogram of which the given line-segments are the diagonals including the angle γ."

In both problems, the data are the same: the line-segments a and b, and the angle γ.   In both problems, the unknown is a parallelogram and so our problems are not *a priori* distinguishable by the nature of the unknown. What makes a difference between our two problems is the condition, the required relation between the unknown and the data: of course, the relation of the parallelogram to its sides differs from its relation to its diagonals.

The *unknown* may be of every imaginable category.   In a problem of geometric construction the unknown is a figure, for instance a triangle. When we are solving an algebraic equation, our unknown is a number, a root of that equation.   When we ask "What did he say?" the unknown may be a word, or a sequence of words, a sentence, or a sequence of sentences, a speech.   A problem clearly stated must specify the category (the set) to which the unknown belongs; we have to know from the start what kind of unknown we are supposed to find: a triangle, or a number, or a word, or....

A problem clearly stated must specify the *condition* that the unknown has to satisfy.   In the set of objects specified by the problem to which the unknown must belong, there is the subset of those objects that satisfy the condition, and any object belonging to this subset is called a *solution*. This subset may contain just one object: then the solution is unique. This subset may be empty: then there is no solution. (For remarks on the term "solution" see ex. 5.13.) We observe here that a problem to find can be taken in various meanings. Taken in a strict sense, the problem demands to find (produce, construct, identify, list, characterize,...) *all* the solutions (the entire subset mentioned above). Taken in a less strict sense, the problem may ask for just one (any one) solution, or some solutions. Sometimes it is enough to decide the existence of a solution, that is, to decide whether the set of solutions is empty or not. It is usual to take mathematical problems in the strict sense unless the contrary is explicitly stated, but for many practical problems the "strict sense" would make little sense.

When we deal with mathematical problems (unless the context hints the contrary) we shall use the phrase "the data" to denote all the given (known, granted,...) objects (or their full set) connected with the unknown by the condition. If the problem is to construct a triangle from its sides a, b, and c, the data are the three line-segments a, b, and c. If the problem is to solve the quadratic equation

$$x^2 + ax + b = 0$$

the data are the two given numbers a and b. A problem may have just one datum, or no data at all. Here is an example: "Find the ratio of the area of a circle to the area of a circumscribed square." The required ratio is independent of the size of the figure and so it is unnecessary to give the length of the radius or some other datum of this kind.

We shall call the unknown, the condition, and the data the *principal parts* of a problem to find. In fact, we can not reasonably hope to solve a problem that we do not understand. Yet, to understand a problem, we should know, and know very well, what is the unknown, what are the data, and what is the condition. And so it is advisable when we are working at a problem to pay especial attention to its principal parts.

5.4. Problems to prove

There is a rumor that, on a certain occasion, Secretary Soandso used an extremely crude expression (which we shall not reproduce here) in referring to Congressman Un Tel. It is just a rumor to which considerable doubt is attached. Yet the question "Did he say that?" has agitated many persons, was debated in the press, mentioned in a congressional committee, and may reach the courts. Whoever takes the question seriously has a "problem to prove" on his hands: he should lift the doubt about the rumor, he should prove the use of the alleged expression or disprove it, and proof or disproof should be supported by the best evidence available.

When we have a mathematical "problem to prove," we should lift the doubt about a clearly stated mathematical assertion A, we should prove A or disprove A. A celebrated unsolved problem of this kind is to prove or disprove Goldbach's conjecture: If the integer n is even and $n > 4$, then n is the sum of two odd primes.[1]

Goldbach's assertion (it is a mere assertion, we do not know yet whether it is true or false) is stated here in the *most usual form* of mathematical propositions: it consists of *hypothesis* and *conclusion*; the first part starting with "If" is the hypothesis, the second part starting with "then" is the conclusion.[2]

When we have to prove or disprove a mathematical proposition stated in the most usual form, the hypothesis and the conclusion of the proposition are appropriately called the *principal parts* of our problem. In fact, these principal parts deserve our especial attention. To prove the proposition we should discover a binding logical link between the principal parts, the hypothesis and the conclusion; to disprove the proposition we should show (by a counter-example, if possible) that one of the principal parts, the hypothesis, does not imply the other, the conclusion. Many mathematicians, great and small, tried to lift the doubt about Goldbach's conjecture, but without success: although very little knowledge is needed to understand the hypothesis and the conclusion, nobody has succeeded yet in linking them with a strict argument, and nobody has been able to produce a counter-example.

5.5. The components of the unknown, the clauses of the condition

If our problem is to construct a circle, we have to find, in fact, two things: the center of the circle and its radius. It may be advantageous to subdivide our task: of the two things wanted, the center and the radius, we may try to find first one then the other.

If our problem is to find a point in space and we use analytic geometry, we have to find, in fact, three numbers: the three coordinates x, y, and z of the point.

---

[1] MPR, vol. 1, pp. 4–5.

[2] There are mathematical propositions which cannot be naturally split into hypothesis and conclusion; see HSI, p. 155, Problems to find, problems to prove, 4. Here is a proposition of this kind: "In the decimal fraction of the number π there are nine consecutive digits 9." To prove or disprove this proposition is a definite mathematical problem—which seems to be, at present, hopelessly difficult. "A fool can ask more questions than nine wise men can answer."

According to the viewpoint which we prefer, we can say that, in our first example, there are two unknowns or just one unknown and, in our second example, there are three unknowns or just one unknown. There is, however, still another viewpoint which is often advantageous: we may say that, in both examples, there is just one unknown, but it is, in a sense, "subdivided." Thus, in our first example, the circle is the unknown, but it is a *bipartite* or *two-component* unknown; its *components* are its center and its radius. Similarly, in our second example, the point is a *tripartite* or *three-component* unknown; its components are its three coordinates x, y, and z. Generally, we may consider a *multipartite* or *multicomponent* unknown x having n components $x_1, x_2, \ldots, x_n$.

One advantage of the terminology we have just introduced is that, in certain general discussions, we need not distinguish between problems with one unknown and problems with several unknowns: in fact, we can reduce the latter case to the former by considering those several unknowns as components of one unknown. For instance, what we have said in sect. 5.3 applies essentially also to problems where we have to find several unknowns, although this case has not been explicitly mentioned in sect. 5.3. We shall see later that our terminology is useful in various contexts.

If our problem is a problem to find, there may be advantage in subdividing the condition into several parts or *clauses* as we have had ample opportunity to observe. In solving a problem of geometric construction, we may split the condition into two parts so that each part yields a locus for the unknown point (chapter 1). In solving a "word problem" by algebra, we split the condition into as many parts as there are unknowns, so that each part yields an equation (chapter 2).

If our problem is a problem to prove, there may be advantage in subdividing the hypothesis, or the conclusion, or both, into appropriate parts or clauses.

5.6. Wanted: a procedure

In constructing a figure in the style of Euclid's Elements, we are not free to choose our tools or instruments: we are supposed to construct the figure with ruler and compasses. Thus, the solution of the problem consists, in fact, in a sequence of *well-coordinated geometric operations* which start from the data and end in the required figure: our operations are drawing straight lines and circles, and determining their points of intersection.

This example may open our eyes and, looking sharper, we may perceive that the solution of many problems consists essentially in a *procedure*, a course of action, a scheme of well-interrelated operations, a *modus operandi*.

Take the problem of solving an equation of the second (or third, or fourth) degree. The solution consists in a scheme of well-coordinated *algebraic* operations which start from the data, the given coefficients of the equation, and end in the required roots: our operations are adding, subtracting, multiplying, or dividing given, or previously obtained, quantities, or extracting roots from such quantities.

Or consider a problem "to prove." The solution of the problem, the result of our efforts, is a proof, that is, a sequence of well-coordinated *logical* operations, of steps which start from the hypothesis and end in the desired conclusion of the theorem: each step infers some new point from appropriately chosen parts of the hypothesis, from known facts, or from points previously inferred.

Nonmathematical problems present a similar aspect. The builder of a bridge has to organize, coordinate, bring into a coherent scheme a tremendous multiplicity of operations: constructing approaches, shipping supplies, erecting scaffoldings, pouring concrete, riveting metallic parts, etc. etc. Moreover, he may be obliged to interrelate these operations with others of a very different nature: with financial, legal, even political transactions. All these operations depend on each other: most of them suppose that certain others have been previously performed.

Or take the case of the mystery story. The unknown is the murderer; the author tries to impress us by the performance of the detective-hero who devises a scheme, a course of action which, starting from the first indications, ends in recognizing and trapping the murderer.

The object of our quest may be an unknown of any nature or the discovery of the truth about any kind of question; our problem may be theoretical or practical, serious or trifling. To solve our problem, we have to devise a well-conceived, coherent scheme of operations, of logical, mathematical, or material operations proceeding from the hypothesis to the conclusion, from the data to the unknown, from the things we have to the things we want.

Examples and Comments on Chapter 5

5.1. Of a right prism with square base, find the volume V, being given the length a of a side of the base and the height h of the prism.

What is the unknown? What are the data? What is the condition?

5.2. Find two real numbers x and y satisfying the equation

$$x^2 + y^2 = 1$$

What is the unknown? What are the data? What is the condition? Describe the set of solutions.

5.3. Find two real numbers x and y satisfying the equation

$$x^2 + y^2 = -1$$

Describe the set of solutions.

5.4. Find two integers x and y satisfying the equation

$$x^2 + y^2 = 13$$

Describe the set of solutions.

5.5. Find three real numbers x, y, and z such that

$$|x| + |y| + |z| < 1$$

(1) Describe the set of solutions.

(2) Modify the problem by substituting $\leq$ for $<$, and describe the set of solutions for the modified problem.

5.6. State the theorem of Pythagoras.
What is the hypothesis? What is the conclusion?

5.7. Let n denote a positive integer and $d(n)$ the number of divisors of n (we mean positive integral divisors including 1 and n). For example

6 has the divisors 1, 2, 3, 6; $d(6) = 4$
9 has the divisors 1, 3, 9 ; $d(9) = 3$

Consider the proposition:
   $d(n)$ is odd or even according as n is, or is not, a square.
What is the hypothesis? What is the conclusion?

5.8. *To prove or to find?* Are the two numbers $\sqrt{3} + \sqrt{11}$ and $\sqrt{5} + \sqrt{8}$ equal? If they are not, which one is greater?

Restated in a generalized form, the problem is concerned with two numbers a and b, well defined by arithmetic operations, and requires of us to decide which one of the three possible cases

$$a = b, \qquad a > b, \qquad a < b$$

is actually true.

We may perceive different aspects of this problem.

(1) First, we have to prove or disprove the proposition $a = b$. If it turned out that this proposition is false, we have to prove or disprove the proposition $a > b$. We may tackle these two tasks also in the reverse order, or perhaps simultaneously; at any rate, we have here two problems to prove, linked with each other.

(2) There is a notation extensively used in various branches of mathematics: sgn x (read "the sign of x" or "signum x") is defined as follows:

$$\text{sgn } x = \begin{cases} 1 & \text{when} \quad x > 0 \\ 0 & \text{when} \quad x = 0 \\ -1 & \text{when} \quad x < 0 \end{cases}$$

The problem stated requires of us to find the number sgn $(a - b)$: this is a problem to find.

There is no formal contradiction (there need not be one if our terminology is carefully devised): in (1), we have a problem A consisting of two linked, simultaneously stated, problems to prove; in (2), we have a problem B, which is a problem to find. We do not regard these two problems A and B, stated in different terms, as *identical*—but they are *equivalent*. (This usage of the term "equivalent" is explained in HSI, Auxiliary problem 6, pp. 53–54, and will be explained again in Chapter 9.)

Moreover, there is no material disadvantage. On the contrary, it may be a good thing to see two different aspects of the same difficulty: one aspect may appeal more to us than the other, it may show a more accessible side and so give us a chance to attack the difficulty from that more accessible side.

5.9. *More problems.* Take any problem (there are many in the foregoing chapters), determine whether it is a problem "to find" or one "to prove," and ask accordingly:

What is the unknown? What are the data? What is the condition?
What is the conclusion? What is the hypothesis?

The aim of these questions here is just to familiarize you with the principal parts of problems. Yet, experience may show you that these questions, if seriously asked and carefully answered, are a great help in problem solving: in focusing your attention upon the principal parts of the problem, they deepen your understanding of the problem and they may start you in the right direction.

5.10. *The procedure of solution may consist of an unlimited sequence of operations.* If we are required to solve the equation

$$x^2 = 2$$

our task can be interpreted in various ways. The interpretation may be: "Find the positive square root of 2 to five significant figures"; in this case we fully discharge our duty in producing the decimal fraction 1.4142. Yet, the interpretation may also be "Extract the square root of 2" without any additional qualification or alleviation, and then we cannot perform our task by producing four, or any other given number of figures after the decimal point: the answer must be a *procedure*, a *scheme* of arithmetical operations that can yield *any* required number of decimal figures.

Here is another example: "Find the ratio of the area of the circle to the area of the circumscribed square." The answer is $\pi/4$ if we take the value of π for granted. Leibnitz gave the answer (expressed the ratio $\pi/4$) in form of an infinite series

$$\frac{1}{1} - \frac{1}{3} + \frac{1}{5} - \frac{1}{7} + \frac{1}{9} - \frac{1}{11} + \cdots$$

This series prescribes, in fact, a never ending sequence of arithmetical operations which can lead us to any number of figures in the decimal fraction of π

(in theory—in practice the procedure is much too slow). "Although this series, as it stands, is not suitable for a rapid approximation, but to present to the mind the ratio of the circle to the circumscribed square, I do not think that anything more suitable or more simple can be imagined," says Leibnitz.[3]

5.11. *Squaring the circle.* In solving a problem "to find" we seek an object, the "unknown" object, and very often we are led to seeking a procedure (a sequence of operations) to obtain that object—let us call the procedure sought, just for the sake of neater distinction, the "operational unknown." That a neat distinction is very desirable here may be illustrated by a historic example.

Being given the radius of a circle, construct *with ruler and compasses* a square having exactly the same area as the circle.

This is the strict form of the celebrated ancient problem of "squaring the circle" which originated with the early Greek geometers. We emphasize that the problem *prescribes the nature of the procedure* (of the "operational unknown"): we should construct a side of the desired square with straightedge and compasses, by drawing straight lines and circles, in using only points given or obtained by the intersection of previously drawn lines. And, of course, starting from the two endpoints of the given radius, we should attain the two endpoints of a side of the desired square in a *finite number* of steps.

After many centuries, in which an uncounted number of persons attempted the solution, it was proved (by F. Lindemann in 1882) that there is no solution: The square having the same area as the given circle undoubtedly "exists" (its side can be approximated to any given precision by various infinite processes known today, one of which is provided by the celebrated series of Leibnitz mentioned in ex. 5.10). Yet, a procedure of the desired kind (consisting of a finite sequence of operations with straightedge and compasses) does not exist. I wonder whether a clear distinction between the desired figure and the desired procedure, between the "unknown object" and the "operational unknown," would have diminished the number of unfortunate circle-squarers.

5.12. *Sequence and consequence.* In the construction of a bridge, fixing a prefabricated metallic piece in its proper place is an important operation. It may be essential that of two such operations one should precede the other (when the second piece cannot be fastened unless the first has been fastened before), but again it may be unessential (when the two pieces are independent). Thus, it may or may not be necessary to observe a definite sequence in the performance of two operations. Similarly, the steps of a proof are presented successively in a lecture or in print. Yet, a step may precede another step in time without preceding it in logic. We must distinguish between sequence and consequence, between succession in time and logical concatenation. (We shall return to this important matter in chapter 7.)

5.13. *An unfortunate ambiguity.* The word "solution" has several different meanings, some of which are very important and would deserve to be designated by an unambiguous term. In default of better ones, I propose a few such terms (adding a German equivalent to each).

[3] *Philosophische Schriften*, edited by Gerhardt, vol. IV, p. 278.

Solving object (Lösungsgegenstand) is an object satisfying the condition of the problem. If the aim of the problem is to solve an algebraic equation, a number satisfying the equation, that is, a root of the equation, is a solving object. Only a problem "to find" can have a solving object. A category (set) to which the solving object belongs must be specified in advance in a clearly stated problem—we must know in advance whether we seek a triangle, or a number, or what not. In fact, such specification (the precise designation of a set to which the unknown belongs) is a principal part of the problem. "To find the unknown" means to find (identify, construct, produce, obtain, . . .) the solving object (the set of all solving objects).

Solving procedure (Lösungsgang) is the procedure (the construction, the scheme of operations, the system of conclusions) that ends in finding the unknown of a problem to find, or in lifting the doubt about the assertion proposed by a problem to prove. Thus, "solving procedure" is a term applicable to both kinds of problems. At the beginning of our work, we do not know the solving procedure, the appropriate scheme of operations, but we are seeking it all the time in the hope that we shall know it at the end: this procedure is the aim of our quest, it is effectively, in a sense, our unknown, it is, let us say, our "operational unknown." (Cf. ex. 5.11.)

We could also talk about the "work of solving" (Lösungsarbeit) and the "result of solving" (Lösungsergebnis), but, in fact, I shall try not to appear too fussy and, except in a few important cases, I shall leave it to the reader to discover from the context what the word "solution" means in a given case: whether it means the object, the procedure, the result of the work, or the work itself.[4]

5.14. *Data and unknown, hypothesis and conclusion.* Euclid's Elements have a peculiar consistent style which some of us may be inclined to call solemn and others pedantic. All propositions are phrased according to a set pattern and, in this phrasing, the data and the unknown of a problem to find are so treated as if they were similar, or parallel, to the hypothesis and the conclusion of a problem, respectively. In fact, there is, as we shall see later, a certain similarity or parallelism between these principal parts of the two kinds of problems, which is of some importance from the problem-solver's viewpoint, and so from the viewpoint of our subject. Yet, it is inadmissible and illiterate to mix up the terms data and hypothesis or the terms unknown and conclusion and to apply any one of these terms to the kind of problem for which it is unfit. It is sad that such inadmissible and illiterate use of these important terms occurs sometimes even in print.

5.15. *Counting the data.* A triangle is determined by three sides, or by two sides and one angle (the included angle), or by one side and two angles, but it is not determined by three angles: 3 independent data are required to determine a triangle. (See also ex. 1.43 and 1.44.) To determine a polynomial of degree n in one variable (in x) $n + 1$ independent data are required: the $n + 1$ coefficients in the expansion of the polynomial in powers of x, or the $n + 1$

---

[4] Cf. HSI, p. 202; Terms, old and new, 8.

values taken by the polynomial at the points $x = 0, 1, 2, \ldots, n$, or at any other $n + 1$ different given points, and so on.   There are many important kinds of mathematical objects to determine which a definite number of independent data is required.   Therefore, when we are solving a problem to find, it is often advantageous to *count the data*, and to count them early.

5.16. To determine a polygon with n sides

$$(n - 1) + (n - 2) = (n - 3) + n = 3 + 2(n - 3) = 2n - 3$$

independent data are required.   What do these four different expressions for the same number suggest to you?

5.17. How many data are required to determine a pyramid the base of which is a polygon with n sides?

5.18. How many data are required to determine a prism (which may be oblique) the base of which is a polygon with n sides?

5.19. How many data are required to determine a polynomial of degree n in v variables?   (Its terms are of the form $cx_1^{m_1}x_2^{m_2}\ldots x_v^{m_v}$ where c is a constant coefficient and $m_1 + m_2 + \cdots + m_v \leq n$.)

CHAPTER 6

WIDENING THE SCOPE

*Divide each problem that you examine into as many parts as you
can and as you need to solve them more easily.*

DESCARTES: *Œuvres*, vol. VI, p. 18; Discours de la Méthode, Part II.

*This rule of Descartes is of little use as long as the art
of dividing...remains unexplained.... By dividing his
problem into unsuitable parts, the unexperienced problem-solver
may increase his difficulty.*

LEIBNITZ: *Philosophische Schriften*, edited by Gerhardt, vol. IV, p. 331.

6.1. Wider scope of the Cartesian pattern

There are important methodical ideas involved in the Cartesian pattern
that are not necessarily connected with the setting up of equations. The
present chapter undertakes to disentangle some such ideas. We shall pass
cautiously from equations to more general concepts. We begin with an
example which is sufficiently general in some respects, but very concrete
in another respect; it indicates the direction of our subsequent work.

(1) A certain problem has been translated into a system of four equations with four unknowns, x_1, x_2, x_3, and x_4. This system has a peculiar
feature: not all of its equations involve all the unknowns. Now, we wish
to emphasize just this feature: our notation will clearly show which equation involves which unknowns, but it will neglect other details. In fact,
we write the four equations as follows:

$$r_1(x_1) = 0$$
$$r_2(x_1, x_2, x_3) = 0$$
$$r_3(x_1, x_2, x_3) = 0$$
$$r_4(x_1, x_2, x_3, x_4) = 0$$

That is, the first equation contains just the first unknown, x_1, whereas the next two equations contain the first three unknowns, x_1, x_2, and x_3, and only the last, fourth, equation contains all the four unknowns.

This situation suggests an obvious plan to deal with the proposed system of equations: We begin with x_1 which we compute from the first equation. Having obtained the value of x_1, we observe that the next two equations form a system from which we can determine the next two unknowns, x_2 and x_3. Having so obtained x_1, x_2, and x_3, we use the last, fourth, equation to compute the last unknown x_4.

(2) Let us now realize that the system of equations considered expresses the *condition* of a problem. This condition is split into four parts and each single equation represents a part (or clause, or proviso) of the full condition: the equation expresses that the unknowns involved are connected with each other and with the data by just such a relation as the corresponding part, or clause, of the condition prescribes. And so the condition has a peculiar feature: not all of its clauses involve all the unknowns. Our notation clearly shows which clause involves which unknowns.

Of course, the condition can be split into clauses in just this peculiar manner (with each clause involving just the indicated particular combination of unknowns) even if we have not yet translated those clauses into equations, or even if we are not able to translate those clauses into equations. We may suspect that the *plan* sketched above, under (1), for a system of equations may *remain valid* in some sense for a system of clauses even if those clauses are not expressed, or are not expressible, algebraically.

This remark opens a broad vista of new possibilities.

(3) In order to see these possibilities more clearly we have to reinterpret our notation.

Till now we have interpreted the symbol $r(x_1, x_2, \ldots, x_n)$ in the usual way as an algebraic expression in (or a polynomial in, or a function of) the unknown (variable) numbers $x_1, x_2, \ldots, x_n$. And so we have interpreted

$$r(x_1, x_2, \ldots, x_n) = 0$$

as an (algebraic) equation linking the unknowns $x_1, x_2, \ldots, x_n$. If we deal with a problem in which $x_1, x_2, \ldots, x_n$ are unknowns, such an equation expresses a part of the condition (a clause or a proviso of the condition), that is, a relation between the unknowns $x_1, x_2, \ldots, x_n$ and the data required by the condition.

We do not repudiate this interpretation but we do extend it: Even if the clause is not translated into an equation, or even if $x_1, x_2, \ldots, x_n$ are not unknown numbers but unknown things of any kind, the symbolic equation

$$r(x_1, x_2, \ldots, x_n) = 0$$

should express a *relation, required by the condition of the problem, which involves the indicated unknowns* $x_1, x_2, \ldots, x_n$. We may also say that such a symbolic equation expresses a part of the condition (a clause, proviso, stipulation, or requirement imposed by the condition).

We need a few examples to understand properly this extended scope of the notation, and still more examples to convince ourselves that this extension is useful.

(4) The notation that we have just introduced can be suitably illustrated by crossword puzzles. Let us look at a (miniature) example.

Across	Down
1. German mathematician	1. Do not write dagre
2. Do not write beaus	5. This is boring as
3. Swiss mathematician	6. Reset reset.

In a crossword puzzle the unknowns are words. Let $x_1, x_2, \ldots, x_6$ stand for the six unknown words of our puzzle. Both words x_1 and x_4 have their initial letter in the same square numbered 1, but x_1 should be written horizontally across and x_4 vertically downward; if $n = 2, 3, 5,$ or $6,$ x_n stands for the word the initial letter of which should be written in the square numbered n. If we spell out pedantically the conditions implied by the square diagram that contains black and white, numbered and unnumbered, smaller squares, we have a system of 21 conditions.

There are the six most conspicuous conditions expressed by the "clues." Let us represent them by

$$r_1(x_1) = 0, \quad r_2(x_2) = 0, \quad \ldots, \quad r_6(x_6) = 0$$

Thus, the symbolic equation $r_1(x_1) = 0$ represents the condition that the word x_1 is the name (we hope the last name) of a German mathematician;

$r_4(x_4) = 0$ expresses the import of the (for the moment, rather cryptic) sentence, "Do not write dagre," and so on.

There are the six conditions, visible from the diagram, concerned with the lengths of the six unknown words:

$$r_7(x_1) = 0, \quad r_8(x_2) = 0, \quad \ldots, \quad r_{12}(x_6) = 0$$

For instance, $r_7(x_1) = 0$ prescribes the length of the word x_1. The meaning in our case is that each of the words, $x_1, x_2, \ldots, x_6$ should be a five-letter word.

The diagram shows which word crosses which other and where, and so implies nine conditions:

$$r_{13}(x_1, x_4) = 0, \quad r_{14}(x_1, x_5) = 0, \quad r_{15}(x_1, x_6) = 0$$
$$r_{16}(x_2, x_4) = 0, \quad r_{17}(x_2, x_5) = 0, \quad r_{18}(x_2, x_6) = 0$$
$$r_{19}(x_3, x_4) = 0, \quad r_{20}(x_3, x_5) = 0, \quad r_{21}(x_3, x_6) = 0$$

For example, $r_{14}(x_1, x_5) = 0$ stipulates that the third letter of the word x_1 is the initial of the word x_5, and so on.

Now, we have listed all conditions; their number is $6 + 6 + 9 = 21$.

(5) In general, if the problem involves n unknowns $x_1, x_2, \ldots, x_n$ and we split the condition into l different parts (requirements, stipulations, clauses, provisos) we have a system of l relations connecting n unknowns which we may express by a system of l symbolic equations connecting those n unknowns as follows:

$$r_1(x_1, x_2, \ldots, x_n) = 0$$
$$r_2(x_1, x_2, \ldots, x_n) = 0$$
$$\cdot \quad \cdot \quad \cdot \quad \cdot \quad \cdot \quad \cdot \quad \cdot$$
$$r_l(x_1, x_2, \ldots, x_n) = 0$$

In chapter 2 we were concerned with the particular case in which the unknowns $x_1, x_2, \ldots, x_n$ are unknown numbers, the equations not merely symbolic, but actual algebraic equations, and $l = n$. In the present chapter we shall often be concerned with special situations, such as that discussed under (1) and (2), in which not all the clauses involve all the unknowns.

(6) It can happen that two problems are expressed by the same system of symbolic equations. Such problems may deal with very different matters, but they have something in common: They are similar to each other in some (rather abstract) respect, we may put them into the same class. In this way we obtain a new, more refined, classification of problems (of problems *to find*). Has this classification some interest for our study? If two problems are expressed by the same system of symbolic equations, is there a procedure of solution applicable to both?

This is a good question, I think. Taken in full generality it may be not very useful, but it helps to understand the special situations which we are going to discuss.

6.2. Wider scope of the pattern of two loci

In the foregoing sect. 6.1 we have outlined a very general picture. How do our former observations fit into this picture? How does the very first pattern that we have discerned fit into it?

(1) We may give the answer in a more striking form if we adapt our terminology.

In dealing with geometric constructions, we considered "loci." Such a locus is really just a set of points. In what follows we shall call a set a *locus* if it intervenes in the solution of a problem in a certain characteristic manner which will be indicated by the following examples. As the term "set," see ex. 1.51, has already so many synonyms (class, aggregate, collection, category) it may seem wanton to add one more. Yet the term "locus" may remind us of our experience with certain elementary geometric problems and so it may suggest, by analogy, useful steps when we are dealing with other, perhaps more difficult, problems.

(2) *Two loci for a point in the plane.* Let us return to the very first example that we have discussed: *Construct a triangle being given its three sides.*

Let us look back at the familiar solution (sect. 1.2). By laying down one side, say a, we locate two vertices, B and C, of the required triangle. Just one more vertex remains to be found; call this third vertex, still unknown at this stage of our work, x. The condition requires two things of this point x:

(r_1) the point x is at the given distance b from the given vertex C,
(r_2) the point x is at the given distance c from the given vertex B.

Using the notation introduced in sect. 6.1, we write these two requirements, (r_1) and (r_2), as two symbolic equations:

$$r_1(x) = 0$$
$$r_2(x) = 0$$

The points x satisfying the first requirement (r_1) (the first of the two symbolic equations) fill the periphery of a circle (with center C and radius b). This circular line forms the *set*, or *locus*, of all points complying with the requirement (r_1). The locus of the points satisfying the second requirement (r_2) (the second symbolic equation) is another circular line. Now, the point x that solves the proposed problem about the triangle has to satisfy both requirements, it must belong to both loci. Therefore, the intersection of these two loci is the set of the solutions of the proposed problem. This set contains two points: there are two solutions, two triangles symmetric to each other with respect to the side BC.

(3) *Three loci for a point in space.* We consider the following problem of solid geometry, which is analogous to the simple problem of plane geometry that we have just discussed, under (2): *Find a tetrahedron being given its six edges.*

By the procedure that we have just recalled, under (2), we construct the base of the tetrahedron, a triangle, from the three edges that are required to surround it. Laying down the base, we locate three vertices of the tetrahedron, say A, B, and C. Just one more vertex remains to be found; call this fourth vertex, still unknown at this stage of our work, x, and call the given distances from the three already located vertices a, b, and c, respectively. The condition requires three things of the point x:

(r_1), x is at the distance a from the point A,
(r_2), x is at the distance b from the point B,
(r_3), x is at the distance c from the point C.

Using the notation introduced in sect. 6.1, we write these three requirements, (r_1), (r_2), and (r_3), as three symbolic equations

$$r_1(x) = 0$$
$$r_2(x) = 0$$
$$r_3(x) = 0$$

The points x satisfying the first requirement (r_1) (the first symbolic equation) fill the surface of a sphere (with center A and radius a). This spherical surface forms the set, or locus, of all points complying with the first requirement (r_1). To each of the other two requirements there corresponds a spherical surface, the locus of the points x satisfying that requirement. Now the point x that solves the proposed problem about the tetrahedron has to satisfy all three requirements, it must belong to all three loci. Therefore, the intersection of these three loci (three spheres) is the set of the solutions of the proposed problem. This set contains two points: there are two solutions, two tetrahedra symmetric to each other with respect to the plane of the triangle ABC.

(4) *Loci for a general object.* The examples discussed under (2) and (3) may remind us of several other problems that we have solved in chapter 1 following the same pattern. Behind these examples we may perceive a general situation.

The unknown of a problem is x.   The condition of the problem is split into l clauses which we express by a system of l symbolic equations:

$$r_1(x) = 0, \quad r_2(x) = 0, \quad \ldots, \quad r_l(x) = 0$$

Those objects x that satisfy the first clause, expressed by the first symbolic equation, form a certain set which we call the first locus.   The objects satisfying the second clause form the second locus, . . . the objects satisfying the last clause form the lth locus.   The object x that solves the proposed problem must satisfy the full condition, that is, all the l clauses of the condition, and so it must belong to all l loci.   On the other hand, any object x that does belong simultaneously to all l loci satisfies all the l clauses, and so the full condition, and is, therefore, a solution of the proposed problem.   In short, the *intersection* of those l loci constitutes the set of solutions, that is, the set of all objects satisfying the condition of the proposed problem.

This suggests a vast generalization of the pattern of two loci, a scheme that could work in an inexhaustible variety of cases, could solve almost any problem: first, split the condition into appropriate clauses, then form the loci corresponding to the various clauses, finally find the solution by taking the intersection of those loci.   Before judging this vast scheme, let us get down to concrete cases.

(5) *Two loci for a straight line.*   Construct a triangle being given r, h_a, and α.

The reader should remember the notation used in chapter 1: r stands for the radius of the inscribed circle, h_a for the height perpendicular to the side a, and α for the angle opposite the side a.

The problem is not too easy, but certain initial steps are rather obvious. *Could you solve a part of the problem?*   We can easily draw a part of the required figure: a circle with radius r and two tangents to it that include the angle α.   (Observe that the two radii drawn to the two points of contact include the angle $180° - \alpha$.)   The vertex of this angle α will be the vertex A of the required triangle.   The problem is reduced to the construction of the (infinite) straight line of which the side opposite A is a segment. This line, say x, is our new unknown, given that part of the figure that we have already drawn.

The condition for the line x consists of two clauses:

(r_1), x is tangent to the circle with radius r already constructed,
(r_2), x is at the given distance h_a from the given point A.

The first locus for x is the set of the tangents of the given circle with radius r.

The second locus for x is again the set of the tangents of a circle the center of which is A and the radius h_a.

The intersection of these two loci consists of the common tangents of the two circles. We can construct these tangents; see sect. 1.6(1) and ex. 1.32.

(In fact, only the exterior common tangents solve the problem as stated; the interior common tangents, which may not exist, would render the circle with radius r an escribed circle.)

Regarding the common tangents of two circles as the *intersection of two loci for straight lines* is a useful idea; it is even more useful if we include in it similar cases, especially the extreme case in which one of the circles degenerates into a point.

(6) *Three loci for a solid.* Design a "multipurpose plug" that fits exactly into three different holes, circular, square, and triangular.

See Fig. 6.1; the diameter of the circle, the side of the square, the base and the altitude of the isosceles triangle are equal to each other.

In geometric terms, three orthogonal projections of the required solid should coincide with the three given shapes. We assume (in fact, this assumption narrows down the question) that the directions of the three projections are perpendicular to each other. Our unknown is a solid, say x, and the condition of our problem consists of three clauses:

(r_1) the projection of x onto the floor is a circle,
(r_2) the projection of x onto the front wall is a square,
(r_3) the projection of x onto the side wall is an isosceles triangle.

It is understood that the solid x is placed into a room having the usual shape of a rectangular parallelepiped, that the projections are orthogonal, and that the measurements of the three shapes in Fig. 6.1 are related as has been explained.

Let us examine the first locus, that is, the set of solids satisfying the requirement (r_1). The given circle is placed on the floor. Consider any infinite vertical straight line passing through the area of this circle and call it a "fiber." These fibers fill an infinite circular cylinder of which the given circle is a cross section. A solid x satisfies the first requirement (r_1)

Fig. 6.1. Three holes for the multipurpose plug.

if it is contained in the cylinder and contains at least one point of each fiber.   The set of all such solids is the first locus.

As the first locus is connected with the infinite vertical cylinder so are the two other loci connected with two infinite horizontal prisms.   The cross section of the prism corresponding to (r_2) is a square.   If this prism lies in the north-south direction, the prism corresponding to (r_3) which has a triangular cross section lies in the east-west direction.

Any solid x that belongs to all three loci solves the problem, is a "multi-purpose plug."   The most extensive solid of this kind is the intersection of the three infinite figures, of the cylinder and the two prisms; it is sketched in Fig. 6.2.

(Why is it the most extensive?   Describe the various parts of its surface. Describe some other solids that solve the problem.)

(7) *Two loci for a word.*   In a crossword puzzle that allows puns and anagrams we find the following clue:

"This form of rash aye is no proof (7 letters)."

This is a vicious little sentence; it almost makes sense: "If you say Yes so rashly, it does not prove a thing."   Yet we suspect that some vague echo of a sense was put into the clue just to lead us astray.   There may be a better lead: the phrase "form of" may mean "anagram of."   And so we may try to interpret the clue as follows.

The unknown x is a word.   The condition consists of two parts:

(r_1), x is an anagram of (has the same seven letters as) RASH AYE;
(r_2), "x is no proof" is a meaningful (probably usual) phrase.

Let us examine this interpretation of the problem.   The condition is neatly split into two clauses: (r_1) is concerned with the spelling of the word, (r_2) with its meaning.   To each clause corresponds a "locus"—but these loci are less "manageable" than in the foregoing cases.

The first locus is quite clear in itself.   We can order the seven letters

A A E Y H R S

in 2520 different ways (the reader need not examine now the derivation of

Fig. 6.2. The best multipurpose plug.

this number which equals, in fact, 7!/2!). If it were absolutely necessary, we could write down the 2520 different arrangements of the 7 given letters without repetition or omission and so exhaust the possibilities left open by clause (r_1), that is, describe or construct completely the corresponding locus. This, however, would be boring and wasteful (many of the arrangements would have combinations of vowels and consonants never arising in English). Moreover, such a mechanical exhaustion of all cases was not intended, it is not in the spirit of the game. And so the locus corresponding to clause (r_1) is, if not in principle, but in practice, inexhaustible, unmanageable.

The locus corresponding to clause (r_2) is not only inexhaustible but somewhat hazy. An English word x is given; does the phrase "x is no proof" make sense? Is it a usual phrase? In many cases, the answer is debatable.

And so, for different reasons, neither of the two loci is manageable, neither can be conveniently described, surveyed, or constructed. And, of course, we have no clear procedure to construct the intersection of the two loci. Still, it may be helpful to realize that the condition has two different clauses and that the required word has to satisfy both. Focusing now one clause of the condition, then the other, thinking of words which almost fulfill one clause, or the other, a stab in this direction, then one in the other—eventually, our memory, our store of words and phrases may be sufficiently stirred, and the desired word may pop up.

(We have insisted on the circumstance that neither of the two clauses, (r_1) and (r_2), is manageable—this point is useful in assessing the general scheme we are considering. In fact, however, one of the two clauses is somewhat more manageable than the other—this point may be useful in solving the little riddle at hand.)

6.3. The clause to begin with

In the foregoing section we have discussed problems of various kinds and solved them following the same pattern which we may call the "pattern of l loci." Yet we did not solve the last problem, in sect. 6.2(7). What was the difficulty? We succeeded in splitting the condition into clauses quite neatly, but we failed to manage the loci corresponding to these clauses: we could not exhaust, could not describe conveniently those loci and so we could not form their intersection.

There are cases in which we have this difficulty but not in its most formidable form, and we may be able to handle such cases.

(1) *Two loci for a word.* In a crossword puzzle that allows puns and anagrams we find the following clue:
"Flat both ways (5 letters)".

After some trials we may be led to the following interpretation: The unknown x is a word. The condition consists of two clauses:

(r_1), x means "flat"

(r_2), x is a word having 5 letters which, read backward, still has the same meaning "flat."

With which clause should we begin? There is a difference. To manage the clause (r_2) efficiently, you should have in your head a list of all five-letter words that can be read backward with some meaning. Now, very few of us have such a list. But most of us can remember words that have more or less the same meaning as "flat"; we have to examine them as they emerge whether they also fulfill the clause (r_2). Here are some such words:

even, smooth, unbroken—plain, dull—horizontal—of course, LEVEL![1]

(2) Let us try to disentangle the essential feature of the foregoing procedure.

The clause (r_1) selects from the vast range of all words a small set of words, one among which is the solution. The clause (r_2) does the same, but there is a difference: the selection is easier in one case than in the other, we can handle (r_1) more efficiently than (r_2). We have used the more manageable clause for a first selection and the less manageable clause for a subsequent second selection. It is more necessary to be efficient in the first selection: we select elements the first time from the immense reservoir of all words, the second time from the very much more restricted first locus, obtained by the first selection.

The moral is simple: To each clause corresponds a locus. *Begin with the clause, for which the locus can be more fully, or more efficiently, formed.* Doing so, you may avoid forming the loci corresponding to the other clauses: you use those other clauses in selecting elements from the first locus.

(3) *Two loci for a tripartite unknown.* How old is the captain, how many children has he, and how long is his boat? Given the product 32118 of the three desired numbers (integers). The length of the boat is given in feet (is several feet), the captain has both sons and daughters, he has more years than children, but he is not yet one hundred years old.

This puzzle demands to find three numbers,

$$x, \qquad\qquad y, \qquad\qquad z$$

which represent the captain's

            number of children,   age,   length of boat

[1] HSI, Decomposing and recombining 8, pp. 83–84, contains a very similar example and anticipates the essential idea of the present section.

respectively. It will be advantageous to conceive the problem thus: We have but one unknown; this unknown, however, is not a number but a tripartite unknown, a triplet (x, y, z) of numbers.

It is very important to split the condition that is expressed by the statement of the problem into appropriate clauses. This needs careful consideration of details and considerable regrouping. After several trials (which we skip to save space) we may arrive at the following two clauses:

(r_1), x, y, and z are positive integers different from 1 and such that

$$xyz = 32118$$

(r_2) $$4 \leqq x < y < 100$$

With which one of the two clauses should we begin? Of course, with (r_1) which leaves only a finite number of possibilities, whereas (r_2), which does not restrict z at all, leaves an infinite number.

Therefore, we examine (r_1). Now, 32118 is divisible by 6, and so we easily decompose it into prime factors:

$$32118 = 2 \times 3 \times 53 \times 101$$

For a decomposition into three factors we have to combine two of the four primes. Therefore, there are only six different ways to decompose the number 32118 into a product of three factors all different from 1:

$$
\begin{array}{rrr}
6 \times & 53 \times & 101 \\
3 \times & 101 \times & 106 \\
3 \times & 53 \times & 202 \\
2 \times & 101 \times & 159 \\
2 \times & 53 \times & 303 \\
2 \times & 3 \times & 5353
\end{array}
$$

Of these six possibilities, the remaining requirement (r_2) rejects all except the first one, and so we obtain

$$x = 6, \qquad y = 53, \qquad z = 101$$

The captain has 6 children, is 53 years old, and the length of his boat is 101 feet.

The essential idea of the solution of this simple puzzle is often applicable, also in more complicated cases: split off from the full condition a "major" clause that leaves open only a small number of possibilities, and choose between these possibilities by using the remaining "minor" part of the condition.[2]

†(4) *Two loci for a function.* There is a very important type of mathematical problem, of daily use in physics and engineering, the condition of

---

[2] Cf. ex. 6.12 to 6.17.

which is naturally split into two clauses: to determine a function by a differential equation and initial, or boundary, conditions. Here is a simple example: the unknown x is a function of the independent variable t; it is required to satisfy

(r_1) the differential equation $\dfrac{d^2x}{dt^2} = f(x, t)$, where $f(x, t)$ is a given

   function,

(r_2) the initial conditions $x = 1$, $\dfrac{dx}{dt} = 0$ for $t = 0$.

Should we begin with the differential equation or with the initial conditions? That depends on the nature of the given function $f(x, t)$.
   First case. Take $f(x, t) = -x$, so that the proposed differential equation is

$$\frac{d^2x}{dt^2} = -x$$

This differential equation belongs to those few privileged types of which we can exhibit the "general integral" explicitly. In fact, the most general function satisfying the differential equation is

$$x = A \cos t + B \sin t$$

where A and B are arbitrary constants (constants of integration). Thus we have obtained the "locus" corresponding to the clause (r_1).
   We proceed now to the clause (r_2) which we use to pick out the solution from the first locus that we have just obtained: setting $t = 0$ in the expressions for x and $\dfrac{dx}{dt}$, we find from the initial conditions that

$$A = 1, \qquad B = 0, \qquad x = \cos t$$

   Second case. We examine the differential equation, but do not succeed in finding its general integral (or any of its integrals) and decide that we shall not make any further efforts in this direction. What should we do next? With which one of the two clauses, (r_1) and (r_2), should we now begin?
   In this situation we may use (r_2) first: we set up x as a power series in t, the initial coefficients of which are determined by the initial conditions whereas the remaining coefficients $u_2, u_3, u_4, \ldots$ appear, at this stage of our work, undetermined (they are, in fact, our unknowns, see ex. 3.81):

$$x = 1 + u_2 t^2 + u_3 t^3 + u_4 t^4 + \cdots$$

Thus, in a sense, the locus corresponding to (r_2) has been obtained. We now proceed to (r_1), the first clause, to determine the remaining coefficients

u_2, u_3,... from the differential equation (by recursion, if possible; see again ex. 3.81).

Observe that, in any case, the differential equation is more "selective" (narrows down the choice of the function much more) than the initial condition. Thus, the proposed (r_2) determines only two coefficients of the power series; the differential equation (the condition (r_1)) has to determine the remaining infinite sequence. This shows that the more selective clause is not always the best to begin with.

6.4. Wider scope for recursion

In the foregoing section we have observed an important difference between clauses and clauses: there may be reasons, and even strong reasons, to begin the work rather with one clause than with the other. It is true, there was a limitation: we have considered the case of one unknown. (This limitation is not really restrictive; see the indication in sect. 5.5.) Let us now consider the case of several unknowns.

(1) There is an important general situation which is suggested by several examples considered in chapter 3. There are n unknowns $x_1, x_2, x_3, \ldots, x_n$ which satisfy n conditions of the following form:

$$r_1(x_1) = 0$$
$$r_2(x_1, x_2) = 0$$
$$r_3(x_1, x_2, x_3) = 0$$
$$. \quad . \quad . \quad . \quad . \quad .$$
$$r_n(x_1, x_2, x_3, \ldots, x_n) = 0$$

This particular system of n relations suggests not only where we should begin, but also how we should go on. In fact, it suggests a full plan of campaign: Begin with x_1, which you should determine from the first relation. Having obtained x_1, determine x_2 from the second relation. Having obtained x_1 and x_2, determine x_3 from the third relation, and so on: determine the unknowns $x_1, x_2, \ldots, x_n$ one at a time, in the order in which they are numbered, using the values of those already obtained in determining the next one. This plan works well if the kth relation is an equation

$$r_k(x_1, x_2, \ldots, x_{k-1}, x_k) = 0$$

from which we can express x_k in terms of $x_1, x_2, \ldots, x_{k-1}$, for $k = 1, 2, 3, \ldots, n$. The situation is particularly favorable if the kth equation is linear with respect to x_k (the coefficient of which should not vanish, of course).

This is the pattern of *recursion*: we determine x_k by recurring, or going back, to the previously obtained $x_1, x_2, \ldots, x_{k-1}$.

Following this pattern we simply proceed step by step, beginning with x_1, tackling x_2 after x_1, x_3 after x_2, and so on, which seems to be the most obvious, the most natural thing to do. At each step we refer to information accumulated by the foregoing steps—and this is perhaps the most significant feature of the pattern. We shall see the point more clearly after a few examples.

(2) In sect. 2.5(3) we obtained a system of 7 equations for 7 unknowns. Let us relabel the unknowns as follows:

$$D = x_7$$
$$a = x_4 \qquad b = x_5 \qquad c = x_6$$
$$p = x_1 \qquad q = x_2 \qquad r = x_3$$

And let us rewrite the system of equations, expressing precisely which unknowns are linked by each equation, but disregarding other details, and numbering the equations so that the order in which they should be treated is clearly visible.

Thus we obtain the following system of relations:

$$r_1(x_2, x_3) = 0$$
$$r_2(x_3, x_1) = 0$$
$$r_3(x_1, x_2) = 0$$
$$r_4(x_2, x_3, x_4) = 0$$
$$r_5(x_3, x_1, x_5) = 0$$
$$r_6(x_1, x_2, x_6) = 0$$
$$r_7(x_4, x_5, x_6, x_7) = 0$$

So written, the system renders the following plan obvious: Let us separate the first three relations from the rest. They contain only the first three unknowns x_1, x_2, x_3 and may be regarded as a system of three equations for these three unknowns. (In fact, we can easily express $x_1 = p$, $x_2 = q$, $x_3 = r$ from the system of the three equations given in sect. 2.5(3) that are indicated here by the first three relations.) Once the first three unknowns x_1, x_2, x_3 have been found, the system "becomes recursive": First, we obtain x_4, x_5, x_6, each unknown from the correspondingly numbered relation. (In fact, the order in which we treat these three unknowns does not matter.) Having found x_4, x_5, x_6, we use the last relation to obtain x_7 (which is the principal unknown in the original problem, see sect. 2.5(3); the others are only auxiliary unknowns).

The reader should compare the system just discussed with the system considered in sect. 6.1(1).

(3) *Solve the equation*

$$(he)^2 = she$$

Of course, *he* and *she* are ordinary numbers (positive integers) written in the ordinary decimal notation, the one a two-digit number, the other a three-digit number, and h, e, and s are digits. We may restate the problem quite fussily: find h, e, and s satisfying

$$(10h + e)^2 = 100s + 10h + e$$

where h, e, and s are integers, $1 \leq h \leq 9$, $0 \leq e \leq 9$, $1 \leq s \leq 9$.

This little puzzle is not difficult. If the reader has solved it by himself, he will be in a better position to appreciate the following scheme. In the initial phase we shall examine just one unknown. In the next phase we shall bring in one more and consider the two unknowns jointly. Only in the last phase shall we deal with all three unknowns.

Phase (e). We begin with e, since there is a *requirement for e alone*: the last decimal of e^2 must be e. We list the squares of the ten digits,

$$0, \quad 1, \quad 4, \quad 9, \quad 16, \quad \mathbf{25}, \quad \mathbf{36}, \quad 49, \quad 64, \quad 81$$

and find that there are only four out of ten that satisfy the requirement, and so

$$e = 0 \quad \text{or} \quad 1 \quad \text{or} \quad 5 \quad \text{or} \quad 6$$

Phase (e, h). There is a requirement that involves just two out of the three digits, e and h:

$$100 \leq (he)^2 < 1000$$

from which we easily conclude that

$$10 \leq he \leq 31$$

Combining this information with the result of the foregoing consideration under (e), we find that the two-digit number he must be one of the following ten numbers:

$$\begin{array}{cccc} 10, & 11, & 15, & 16, \\ 20, & 21, & 25, & 26, \\ 30, & 31 & & \end{array}$$

Phase (e, h, s). Now we list the squares of the ten numbers just obtained

$$\begin{array}{cccc} 100, & 121, & 225, & 256 \\ 400, & 441, & \mathbf{625}, & 676 \\ 900, & 961 & & \end{array}$$

and we find just one that satisfies the full condition. Hence

$$e = 5, \quad h = 2, \quad s = 6$$

$$(25)^2 = 625$$

(4) In the foregoing subsection (3) we have split the condition of the

proposed problem into three clauses which we may represent (using the notation introduced in sect. 6.1) by a system of three symbolic equations

$$r_1(e) = 0$$
$$r_2(e, h) = 0$$
$$r_3(e, h, s) = 0$$

Let us compare this system of three clauses with the following system of three linear equations:

$$a_1 x_1 = b_1$$
$$a_2 x_1 + a_3 x_2 = b_2$$
$$a_4 x_1 + a_5 x_2 + a_6 x_3 = b_3$$

x_1, x_2, x_3 are the unknowns, a_1, $a_2, \ldots$, a_6, $b_1, \ldots$, b_3 are given numbers, a_1, a_3, and a_6 are supposed to be different from 0.

The similarity of these two systems is more obvious than their difference; let us compare them carefully.

Let us look first at the system of three linear equations for x_1, x_2, and x_3. The first equation determines the first unknown x_1 completely: the later equations will in no way influence or modify the value of x_1 obtained from the first.   Based on this value of x_1, the second equation completely determines the second unknown x_2.

The system of the three clauses into which we have split the condition for the unknowns e, h, and s is formally similar to, but materially different from, the system of three equations for x_1, x_2, x_3.   The first clause does not completely determine the first unknown e; it just narrows down the choice for e; it yields (this is the most appropriate expression) a *locus* for e. Similarly, the second clause does not completely determine the second unknown h; it yields a locus for the couple of two unknowns (e, h).   Only the last clause achieves a definitive determination: it picks out from the locus previously established the only triplet (e, h, s) that satisfies the full condition.

6.5. Gradual conquest of the unknown

Considering n numerical unknowns x_1, $x_2, \ldots$, x_n, we may regard them as the successive components of one multipartite unknown x (see sect. 5.5). Let us so view the n unknowns which we determine successively from a recursive system of equations such as we have considered in sect. 6.4(1). The recursive procedure of solution unveils our multipartite unknown gradually, step by step.   At first we obtain little information about the unknown, the value of just one component, x_1.   Yet, using this initial information to advantage, we obtain more: we add the knowledge of the second component x_2 to that of the first.   At each stage of our work we add the knowledge of one more component to our previously acquired knowledge, at each stage we use the information already obtained to obtain additional information. We conquer an empire province by province, using at each stage the provinces already won as *base of operations* to win the next province.

We have seen cases in which this procedure is more or less modified. The provinces may not be conquered exactly one at a time, but the empire builder sometimes takes a bigger bite, two or three provinces at the same time; cf. sect. 6.4(2) and sect. 6.1(1). Or a province is not conquered fully at one stroke; first one, then another province is made partially dependent, and a final successful move acquires all at once; cf. sect. 6.4(3).

We may have met with still other variations of the procedure in our past experience. We certainly had opportunity to be impressed with the peculiar expanding pattern of the work for the solution; cf. sect. 2.7. If the unknown has many components (as in a crossword puzzle), we may advance along several lines simultaneously: we need not thread all our beads on one string, but may use several strings. Yet the essential thing is to *use the information already gathered as a base of operations to gather further information.* Perhaps all rational procedures of problem solving and learning are recursive in this sense.

Examples and Comments on Chapter 6

6.1. *A condition with many clauses.* In a *magic square* with n rows, n^2 numbers are so arranged that the sum of the numbers in each of the n rows, in each of the n columns and in each of the two diagonals is the same; this sum is called the "constant" of the magic square. The simplest and best-known magic square has $n = 3$ rows and is filled with the first nine natural numbers $1, 2, \ldots, 9$. Let us state in minute detail the problem that requires us to find this simple magic square.

What is the unknown? There are nine unknowns; let x_{ik} denote the desired number in the ith row and the kth column; $i, k = 1, 2, 3$.

What is the condition? The condition has four different kinds of clauses:

(1) x_{ik} is an integer
(2) $1 \leq x_{ik} \leq 9$
(3) $x_{ik} \neq x_{jl}$ unless $i = j$ and $k = l$
(4) $x_{11} + x_{12} + x_{13} = x_{11} + x_{22} + x_{33}$ for $i = 1, 2, 3$
    $x_{1k} + x_{2k} + x_{3k} = x_{11} + x_{22} + x_{33}$ for $k = 1, 2, 3$
    $x_{13} + x_{22} + x_{31} = x_{11} + x_{22} + x_{33}$

State the number of clauses of each kind and the total number of clauses. Which shape have these clauses in the notation of sect. 6.1(3)?

6.2. By introducing a multipartite unknown reduce the general system considered in sect. 6.1(5) to the (apparently more particular) system considered in sect. 6.2(4).

6.3. By introducing a multipartite unknown reduce the system considered in sect. 6.1(1) to a particular case of the system considered in sect. 6.4(1).

6.4. Reduce the system considered in sect. 6.4(2) in the manner of ex. 6.3.

6.5. Devise a plan to solve the system

$$r_1(x_1, x_2, x_3) = 0$$
$$r_2(x_1, x_2, x_3) = 0$$
$$r_3(x_1, x_2, x_3) = 0$$
$$r_4(x_1, x_2, x_3, x_4) = 0$$
$$r_5(x_1, x_2, x_3, x_5) = 0$$
$$r_6(x_1, x_2, x_3, x_6) = 0$$
$$r_7(x_1, x_2, x_3, x_4, x_5, x_6, x_7) = 0$$

6.6. The system of relations

$$r_1(x_1) = 0$$
$$r_2(x_1, x_2) = 0$$
$$r_3(x_2, x_3) = 0$$
$$r_4(x_3, x_4) = 0$$
$$\cdot \quad \cdot \quad \cdot \quad \cdot$$
$$r_n(x_{n-1}, x_n) = 0$$

is a particularly interesting particular case of a system considered in the text: of which one?

Have you seen it before? Where have you had opportunity to compare two systems analogously related to each other?

6.7. Through a given interior point of a circle construct a chord of given length.
Classify this problem.

6.8. Two straight lines, *a* and *b*, and a point *C* are given in position; moreover, a length *l* is given. Draw a straight line *x* through the point *C* so that the perimeter of the triangle formed by the lines *a*, *b*, and *x* is of length *l*.
Classify this problem.

6.9. *Keep only a part of the condition.* Of the two clauses of the problem considered in sect. 6.2(7), (r_1) is somewhat more manageable: in trying to satisfy this requirement, we can map out some sort of a plan. To find an anagram of a given set of letters, such as RASH AYE, we have to find a word that has only letters of this set and has all of them. A procedure that may help is the following: drop the last part of the condition "and has all of them" and try to find words, or usual word parts, word endings, formed with the letters of the given set. Short words of this kind are easy to form, and proceeding to

longer ones we may hope to arrive eventually at the desired anagram.   In our
case, we may hit upon the following:

ASH, YES, SAY, SHY, RYE, EAR
HEAR, HARE, AREA
SHARE
RE- (beginning)
-ER, -AY, -EY (endings).

To solve the problem of sect. 6.2(7), we look at these bits of words, having in
mind not only the anagram or clause (r_1), but also clause (r_2).   Some of the
bits may combine into a full anagram—but SHY AREA is not an acceptable
solution.

6.10. *The thread of Ariadne.*   The daughter of King Minos, Ariadne, fell in
love with Theseus and gave him a clue of thread which he unwound when
entering the Labyrinth and found his way out of its mazes by following back
the thread.

Did some prehistoric heuristic genius contribute to the formation of this
myth?   It suggests so strikingly the nature of certain problems.

In trying to solve a problem, we often run into the difficulty that we see no
way to proceed farther from the last point that we have attained.   The
Labyrinth suggests another kind of problem in which there are many ways proceeding from each point attained, but the difficulty is to choose between them.
To master such a problem (or to present its solution when we have succeeded
in mastering it) we should try to treat the various topics involved *successively*,
in the most appropriate, the most *economical* order: whenever an alternative
presents itself, we should choose *the next topic so that we can derive the maximum help from our foregoing work.*   The "most appropriate choice at the
crossroads" is strongly suggested by the "thread of Ariadne" which was, by
the way, one of Leibnitz's pet expressions.

Problems involving several unknowns, several interrelated tasks and conditions, are often of such a labyrinthine nature; crossword puzzles and constructions of complex geometric figures may yield good illustrations.   Having
to solve such a problem, we have a choice at each stage: to which task (to
which word, to which part of the figure) should we turn next?   At first, we
should look for the weakest spot, for the clause to begin with, for the most
accessible word of the puzzle, or the most easily constructible part of the
figure.   Having found a first word or having constructed a first bit of the
figure, we should carefully select our second task: that word (or that part of
the figure) different from the first to find which the first word (or part) already
found offers the most help.   And so on, we should always try to select our
next task, the next unknown to find, so that we get the maximum help from
the unknowns previously found.   (This spells out an idea already voiced in
sect. 6.5.)

There follow a few problems which give an opportunity to the reader to try
out the preceding advice.

6.11. Find the magic square with three rows minutely described in ex. 6.1. (You may know one solution, but you should find all solutions. The order in which you examine the various unknowns is very important. Especially, try to spot those unknowns whose values are uniquely determined and begin with them.)

6.12. Multiplication by 9 reverses a four-digit number (produces a four-digit number with the same digits in reverse order). What is the number? (Which part of the condition will you use first?)

6.13. Find the digits a, b, c, and d, being given that

$$ab \times ba = cdc$$

It is assumed that the digits a and b of the two-digit number ab (that is $10a + b$) are different.

6.14. A triangle has six "parts": three sides and three angles. Is it possible to find two such *noncongruent* triangles that five parts of the first are identical with five parts of the second? (I did *not* say that those five identical parts are *corresponding* parts.)

6.15. Al, Bill, and Chris planned a big picnic. Each boy spent 9 dollars. Each bought sandwiches, ice cream, and soda pop. For each of these items the boys spent jointly 9 dollars, although each boy split his money differently and no boy paid the same amount of money for two different items. The greatest single expense was what Al paid for ice cream; Bill spent twice as much for sandwiches as for ice cream. How much did Chris pay for soda pop? (All amounts are in round dollars.)

6.16. In preparation for Hallowe'en, three married couples, the Browns, the Joneses, and the Smiths, bought little presents for the neighborhood youngsters. Each bought as many identical presents as he (or she) paid cents for one of them. Each wife spent 75 cents more than her husband. Ann bought one more present that Bill Brown, Betty one less than Joe Jones. What is Mary's last name?

6.17. It was a very hot day and the 4 couples drank together 44 bottles of coca-cola. Ann had 2, Betty 3, Carol 4, and Dorothy 5 bottles. Mr. Brown drank just as many bottles as his wife, but each of the other men drank more than his wife: Mr. Green twice, Mr. White three times, and Mr. Smith four times as many bottles. Tell the last names of the four ladies.

6.18. *More problems.* Try to consider further examples from the viewpoint of this chapter. Pay attention to the division of the condition into clauses and weigh the advantages and the disadvantages of beginning the work with this or that clause. Review a few problems you have solved in the past with this viewpoint in mind and seek new problems in solving which this viewpoint has a chance to be useful.

6.19. *An intermediate goal.* We have started already working at our problem, but we are still in an initial phase of our work. We have understood our problem as a whole; it is a problem to find.   We have answered the question "What is the unknown?"; we know what kind of thing we are looking for. We have also listed the data and have understood the condition as a whole, and now we want to *split the condition into appropriate parts*.

Observe that this task need not be trivial: there may be several possibilities to subdivide the condition and we want, of course, the most advantageous subdivision.   For instance, in solving a geometric problem by algebra, we express each clause of the condition by an equation; different subdivisions of the condition into clauses yield different systems of equations and, of course, we want to pick the system that is most convenient to handle (cf. sect. 2.5(3) and 2.5(4)).

In the statement of the proposed problem, the condition may appear as an undivided whole or it may be divided into several clauses.   In either case we are facing a task: to split the condition into appropriate clauses in the first case and, in the second case, to split the condition into more appropriate clauses.   The subdivision of the condition may bring us nearer to the solution: it is an *intermediate goal*, very important in some cases.

6.20. *Graphical representation.*   We have expressed a relation, required by the condition of the problem, which involves specified unknowns by a symbolic equation [introduced in sect. 6.1(3)].   We may express such relations also graphically, by a diagram, and the graphical representation may contribute to a clearer conception of a system of such relations.

We represent an unknown by a small circle, a relation between unknowns by a small square, and we express the fact that a certain relation involves a certain unknown by joining the square representing the relation to the circle representing the unknown.   Thus diagram (*a*) in Fig. 6.3 represents a system of four relations between four unknowns; we see from it, for instance, that there is just one unknown involved in all four relations and just one relation involving all four unknowns; in fact, the diagram (*a*) and the system of four equations in sect. 6.1(1) express exactly the same state of affairs, the one in geometric language and the other in the language of formulas.   The crossing of lines in a point which lies outside the little circles and squares [as it happens once in diagram (*a*)] is immaterial; we can imagine, in fact, that only the little circles and squares lie in the plane of the paper and the connecting lines are drawn through space and have no point in common, although their projections on the plane of the paper may cross accidentally.

As the diagram (*a*), also the diagrams (*b*), (*c*), (*d*), and (*e*) of Fig. 6.3 represent systems of relations considered before: point out the section or example where they have been considered.

(Fig. 6.4 exemplifies another kind of diagrammatic representation which is "dual" to the foregoing: both relations and unknowns are represented by lines, a relation by a horizontal, an unknown by a vertical, line; iff a relation contains an unknown, the lines have a common point.   The same fact is expressed by (*c*) in Fig. 6.3 and Fig. 6.4, and the same holds for (*d*).

†An algebraic representation is suggested by Fig. 6.4: a matrix in which

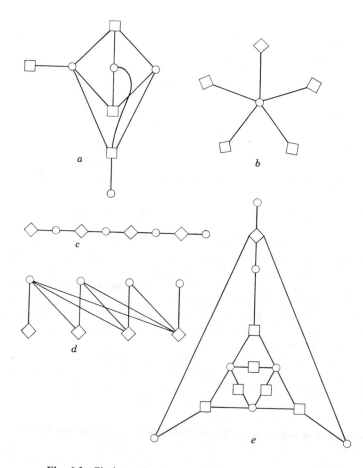

Fig. 6.3. Circles and squares, unknowns and relations.

each row corresponds to a relation and each column to an unknown of the system; an element of the matrix is 1 or 0 according as the relation concerned does or does not involve the unknown concerned.)

6.21. *Some types of nonmathematical problems.*   Which clause of the condition should we try to satisfy first?   This question arises typically in various situations.   Having chosen a clause which appears to be of major importance, and having listed the objects (or some objects) satisfying this "major" clause, we bring into play the remaining "minor" clauses which remove most of the objects from the list and leave eventually one that satisfies also the minor clauses and so the full condition.   This pattern of procedure which we had opportunity to observe in the foregoing [sect. 6.3(3), ex. 6.15, ex. 6.16] is

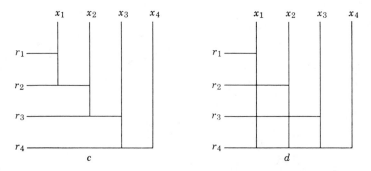

Fig. 6.4. Unknowns and relations, vertical and horizontal lines.

suitable for, and arises naturally in, various types of nonmathematical problems.

The translator's problem. In translating a French text into English, we have to find the correct English equivalent of a French word, for instance of the word "confiance." A French–English dictionary yields a list of English words (confidence, trust, reliance, assurance) which satisfy only a first, rather rough, clause of the full condition of our problem: we have to look carefully into the context to discover further, more subtle clauses hidden in it and bring these clauses into play to remove the less fitting words and choose the most appropriate one from the list.

Checkmate in two moves. There is given an arrangement of white and black chessmen on the chessboard, consistent with the rules of the game. The unknown is a move of white. The condition requires this move of white to be such that whatever move of black may follow there is a subsequent move of white that checkmates the black king.

The desired move of white has to "ward off" each possible move of black (prevent it from happening or prepare for answering it with a checkmate). And so the condition has as many clauses, we may say, as there are possible moves of black to be warded off.

A workable strategy is to begin with a crucial move of black which appears to involve a major threat and to list the moves of white capable of warding off this major threat. Then we consider other "minor" moves of black and remove from the list such moves of white which are not able to ward off one or the other "minor" move of black; the true solution should eventually remain alone on the list.

Engineering design. An engineer wants to design a new gadget. To be put into production, the new gadget has to fulfill a host of requirements; some are "technical" requirements such as smooth working, no danger for the user, durability, and so on, others are "commercial" requirements such as low price of manufacturing, sales appeal, and so on. The engineer retains at first the technical requirements (or some of them) which we may regard as constituting the "major" condition so that he has a clear-cut technical (physical) problem to solve.   This problem has usually several solutions which the engineer lists and examines.   This being done, the commercial requirements (which we have heretofore regarded as "minor") enter into play; they may discard many a smoothly operating gadget and may leave one the production of which appears the most profitable.

6.22. Without using paper and pencil, just by looking at it, solve the following system of three equations with three unknowns:

$$3x + y + 2z = 30$$
$$2x + 3y + z = 30$$
$$x + 2y + 3z = 30$$

Prove your solution.

6.23. Given a, b, and c, the lengths of the three sides of a triangle.   Each vertex of the triangle is the center of a circle; these three circles are exterior to each other and touch each other.   Find the three radii x, y, and z.

6.24. Find x, y, u, and v, satisfying the system of four equations:

$$y + u + v = -5$$
$$x + u + v = 0$$
$$x + y + v = -8$$
$$x + y + u = 4$$

Don't you see a short cut?

6.25. *A more refined classification.*   The foregoing examples 6.22, 6.23, and 6.24 illustrate an important point: The circumstance that the condition of a problem involving several unknowns is *symmetric* with respect to these unknowns may, if recognized, influence the course of, and greatly facilitate, the solution.   (Cf. also ex. 2.8 and MPR, vol. 1, pp. 187–188, ex. 41.   Sometimes, as in ex. 6.23, we should consider not only permutations of the unknowns, but permutations of the unknowns *and* the data.)   There are other cases, of less common occurrence but nevertheless interesting, in which the condition remains unchanged, not by all, but only by some permutations (by a certain group of permutations) of the unknowns (and the data).   By following up this remark systematically, we would arrive at a still more refined classification of problems to find than the one implied by the basic remark of the present chapter [see sect. 6.1(6)] and we can foresee that such a classification would be of some interest for our study.

SOLUTIONS

Chapter 1

1.1. Circle with the given point as center and the given distance as radius.

1.2. Two straight lines parallel to the given line.

1.3. Straight line, the perpendicular bisector of the segment the endpoints of which are given.

1.4. Straight line parallel to the given parallels midway between them, that is, bisecting their distance.

1.5. Two straight lines, perpendicular to each other, bisectors of the angles included by the given lines.

1.6. Two circular arcs, symmetric to each other with respect to the line AB; they have the same endpoints, A and B.

1.8. Two loci, ex. 1.1.

1.9. Two loci, exs. 1.1, 1.2.

1.10. Two loci, exs. 1.2, 1.6.

1.11. Two loci, exs. 1.1, 1.6.

1.12. Two loci, ex. 1.5.

1.13. Two loci, ex. 1.2.

1.14. Two loci, exs. 1.1, 1.2.

1.15. Two loci, ex. 1.6.

1.16. By symmetry reduces to sect. 1.3(2), or to ex. 1.12.

1.17. Two loci, ex. 1.6.

1.18. (*a*) If X varies so that the areas of the two triangles $\triangle XCA$ and $\triangle XCB$ remain equal, the locus of X is the median passing through C. (Prove it!) The required point is the intersection of the medians. (*b*) If X varies so that the area of $\triangle ABX$ remains one third of the area of the given $\triangle ABC$, the locus

154

of X is a parallel to the side AB, at a distance equal to one third of the altitude dropped from C; see ex. 1.2.   The required point is the intersection of such parallels to the sides.   Both solutions use "two loci."

1.19. Join the center of the inscribed circle to both endpoints of the side a; in the triangle so obtained the angle at the center of the inscribed circle is

$$180° - \frac{\beta + \gamma}{2} = 90° + \frac{\alpha}{2}$$

Two loci, exs. 1.2, 1.6.

1.20. Auxiliary figure: the right triangle with hypotenuse a and leg h_b.

1.21. Auxiliary figure, see ex. 1.20.

1.22. Auxiliary figures, see ex. 1.20.

1.23. Auxiliary figure: right triangle with leg h_a and opposite angle β.

1.24. Auxiliary figures, see ex. 1.23.   Other solution: see ex. 1.34.

1.25. Auxiliary figure: right triangle with hypotenuse d_α and height h_a.

1.26. Auxiliary figure: triangle from three sides.

1.27. Assume that a is longer than c.   Auxiliary figure: triangle from sides $a - c$, b, d; see HSI, Variation of the problem 5, pp. 211–213.

1.28. Generalization of ex. 1.27 which corresponds to the case $\epsilon = 0$. Auxiliary figure: triangle from a, c, ϵ; see MPR, vol. 2, pp. 142–145.

1.29. Auxiliary figure: triangle from a, $b + c$, $\alpha/2$.

1.30. Auxiliary figure: triangle from a, $b + c$, $90° + (\beta - \gamma)/2$.

1.31. Auxiliary figure: triangle from $a + b + c$, h_a, $\alpha/2 + 90°$.   See HSI Auxiliary elements 3, pp. 48–50, and Symmetry, pp. 199–200.

1.32. Appropriate modification of the approach in sect. 1.6(1): let one of the two radii shrink at the same rate as the other expands.   Auxiliary figure: tangents to a circle from an outside point, followed by the construction of two rectangles.

1.33. Cf. sect. 1.6(1).   Auxiliary figure: circle circumscribed about the triangle the vertices of which are the centers of the three given circles.

1.34. Similar triangle from α, β; obtain afterwards the required size by using the given length d_γ.   Essentially the same for ex. 1.24.

1.35. Similar figures: center of similarity is the vertex of the right angle in the given triangle.   The bisector of this right angle intersects the hypotenuse in a vertex of the desired square.

1.36. Generalization of ex. 1.35.   Center of similarity is A (or B).   Cf. HSI, section 18, pp. 23–25.

1.37. Similar figures: center of similarity is the center of the circle.   The required square is symmetric with respect to the same line as the given sector.

1.38. Similar figure: any circle touching the given line, the center of which is on the perpendicular bisector of the segment joining the two given points. The point of intersection of this bisector and of the given line is center of similarity.   Two solutions.

1.39. Symmetry with respect to the bisector of the appropriate angle included by the given tangents yields one more point through which the circle must pass, and so reduces the problem to ex. 1.38.

1.40. The radii of the inscribed circle drawn to the points of tangency include the angles $180° - \alpha$, $180° - \beta, \ldots$; hence a similar figure is immediately obtainable. Applicable to circumscribable polygons with any number of sides.

1.41. Let A denote the area and a, b, c the sides of the required triangle (ex. 1.7) so that

$$2A = ah_a = bh_b = ch_c$$

Construct a triangle from the given sides h_a, h_b, h_c and let A' denote its area and a', b', c' its corresponding altitudes so that

$$2A' = h_a a' = h_b b' = h_c c'$$

Therefore,

$$\frac{a}{a'} = \frac{b}{b'} = \frac{c}{c'}$$

and so the triangle with the easily obtainable sides a', b', c' is *similar* to the required triangle.

1.42. The foregoing solution of ex. 1.41 is imperfect: if $h_a = 156$, $h_b = 65$, $h_c = 60$, the required triangle does exist, but the auxiliary triangle with the given sides h_a, h_b, h_c does not.

One possible remedy is a *generalization*: let k, l, m be any three positive integers, and (the notation is *not* the same as in ex. 1.41) a', b', c' the altitudes in the triangle with sides kh_a, lh_b, mh_c; then

$$\frac{a}{ka'} = \frac{b}{lb'} = \frac{c}{mc'}$$

For example, a triangle with sides 156, 65, and $120 = 2 \times 60$ does exist.

1.43. From the center of the circumscribed circle, draw a line to one of the endpoints of the side a and a perpendicular to this side. You so obtain a right triangle with hypotenuse R, angle α, and opposite leg $a/2$. This yields a relation between a, α, and R: you can construct any one of the three if the two others are given. (The relation can also be expressed by the trigonometric equation $a = 2R \sin \alpha$.) If the data of the proposed problem do not satisfy this relation, the problem is impossible; if they do satisfy it, the problem is indeterminate.

1.44. (a) Triangle from α, β, γ: the problem is either impossible or indeterminate. (b) The general situation behind ex. 1.43 and (a): the existence of the solution implies a relation between the data and, therefore, the solution is either indeterminate or nonexistent according as the relation is, or is not, satisfied by the data. (c) By the solution of ex. 1.43 reduce to: triangle from a, β, α. (d) By the solution of ex. 1.43 reduce to ex. 1.19.

1.45. We disregard disturbances influencing the velocity of sound which we cannot control (as wind and varying temperatures). Then, from the time difference of the observations at the listening posts A and B, we obtain the difference of two distances, $AX - BX$ which yields a locus for X: a hyperbola. We obtain another hyperbola comparing C with A (or B) and the intersection of the two hyperbolas yields X. Main analogy with ex. 1.15: the observations yield two loci. Main difference: the loci are circular arcs there, but hyperbolas here. We cannot describe a hyperbola with ruler and compasses, but we can describe it with some other gadget and a machine could be constructed to evaluate conveniently the observations of the three listening posts.

1.46. Those loci would not be usable if we took the statement of the pattern in sect. 1.2 literally. In fact, those loci are useful and have been used several times in the foregoing examples, and it is the statement in sect. 1.2 that needs extension: we should admit a locus when it is a union of a finite number of straight lines, or circles, or segments of straight lines, or arcs of circles.

1.48. If the parts into which the condition is split are jointly equivalent to the condition, the various manners of splitting must be equivalent to each other. Hence a theorem for the triangle: the perpendicular bisectors of the sides (there are three) pass through the same point. And for the tetrahedron: the perpendicular bisecting planes of the edges (there are six) pass through the same point.

1.50. (1) Avoiding certain exceptional cases (see exs. 1.43 and 1.44) take any three different constituent parts of a triangle listed in ex. 1.7 as data and propose to construct a triangle. Here are a few more combinations with which the construction is easy:

$a,$	$h_b,$	R
$a,$	$h_b,$	m_b
$a,$	$h_b,$	m_a
$h_a,$	$d_\alpha,$	b
$h_a,$	$m_a,$	m_b
$h_a,$	m_b	m_c
$h_a,$	$h_b,$	m_a
$a,$	$b,$	R

Also α, β, and any line not yet mentioned in ex. 1.24 or ex. 1.34. Less easy

$$\alpha, \quad r, \quad R$$

(2) There are several problems about trihedral angles, similar to that discussed in sect. 1.6(3), which are important and can be solved without invoking explicitly the help of descriptive geometry. Here is one: "Being given a, a face angle, and β and γ, the adjacent dihedral angles of a trihedral angle, construct b and c, the remaining face angles." The solution is not difficult, but it would take up too much space to explain it here.

(3) Ex. 1.47 is the space analogue of sect. 1.3(1). Discuss the space analogues of sect. 1.3(2), ex. 1.18, sect. 1.3(3), ex. 1.14.

No solution: **1.7, 1.47, 1.49, 1.51.**

Chapter 2

2.1. If Bob has x nickels and y dimes, we can translate the condition into the system of two equations

$$5x + 10y = 350$$
$$x + y = 50$$

which, after an obvious simplification, precisely coincides with the system in sect. 2.2(3).

2.2. There are m pipes to fill, and n pipes to empty, a tank. The first pipe can fill the tank in a_1 minutes, the second pipe in a_2 minutes,... the pipe number m in a_m minutes. Of the other kind of pipes, the first can empty the tank in b_1 minutes, the second in b_2 minutes,...the pipe number n in b_n minutes. With all pipes open, how long will it take to fill the empty tank?

The required time t satisfies the equation

$$\frac{t}{a_1} + \frac{t}{a_2} + \cdots + \frac{t}{a_m} - \frac{t}{b_1} - \frac{t}{b_2} - \cdots - \frac{t}{b_n} = 1$$

(If the solution t turns out negative, how do you interpret it? Possibly, there is no solution. How do you interpret this case?)

2.3. (*a*) Mr. Vokach (his name means "Smith" in Poldavian) spends one-third of his income on food, one-fourth on housing, one-sixth on clothing, and has no other expenses (there is no income tax in lucky Poldavia). He wonders how long he could live on one year's pay.

(*b*) What voltage should be maintained between two points connected by three parallel wires, the resistance of which is 3, 4, and 6 ohms, respectively, in order that the total current carried jointly by the three wires should be of intensity 1?

And so on.

2.4. (*a*) x remains unchanged if we substitute $-w$ for w: starting with the wind and returning against it the plane attains the same extreme point in a given time.

(*b*) Test by dimension; see HSI, pp. 202–205.

2.5. The system

$$x + y = v$$
$$ax + by = cv$$

agrees fully with that obtained in sect. 2.6(2).

2.6. Choose the coordinate system in the same relative position to the line AB as in sect. 2.5(1) and set $AB = a$. The required center (x, y) of the circle touching the four given arcs satisfies the two equations

$$a - \sqrt{x^2 + y^2} = \sqrt{\left(x - \frac{a}{4}\right)^2 + y^2} - \frac{a}{4}$$

$$x = \frac{a}{2}$$

from which follows

$$y = a\sqrt{6}/5$$

2.7. Heron's formula appears rather formidable, but is, in fact, quite manageable if you observe the combination "sum times difference" often enough:

$$
\begin{aligned}
16D^2 &= (a + b + c)(-a + b + c)(a - b + c)(a + b - c) \\
&= [(b + c)^2 - a^2][a^2 - (b - c)^2] \\
&= (2bc - a^2 + b^2 + c^2)(2bc + a^2 - b^2 - c^2) \\
&= 4b^2c^2 - (b^2 + c^2 - a^2)^2 \\
&= 4(p^2 + q^2)(p^2 + r^2) - (2p^2)^2
\end{aligned}
$$

2.8. (a) *Relevant knowledge.* Approach (3) supposes more knowledge of plane geometry (Heron's formula is less familiar than the expression of the area in terms of base and height). Yet approach (4) needs more knowledge of solid geometry (we have to see, and then to prove, that k is perpendicular to a).

(b) *Symmetry.* The three data A, B, and C play the same role, the problem is symmetric in A, B, and C. Approach (3) respects this symmetry, but approach (4) breaks with it and prefers A to B and C.

(c) *Planning.* Approach (3) proceeds more "methodically," we can follow it with some confidence from the start. And, in fact, it leads quite clearly to that system of seven equations which appeared to us, at the first blush, too formidable. [This is not the fault of the approach which hints, in fact, a procedure to solve them; see sect. 6.4(2).] It is less visible in advance that approach (4) will be helpful, but it "muddles through" somehow (thanks to a lucky remark) and attains the final result with a much shorter computation.

2.9. $V^2 = p^2q^2r^2/36 = 2ABC/9.$

2.10. From the three equations in sect. 2.5(3) that express a^2, b^2, and c^2 in terms of p, q, and r, we obtain

$$p^2 + q^2 + r^2 = S^2$$

$$p^2 = S^2 - a^2, \qquad q^2 = S^2 - b^2, \qquad r^2 = S^2 - c^2$$

and so ex. 2.9 yields

$$V^2 = (S^2 - a^2)(S^2 - b^2)(S^2 - c^2)/36$$

2.11. $d^2 = p^2 + q^2 + r^2.$ This problem is broadly treated in HSI, Part I; see pp. 7–8, 10–12, 13–14, 16–19.

2.12. The notation chosen agrees both with ex. 2.11 and with sect. 2.5(3)— pay attention to both diagonals of the same face. Repeating a computation already done in ex. 2.10, we find

$$d^2 = (a^2 + b^2 + c^2)/2$$

2.13. A tetrahedron is determined by the lengths of its six edges—this results from the space analogue of the very first problem we have discussed in sect. 1.1. Yet we obtain the required configuration of the six edges, and so the proposed tetrahedron, in choosing one appropriate diagonal in each face of the box considered in ex. 2.11 and 2.12. The volume of this box is pqr. Cut off from the box four congruent tetrahedra, each with a trirectangular vertex and with volume $pqr/6$, see ex. 2.9; you obtain so the proposed tetrahedron whose volume is, therefore,

$$V = pqr - 4pqr/6 = pqr/3$$

Now, see ex. 2.10, $p^2 = S^2 - a^2$ and so on; hence

$$V^2 = (S^2 - a^2)(S^2 - b^2)(S^2 - c^2)/9$$

2.14. Ex. 2.10: If $V = 0$, one of the factors, for instance $S^2 - a^2 = p^2$ vanishes, and so two faces degenerate into line segments; the two other faces become coincident right triangles.

Ex. 2.13: If $V = 0$, the tetrahedron degenerates into a (doubly covered) rectangle; all four faces become congruent right triangles; in fact, $S^2 - a^2 = 0$ involves $a^2 = b^2 + c^2$.

2.15. As the last equation sect. 2.7 shows, the side x of the desired rectangle is the hypotenuse of a right triangle with legs $3a$ and a. This segment x can be fitted into the cross in four different (but not essentially different) ways; its midpoint must coincide with the center of the cross, which divides it into two parts, each of the same length $x/2$ as the other side of the desired rectangle. All this suggests strongly the solution exhibited by Fig. S2.15.

2.16. (a) $x^2 = 12 \cdot 9 - 8 \cdot 1, x = 10$.

(b) Shift two units to the left and one unit upward, since

$$10 = 12 - 2 = 9 + 1$$

(c) Retention of the central symmetry is more likely.

All this suggests Fig. S2.16.

2.17. Let x and y be the loads carried by the mule and the ass, respectively. Then

$$y + 1 = 2(x - 1), \quad x + 1 = 3(y - 1); \qquad x = 13/5, \quad y = 11/5$$

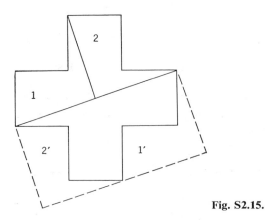

Fig. S2.15.

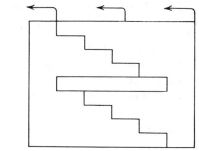

Fig. S2.16.

2.18. Mr. has h pounds, Mrs. w pounds, x pounds are free:

$$h + w = 94, \quad \frac{h - x}{1.5} = \frac{w - x}{2} = \frac{94 - x}{13.5}; \quad x = 40$$

2.19. 700, 500, $x = 400$ from

$$x + (x + 100) + (x + 300) = 1600$$

2.20. Each son receives 3000.

2.21. If the share of each child is x, and the whole fortune y, the shares of the children are

first: $\qquad x = 100 + \dfrac{y - 100}{10}$

second: $\qquad x = 200 + \dfrac{y - x - 200}{10}$

third: $\qquad x = 300 + \dfrac{y - 2x - 300}{10}$

and so on.

The difference of any two consecutive right hand sides is

$$100 - \frac{x + 100}{10}$$

If this difference equals 0 (as it should) $x = 900$, then (from the first equation) $y = 8100$: there were 9 children.

2.22. Let the three players initially have the amounts x, y, and z, respectively; it will be advantageous to consider

$$x + y + z = s$$

($s = 72$). We have to consider the amounts owned by the players at four different instants; any two consecutive instants are separated by a game; the total sum owned by the three is always s:

First	*Second*	*Third*
x	y	z
$2x - s$	$2y$	$2z$
$4x - 2s$	$4y - s$	$4z$
$8x - 4s = 24$	$8y - 2s = 24$	$8z - s = 24$

Hence $x = 39$, $y = 21$, $z = 12$.

2.23. Analogous to sect. 2.4(1) and 2.4(2), particular case of ex. 2.2 with

$$m = 3, \quad n = 0, \quad a_1 = 3, \quad a_2 = 8/3, \quad a_3 = 12/5$$

Hence $t = \frac{8}{9}$ of a week.

2.24. Newton means a generalization tending in the direction of ex. 2.2, but going less far: without "emptying pipes," there are no b's, $n = 0$.

2.25. Wheat, barley, and oats cost 5, 3, and 2 shillings a bushel, respectively. See ex. 2.26.

2.26. Let

$$x, \quad y, \quad z$$

be the prices of three commodities, and let p_ν be the price of the mixture in which

$$a_\nu, \quad b_\nu, \quad c_\nu$$

units of these commodities are contained, respectively, $\nu = 1, 2, 3$. We have thus a system of three equations

$$a_\nu x + b_\nu y + c_\nu x = p_\nu$$

$\nu = 1, 2, 3$. We obtain this generalization from the foregoing ex. 2.25 in passing from the array of numbers

40	24	20	312
26	30	50	320
24	120	100	680

to the array of letters

a_1	b_1	c_1	p_1
a_2	b_2	c_2	p_2
a_3	b_3	c_3	p_3

There is no difficulty in passing from 3 to n different commodities.

2.27. Let

α denote the quantity of grass per acre when the pasture starts to be used,
β the quantity of grass eaten by one ox in one week,
γ the quantity of grass that grows on one acre in one week,
a_1, a_2, a the number of oxen,
m_1, m_2, m the numbers of acres,
t_1, t_2, t the numbers of weeks in the three cases considered, respectively.
a, α, β, and γ are unknown, the remaining eight quantities are numerically
   given.

The conditions are

$$m_1(\alpha + t_1\gamma) = a_1 t_1 \beta$$
$$m_2(\alpha + t_2\gamma) = a_2 t_2 \beta$$
$$m(\alpha + t\gamma) = at\beta$$

a system of 3 equations for the 3 unknowns α/β, γ/β, a, which yields

$$a = \frac{m[m_1 a_2 t_2(t - t_1) - m_2 a_1 t_1(t - t_2)]}{m_1 m_2 t(t_2 - t_1)}$$

and, with the numerical data, $a = 36$.

2.28. Of an arithmetic progression
with five terms $\qquad$ $a, a + d, \ldots, a + 4d$
find the first term $\qquad$ a
and the difference $\qquad$ d
being given that
the sum of all terms equals 100 $\qquad$ $a + (a + d) + \cdots$
$$+ (a + 4d) = 100$$
and the sum of the last three terms $\quad (a + 2d) + (a + 3d) + (a + 4d)$
equals 7 times the sum of the
first two terms $\qquad$ $= 7[a + (a + d)]$

From the equations

$$5a + 10d = 100, \qquad 11a - 2d = 0$$

$a = 5/3, d = 55/6$ and so the progression is

$$10/6, \quad 65/6, \quad 120/6, \quad 175/6, \quad 230/6$$

2.29.
$$\frac{m}{r} + m + mr = 19$$

$$\frac{m^2}{r^2} + m^2 + m^2 r^2 = 133$$

Set
$$r + \frac{1}{r} = x$$

This changes the system into

$$m(x + 1) = 19, \qquad m^2(x^2 - 1) = 133$$

Division yields two *linear* equations for mx and m. Hence $m = 6$, $x = 13/6$, $r = 3/2$ or $2/3$; there are two (only trivially different) progressions: 4, 6, 9 and 9, 6, 4.

2.30.
$$a(q^3 + q^{-3}) = 13, \qquad a(q + q^{-1}) = 4$$

Division yields a quadratic for q^2. The progression is

$$1/5, \quad 4/5, \quad 16/5, \quad 64/5$$

or the same terms in reverse order.

2.31. Let x be the number of the partners. Express the profit of the partnership in two different ways (as received and as distributed):

$$(8240 + 40x \cdot x)\frac{x}{100} = 10x \cdot x + 224$$

The equation

$$x^3 - 25x^2 + 206x - 560 = 0$$

has no negative roots (substitute $x = -p$). If there is a rational root, it must be a positive integer, a divisor of 560. This leads to trying successively $x = 1, 2, 4, 5, 7, 8, 10, 14, 16, \ldots$. In fact, the roots are 7, 8, and 10. (Of course, Euler first made up the equation, then the story—you could try to imitate him.)

2.32. The centers of the four circles nonconcentric with the given square are vertices of another square of which we express the diagonal in two different ways:

$$(4r)^2 = 2(a - 2r)^2$$

and so

$$r = (\sqrt{2} - 1)a/2$$

2.33. Let $x + (d/2)$ stand for the height of the isosceles triangle perpendicular to the base.   Then

$$\left(x + \frac{d}{2}\right)^2 + \left(\frac{b}{2}\right)^2 = s^2, \qquad x^2 + \left(\frac{b}{2}\right)^2 = \left(\frac{d}{2}\right)^2$$

Elimination of x yields

$$4s^4 - 4d^2s^2 + b^2d^2 = 0$$

2.34. (*a*) The equation is of the first degree in d^2 as well as in b^2, but of the second degree in s^2: Hence, the problem to find s may be reasonably regarded as more difficult than the other two.

(*b*) d has a positive value iff $4s^2 > b^2$.

b has a positive value iff $d^2 > s^2$.

s has two different positive values iff $d^2 > b^2$.

The reader can learn here several things.   Newton comments on the solution of ex. 2.33 as follows: "And hence it is that Analysts order us to make no Difference between the given and sought Quantities.   For since the same Computation agrees to any Case of the given and sought Quantities, it is convenient that they should be conceived and compared without any Difference... or rather it is convenient that you should imagine, that the Question is proposed of those *Data* and *Quaesita*, given and sought Quantities, by which you think it is most easy to make out your Equation."   He adds a little later: "Hence, I believe, it will be manifest what Geometricians mean, when they bid you imagine that to be already done which is sought."

("Take the problem as solved"; cf. sect. 1.4.)

2.35. In setting up our equations we proceed in the direction just opposite to the one suggested by the surveyor's situation: We regard x and the angles $\alpha, \beta, \gamma, \delta$ as given, and l as the unknown.   From $\triangle UVG$ we find GV in terms of x, $\alpha + \beta$, and γ (law of sines).   From $\triangle VUH$ we find HV in terms of x, β, and $\gamma + \delta$ (law of sines).   From $\triangle GHV$ we find l in terms of GV, HV, and δ (law of cosines) and, by using the expressions for GV and HV, we obtain

$$l^2 = x^2\left[\frac{\sin^2(\alpha + \beta)}{\sin^2(\alpha + \beta + \gamma)} + \frac{\sin^2 \beta}{\sin^2(\beta + \gamma + \delta)} - \frac{2\sin(\alpha + \beta)\sin\beta\cos\delta}{\sin(\alpha + \beta + \gamma)\sin(\beta + \gamma + \delta)}\right]$$

Hence, express x^2 in terms of l, α, β, γ, and δ.

2.36. Let

| A, | 2s, | a, | b, | c |

stand for

| area, | perimeter, | hypotenuse, | remaining sides, |

respectively; A and s are given, a, b, and c unknown. To solve the system

$$a + b + c = 2s, \qquad bc = 2A, \qquad a^2 = b^2 + c^2$$

express $(b + c)^2$ in two different ways:

$$(2s - a)^2 = a^2 + 4A$$

$$a = s - \frac{A}{s}$$

2.37. Lengths of the sides of the triangle $2a$, u, v; $u + v = 2d$; the altitude perpendicular to side of length $2a$ is of length h.

Given a, h, d, find u, v.

Introduce x and y, orthogonal projections of the sides u and v on $2a$, respectively, and z, where

$$x - y = 2z$$

Also

$$x + y = 2a$$

$$u^2 = h^2 + x^2 \qquad\qquad v^2 = h^2 + y^2$$

Hence,

$$u^2 - v^2 = x^2 - y^2$$

or

$$2d(u - v) = 2a \cdot 2z$$

$$u = d + \frac{a}{d} z \qquad\qquad v = d - \frac{a}{d} z$$

$$x = a + z \qquad\qquad y = a - z$$

$$\left(d + \frac{a}{d} z \right)^2 = h^2 + (a + z)^2$$

$$z^2 = d^2 \left(1 - \frac{h^2}{d^2 - a^2} \right)$$

2.38. If a and b are the lengths of two nonparallel sides, and c and d the lengths of the two diagonals, then

$$2(a^2 + b^2) = c^2 + d^2$$

In fact, the diagonals dissect the parallelogram into four triangles: Apply the law of cosines to two neighboring triangles.

2.39. $\qquad\qquad (2b - a)x^2 + (4a^2 - b^2)(2x - a) = 0$

If $a = 10$, $b = 12$, then $x = 16(-8 + 3\sqrt{11})/7$, very nearly 32/7. Interpret the case $a = 2b$.

2.40. $\qquad\qquad a^2(3 + \sqrt{3})/2$

2.41. $1/3$, $2/9$, $2/9$, $2/9$. In fact, the sides of the larger triangle are divided by the vertices of the inscribed triangle in the proportion 2 to 1.

2.42. (Stanford 1957.) Consider the simplest case first, that of the equilateral triangle. Symmetry may lead us to suspect that in this case the four triangular pieces will also be equilateral. If this is so, however, the sides of the triangular pieces must be *parallel* to the sides of the given triangle: with this remark, we have discovered the essential feature of a configuration that solves the problem not only in the particular case examined, but also in the general case. (We pass from the equilateral triangle to the general triangle by "affinity.") By four parallels to a side of the given triangle, dissect each of the other two sides into five equal segments. Performing this construction three times, with respect to each side of the given triangle, we divide it into 25 congruent triangles similar to it. From these 25 triangular pieces, we easily pick out the four mentioned in the problem: the area of each of them is 1/25 of the given triangle's area. (This solution is not uniquely determined.

2.43. (Stanford 1960.) Generalize: The point P lies in the interior of a rectangle, its distances from the four corners are a, b, c, and d, from the four sides x, y, x', y', in cyclical order (as they are met by the hands of a watch). Then, with appropriate notation

$$a^2 = y'^2 + x^2, \quad b^2 = x^2 + y^2, \quad c^2 = y^2 + x'^2, \quad d^2 = x'^2 + y'^2$$

and so

$$a^2 - b^2 + c^2 - d^2 = 0$$

In our case $a = 5$, $b = 10$, $c = 14$, and so

$$d^2 = 25 - 100 + 196 = 121, \quad d = 11$$

Observe that the data a, b, and c which determine d are insufficient to determine the sides $x + x'$ and $y + y'$ of the rectangle.

2.44. (I) Let s stand for the side of the square. Then, by ex. 2.43, $x + x' = y + y' = s$, and we have three equations for the three unknowns x, y, and s:

$$x^2 + (s - y)^2 = a^2, \quad x^2 + y^2 = b^2, \quad y^2 + (s - x)^2 = c^2$$

Hence,

$$2sy = s^2 + b^2 - a^2, \quad 2sx = s^2 + b^2 - c^2$$

and, by squaring and adding, we find

$$s^4 - (a^2 + c^2)s^2 + [(b^2 - a^2)^2 + (b^2 - c^2)^2]/2 = 0$$

a quadratic equation for s^2.

(II) Check the geometric meaning of the particular cases:

(1) $s^2 = 2a^2$ or $s = 0$.
(2) $s = a$.
(3) s imaginary unless $c^2 = 2b^2 = 2s^2$.
(4) s imaginary unless $a^2 = c^2 = s^2$.

2.45. (Stanford 1959.) $100\pi/4$ and $100\pi/(2\sqrt{3})$ or approximately 78.54% and 90.69%, respectively. The transition from a large (square) table to the infinite plane involves, in fact, the concept of limit on which, however, we do not insist as the result is intuitive.

2.46. In following the procedure of ex. 2.32, we express the diagonal of an appropriate cube in two different ways:

$$(4r)^2 = 3(a - 2r)^2$$
$$r = (2\sqrt{3} - 3)a/2$$

2.47. The four vertices of a rectangle, taken in cyclical order, are at the distance a, b, c, and d, respectively, from a point P (which may be located anywhere in space). Being given three of these distances, find the fourth one.

The relation

$$a^2 - b^2 + c^2 - d^2 = 0$$

found in ex. 2.43 remains valid in the present more general situation, and the solution follows from it immediately. This can be applied, for instance, to a point P and four appropriately chosen vertices of a box (rectangular parallelepiped) since any two diagonals of a box are also the diagonals of a certain rectangle.

2.48. The solution of a problem of solid geometry often depends on a "key plane figure" which opens the door to the essential relations.

Through the altitude of the pyramid, pass a plane that is parallel to two sides of the base (and perpendicular to its two other sides). The intersection of this plane with the pyramid is an isosceles triangle which can be used as key figure: its height is h, its base, say a, is equal in length to a side of the base of the pyramid, its legs are of length $2a$, since each one is the height of a lateral face. Hence,

$$(2a)^2 = \left(\frac{a}{2}\right)^2 + h^2$$

and so the desired area is

$$5a^2 = 4h^2/3$$

2.49. For instance: The area of the surface of a regular pyramid equals four times the area of its hexagonal base. Given a, the length of a side of this base, find h, the height of the pyramid ($h = \sqrt{6}a$).

See also ex. 2.52.

2.50. In a parallelogram the sum of the squares of the 2 diagonals equals the sum of the squares of the 4 sides. (Restatement of the result of ex. 2.38.)

In a parallelepiped, let

| $D,$ | $E,$ | F |

stand for the sum of the squares of the

| 4 diagonals, | 12 edges, | 12 face-diagonals |

respectively. Then

$$D = E = F/2$$

(Follows from the result of ex. 2.38 by repeated application.)

2.51. The square of the desired area is

$$16s(s - a)(s - b)(s - c)$$

This may be regarded as an analogue to Heron's theorem, but is too close to it to be interesting.

2.52. (Stanford 1960.) Let a stand for the side of a triangle, T for the volume of the tetrahedron, and O for the volume of the octahedron.

First solution.   The octahedron is divided by an appropriate plane into two congruent regular pyramids with a common square base the area of which is a^2. The height of one of these pyramids is $a/\sqrt{2}$ (the "key plane figure" passes through a diagonal of the base) and so

$$O = 2\frac{a^2}{3}\frac{a}{\sqrt{2}} = \frac{a^3\sqrt{2}}{3}$$

Pass a plane through the altitude (of length h) of the tetrahedron and through a coterminal edge; the intersection (the key plane figure) shows two right triangles from which

$$h^2 = a^2 - \left(\frac{2a\sqrt{3}}{6}\right)^2 = \left(\frac{a\sqrt{3}}{2}\right)^2 - \left(\frac{a\sqrt{3}}{6}\right)^2$$

$$= \frac{2a^2}{3}$$

and so

$$T = \frac{1}{3}\frac{a}{2}\frac{a\sqrt{3}}{2}\frac{a\sqrt{2}}{\sqrt{3}} = \frac{a^3\sqrt{2}}{12}$$

Finally

$$O = 4T$$

Second solution.   Consider the regular tetrahedron with edge $2a$; its volume is 2^3T; four planes, each of which passes through the midpoints of three of its edges terminating in the same vertex, dissect it into four regular tetrahedra, each of volume T, and a regular octrahedron of volume O.   Hence,

$$4T + O = 8T$$

which yields again $O = 4T$.

2.53. The volumes are as

$$\frac{1}{a} : \frac{1}{b} : \frac{1}{c}$$

the surface areas as

$$\frac{b + c}{a} : \frac{c + a}{b} : \frac{a + b}{c}$$

2.54. (Stanford 1951.) The difference of the volumes, frustum minus cylinder,

$$\pi h\left[\frac{a^2 + ab + b^2}{3} - \left(\frac{a + b}{2}\right)^2\right] = \frac{\pi h(a - b)^2}{12}$$

is positive unless $a = b$ and the solids coincide.

MPR, vol. I, chapter VIII, contains several applications of algebraic inequalities to geometry.

2.55. Let r be the radius of the circle circumscribed about $\triangle ABC$. Then

$$r^2 = h(2R - h), \qquad r = \frac{2}{3}\frac{\sqrt{3}a}{2}$$

and so

$$R = \frac{a^2}{6h} + \frac{h}{2}$$

The term $h/2$ is often negligible in practice.

2.57. 35 miles; see ex. 2.58.

2.58. We list each given numerical value in parenthesis following the letter that generalizes it:

a (7/2) is the velocity of A,

b (8/3) the velocity of B,

c (1) the number of hours that pass between the two starts,

d (59) the distance between the two starting points. Then

$$x + y = d, \qquad \frac{x}{a} - \frac{y}{b} = c, \qquad x = \frac{a(bc + d)}{a + b}$$

Newton formulates the generalized problem as follows: "Having given the Velocities of two moveable Bodies, A and B, tending to the same Place, together with the Interval or Distance of the Places and Times from, and in which, they begin to move; to determine the Place they shall meet in."

2.59. (Stanford 1959.)   We use the notation

u for Al's speed,

v for Bill's speed,

t_1 for the time (counted from the start) the boys meet the first time,

t_2 for the time they meet the second time,

d the desired distance of the two houses.   Then

$$ut_1 = a, \qquad ut_2 = d + b$$
$$vt_1 = d - a, \qquad vt_2 = 2d - b$$

(1) By expressing u/v in two different ways, we obtain

$$\frac{a}{d - a} = \frac{d + b}{2d - b}$$

Hence, after discarding the vanishing root, we find $d = 3a - b$.

(2) Of course, Al.   Numerically: $u/v = 3/2$.

2.60. (Stanford 1955.) See ex. 2.61; see also HSI, pp. 236, 239–240, 247: problem 12.

2.61. Between the start and the first point where all $n + 1$ friends meet again, there are $2n - 1$ different phases:

(1) Bob rides with A
(2) Bob rides alone
(3) Bob rides with B
(4) Bob rides alone

. . . . . .

$(2n - 1)$ Bob rides with L

Fig. S2.61, where $n = 3$, exhibits 5 phases; the lines representing the travels of A, B, or C are marked with the proper letters; the steeper slope renders the line

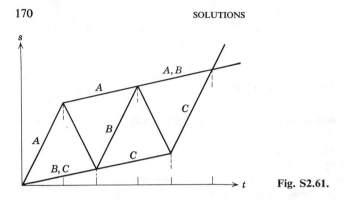

Fig. S2.61.

that represents the itinerary of the car easily recognizable. We see from the symmetry of the arrangement (especially clear from Fig. S2.61) that all n odd-numbered phases are of the same duration, say T, and all $n - 1$ even-numbered phases of the same duration, say T'. Express the total progress through the $2n - 1$ phases [in $nT + (n - 1)T'$ units of time] in two different ways (look first at Bob, then at one of his friends):

$$nTc - (n - 1)T'c = Tc + (n - 1)(T + T')p$$

whence

$$\frac{T}{T'} = \frac{c + p}{c - p}$$

(1) Rate of progress of the company is

$$\frac{nTc - (n - 1)T'c}{nT + (n - 1)T'} = c \, \frac{c + (2n - 1)p}{(2n - 1)c + p}$$

(2) The fraction of time when the car carries Bob alone

$$\frac{(n - 1)T'}{nT + (n - 1)T'} = \frac{(n - 1)(c - p)}{(2n - 1)c + p}$$

(3) Results (1) and (2) become intuitive in the extreme cases [just (2) for $n = \infty$ is less immediate]:

	$p = 0$	$p = c$	$n = 1$	$n = \infty$
(1) Rate of progress	$c/(2n - 1)$	c	c	p
(2) Fraction when Bob alone	$(n - 1)/(2n - 1)$	0	0	$(c - p)/2c$

2.62. Let t_1 be the time of descent of the stone and t_2 the time of ascent of the sound. From

$$T = t_1 + t_2, \qquad d = gt_1{}^2/2, \qquad d = ct_2$$

we find

$$d = \{-c(2g)^{-1/2} + [c^2(2g)^{-1} + cT]^{1/2}\}^2$$

Cf. MPR, vol. I, p. 165 and p. 264, ex. 29.

2.63. Introduce $\beta' = \angle ACO$. Since

$$\frac{\sin \omega}{\sin \beta} = \frac{AB}{AO}, \qquad \frac{\sin \omega'}{\sin \beta'} = \frac{AC}{AO}$$

$$\frac{\sin \omega}{\sin \omega'} \frac{\sin \beta'}{\sin \beta} = \frac{t}{t'}$$

On the other hand, $\beta' = \beta - (\omega' - \omega)$. Expressing $\sin \beta' / \sin \beta$ in two different ways, we obtain

$$\cot \beta = \cot(\omega' - \omega) - \frac{t \sin \omega'}{t' \sin \omega \sin (\omega' - \omega)}$$

2.64. Adding the three equations, we obtain

$$0 = a + b + c$$

If this relation is *not* satisfied by the data a, b, and c, "the problem is impossible," that is, there are no numbers x, y, z fulfilling the three simultaneous equations. If that relation *is* satisfied, the problem is *indeterminate*, that is, there are infinitely many solutions: from the first two equations

$$x = z + (3a + b)/7$$
$$y = z + (2a + 3b)/7$$

where z remains arbitrary.

Cf. ex. 1.43 and 1.44.

2.65. (Stanford 1955.) Comparing coefficients of like powers on both sides of the identity, we obtain 5 equations

$$1 = p^2, \qquad 4 = 2pq, \qquad -2 = q^2 + 2pr, \qquad -12 = 2qr, \qquad 9 = r^2$$

for our 3 unknowns p, q and r. The first equation yields $p = \pm 1$, whence the following 2 equations, used successively, determine two systems of solutions

$$p = 1, \quad q = 2, \quad r = -3, \qquad \text{and} \qquad p = -1, \quad q = -2, \quad r = 3$$

both of which happen to satisfy also the remaining two equations.

Usually, it will not be possible to extract the square root, since usually it is impossible to satisfy a system with more equations than unknowns.

2.66. (Stanford 1954.) Expanding the right-hand side of the hypothetical identity and equating corresponding coefficients, we obtain

(1) $aA = bB = cC = 1$
(2) $bC + cB = cA + aC = aB + bA = 0$

We derive from (2) that

$$bC = -cB, \qquad cA = -aC, \qquad aB = -bA$$

and multiplying these three equations, we derive further that

$$abcABC = -abcABC$$
$$abcABC = 0$$

Yet we derive from (1) that

$$abcABC = 1$$

This contradiction shows that the hypothetical identity from which we have started is impossible.

We have shown here that a system of 6 equations with the 6 unknowns a, b, c, A, B, and C is inconsistent.

2.67. $x = 5t,$ $y = 60 - 18t,$ $z = 40 + 13t$

are positive iff $0 < t < 60/18$. This leaves for t the values 1, 2, 3 and for (x, y, z) the three systems

$$(5, 42, 53),   (10, 24, 66),   (15, 6, 79)$$

2.68. Follow ex. 2.67: the system

$$x + y + z = 30$$
$$14x + 11y + 9z = 360$$

is satisfied by

$$x = 2t,   y = 45 - 5t,   z = 3t - 15$$
$$t = 5, 6, 7, 8, \text{ or } 9$$

2.69. $100 + x = y^2,$ $168 + x = z^2$

Subtracting we obtain

$$(z - y)(z + y) = 68$$

Since $68 = 2^2 17$ can be decomposed into a product of two factors in just three ways:

$$68 = 1 \cdot 68 = 2 \cdot 34 = 4 \cdot 17$$

and y and z must be both odd or both even, there is just one solution:

$$z - y = 2,   z + y = 34,   z = 18,   y = 16,   x = 156$$

2.70. (Stanford 1957.) Bob has x stamps of which y sevenths are in the second book; x and y are positive integers,

$$\frac{2x}{10} + \frac{yx}{7} + 303 = x$$

and hence

$$x = \frac{3 \cdot 5 \cdot 7 \cdot 101}{28 - 5y}$$

The denominator on the right-hand side must be positive and *odd* since it must divide the numerator which is odd. This leaves three possibilities: $y = 1, 3,$ and 5, and only the last one is suitable: $y = 5$ and $x = 3535$ are uniquely determined.

2.71. (Stanford 1960.) If the reduced price is x cents and there are y pens in the remaining stock, $x < 50$ and

$$xy = 3193$$

Now, $3193 = 31 \times 103$ is a product of two prime factors, and so it has precisely four different factors, 1, 31, 103, and 3193. If we *assume* that x is an integer, $x = 1$, or 31. If we *assume also* that $x > 1$, then $x = 31$.

2.75. (1) *Inconsistence*: Either there are, among the three planes, two which are different and parallel; or any two planes intersect and the three lines of intersection are different and parallel.

(2) *Dependence*: The plans possess a common straight line; two, or even all three, may coincide.

(3) *Consistence and independence*: There is just one common point, *the* point of intersection.

2.78. The current textbooks for secondary schools contain "word problems" in great number, although not in great variety. Just such applications and such kinds of questions are usually lacking as could shed light on the general interest of the "Cartesian pattern."

From the foregoing examples, the reader can learn to attach useful questions to a problem that he has just solved. I list a few such questions, referring to one illustrative example for each (the reader should look for further illustrations):

Can you check the result? (Ex. 2.4.)

Check extreme (degenerate, limiting) cases. (Ex. 2.14.)

Can you derive the result differently? Compare the different approaches. (Ex. 2.8.)

Could you devise another interpretation of the result? (Ex. 2.3.)

Generalize the problem. (Ex. 2.2.)

Devise an analogous problem. (Ex. 2.47.)

Starting from any problem and asking the foregoing and similar questions, the reader may evolve new problems and find perhaps some interesting and not too difficult problem. At any rate, in so asking, he has a good chance to deepen his understanding of the problem he started from and to improve his problem-solving ability.

Here are just two (not too easy) problems evolved from the foregoing.

(I) Check the result of ex. 2.35

(1) in supposing $\alpha = \delta$, $\beta = \gamma$, $\alpha + \beta = 90°$;

(2) in supposing $\alpha = \delta$, $\beta = \gamma$, but without prescribing a value for $\alpha + \beta$;

(3) by substituting δ, γ, β, and α for α, β, γ, and δ, respectively.

(II) Consider problems of solid geometry analogous to ex. 2.45. (There is a hint in ex. 3.39.)

No solution: **2.56, 2.72, 2.73, 2.74, 2.76, 2.77.**

Chapter 3

3.1. For $n = 0$ and $n = 1$ the assertion is obvious. Assume that it is valid for some value of n:

$$(1 + x)^n = 1 + \cdots + \binom{n}{r-1} x^{r-1} + \binom{n}{r} x^r + \cdots + x^n$$

Multiplying both sides by $1 + x$ you obtain

$$(1 + x)^{n+1} = 1 + \cdots + \left[\binom{n}{r} + \binom{n}{r-1}\right] x^r + \cdots + x^{n+1}$$

By virtue of the *recursion formula* of sect. 3.6(2), the coefficient of x^r in $(1 + x)^{n+1}$ turns out equal to

$$\binom{n+1}{r}$$

and so the binomial theorem, supposed to be valid for n, turns out to be valid also for $n + 1$. Observe that we have also used the *boundary condition* of sect. 3.6(2). Where?

3.2. Assume the result of ex. 3.1, set

$$x = \frac{b}{a}$$

and consider

$$a^n(1 + x)^n = (a + b)^n$$

3.3. Consider the assertion "S_p is a polynomial of degree $p + 1$" as a conjecture (what it originally was). This conjecture is certainly true in the first particular cases, $p = 0$, 1, and 2 (which suggested it; see the beginning of sect. 3.3). Let us *assume* that the conjecture is verified up to the case $p = k - 1$, that is, for $p = 0, 1, 2, \ldots, k - 1$ (for $S_0, S_1, S_2, \ldots, S_{k-1}$.) *Then* we can conclude (look at the final equation of sect. 3.4) that

$$\binom{k+1}{2} S_{k-1} + \binom{k+1}{3} S_{k-2} + \cdots + S_0 = P$$

(we introduce P as an abbreviation) is a polynomial of degree k. Now, we derive from that equation that

(1)
$$S_k = \frac{(n+1)^{k+1} - 1 - P}{k+1}$$

Since P is of degree k in n, the highest term of $(n + 1)^{k+1}$ which is n^{k+1} cannot be canceled, and so our formula shows that S_k is a polynomial of degree $k + 1$ in n. We reached this conclusion assuming that S_0 is of degree 1, S_1 of degree 2, ... and S_{k-1} of degree k.

To put it intuitively, the property considered of S_k (that it is of degree $k + 1$) has an "unquenchable tendency to propagate itself." We knew pretty early that S_0, S_1, and S_2 have this property; therefore, by our foregoing proof, also S_3 must have it; by the same proof also S_4 must have it, then S_5, and so on.

That the highest term in S_k is of the asserted form, is also obvious now from the formula (1).

Some of the following problems provide other approaches to the result just proved; see also ex. 4.2–4.7.

3.4.
$$S_4 = S_2 \frac{6S_1 - 1}{5}$$
$$= \frac{n(n+1)(2n+1)(3n^2 + 3n - 1)}{30}$$

The proof by mathematical induction follows the standard pattern; see MPR, vol. 1, pp. 108–120.

3.5. Pattern suggested by sect. 3.2, 3.3, and 3.4, and by ex. 3.3.

3.6. Analogous to sect. 3.4:

$$n^k - (n - 1)^k = \binom{k}{1}n^{k-1} - \binom{k}{2}n^{k-2} + \cdots + (-1)^{k-1}\binom{k}{k}$$

3.7. Analogous to sect. 3.4:

$$[n(n + 1)]^k - [(n - 1)n]^k = n^k[(n + 1)^k - (n - 1)^k]$$

$$= 2\binom{k}{1}n^{2k-1} + 2\binom{k}{3}n^{2k-3} + 2\binom{k}{5}n^{2k-5} + \cdots$$

3.8. Analogous to sect. 3.4:

$$(2n + 1)[n(n + 1)]^k - (2n - 1)[(n - 1)n]^k$$
$$= n^k[(n + 1)^k + (n - 1)^k] + 2n^{k+1}[(n + 1)^k - (n - 1)^k]$$

$$= 2\left[\binom{k}{0} + 2\binom{k}{1}\right]n^{2k} + 2\left[\binom{k}{2} + 2\binom{k}{3}\right]n^{2k-2} + \cdots$$

3.9. From ex. 3.7 by recursion and mathematical induction.

3.10. From ex. 3.8 by recursion and mathematical induction.

3.11. By virtue of ex. 3.9 and 3.10, it is enough to verify the assertion for $S_1(x)$ and

$$S_2(x) = S_1(x)\frac{2x + 1}{3}$$

3.12. (a) By "little Gauss's method" (first approach of sect. 3.1):

$$[1 + (2n - 1)] + [3 + (2n - 3)] + \cdots = 2n\cdot\frac{n}{2} = n^2$$

(b) By the second approach of sect. 3.1; see the following.

(c) Generalize: consider the sum of an arithmetic series with initial term a, difference d, and n terms:

$$S = a + (a + d) + (a + 2d) + \cdots + (a + (n - 1)d)$$

Set the last term $a + (n - 1)d = b$; then (this is the second approach of sect. 3.1)

$$S = a + (a + d) + (a + 2d) + \cdots + (b - 2d) + (b - d) + b$$
$$S = b + (b - d) + (b - 2d) + \cdots + (a + 2d) + (a + d) + a$$

By adding and dividing by 2

$$S = \frac{a + b}{2}n$$

Specialize: $a = 1$, $b = 2n - 1$; then

$$S = \frac{1 + (2n - 1)}{2}n = n^2$$

(4) Look at Fig. 3.9.

(5) See ex. 3.13.

3.13. $1 + 4 + 9 + 16 + \cdots + (2n - 1)^2 + (2n)^2 - 4(1 + 4 + \cdots + n^2)$

$$= \frac{2n(2n + 1)(4n + 1)}{6} - 4 \cdot \frac{n(n + 1)(2n + 1)}{6}$$

$$= \frac{n(4n^2 - 1)}{3}$$

3.14. Follow the pattern of ex. 3.13:

$$\frac{4n^2(2n + 1)^2}{4} - 8\frac{n^2(n + 1)^2}{4} = n^2(2n^2 - 1)$$

3.15. Use the notation of ex. 3.11:

$$1^k + 3^k + \cdots + (2n - 1)^k = S_k(2n) - 2^k S_k(n)$$

3.16. *More questions may be easier to answer than just one question.* (This is the "inventor's paradox"; see HSI, p. 121.) Along with the proposed

$$2^2 + 5^2 + 8^2 + \cdots + (3n - 1)^2 = U$$

consider

$$1^2 + 4^2 + 7^2 + \cdots + (3n - 2)^2 = V$$

Then (suggested by ex. 3.15)

$$U + V + 9S_2(n) = S_2(3n)$$

Moreover

$$U - V = 3 + 9 + 15 + \cdots + (6n - 3) = 3n^2$$

We have here a system of two linear equations for the two unknowns U and V which yields not only the required

$$U = n(6n^2 + 3n - 1)/2$$

but also

$$V = n(6n^2 - 3n - 1)/2$$

For another method see ex. 3.17.

3.17. (See Pascal, l.c. footnote 3 in chapter 3.) Generalizing the notation of sect. 3.3 (where we dealt with the particular case $a = d = 1$) we set

$$S_k = a^k + (a + d)^k + (a + 2d)^k + \cdots + [a + (n - 1)d]^k$$

Obviously, $S_0 = n$. Substituting $1, 2, 3, \ldots, n$ for n in the relation

$(a + nd)^{k+1} - [a + (n - 1)d]^{k+1}$

$$= \binom{k + 1}{1}[a + (n - 1)d]^k d + \binom{k + 1}{2}[a + (n - 1)d]^{k-1}d^2 + \cdots$$

and adding, we obtain

$$(a + dn)^{k+1} - a^{k+1} = \binom{k + 1}{1}S_k d + \binom{k + 1}{2}S_{k-1}d^2 + \cdots + S_0 d^{k+1}$$

Hence we find $S_1, S_2, \ldots, S_k$ one after the other, recursively. Do in detail the case $a = 2, d = 3, k = 2$; see ex. 3.16.

3.18. The sum required is

$$\frac{1\cdot 2}{2}2 + \frac{2\cdot 3}{2}3 + \frac{3\cdot 4}{2}4 + \cdots + \frac{(n-1)n}{2}n$$

$$= \tfrac{1}{2}[(2^3 - 2^2 + 3^3 - 3^2 + 4^3 - 4^2 + \cdots + n^3 - n^2)]$$

$$= \tfrac{1}{2}(S_3 - S_2) = \frac{(n-1)n(n+1)(3n+2)}{24}$$

by the results of sect. 3.2 and 3.3.

3.19. (a) $\dfrac{n(n^2 - 1)}{6}$;   (b) $1^{n-1}2^{n-2}3^{n-3}\ldots(n-1)^1$;   (c) $\dfrac{n^2(n^2-1)}{12}$

3.20. We have already computed E_1 in sect. 3.1 and E_2 in ex. 3.18.   A more efficient procedure is based on a classical fact of algebra: the elementary symmetric functions can be expressed in terms of the sums of like powers:

$$E_1 = S_1$$
$$E_2 = (S_1{}^2 - S_2)/2$$
$$E_3 = (S_1{}^3 + 2S_3 - 3S_1S_2)/6$$
$$E_4 = (S_1{}^4 + 3S_2{}^2 + 8S_1S_3 - 6S_1{}^2S_2 - 6S_4)/24$$

Combine these with our former results (sect. 3.1, 3.2, and 3.3, ex. 3.4).   Combining certain properties of the general expression of E_k in terms of $S_1, S_2, \ldots,$ S_k ("isobaric") with ex. 3.9 and 3.10, we can obtain not only the degree but also the coefficient of the highest term,

$$E_k(n) = \frac{n^{2k}}{k!2^k} + \cdots$$

and we can derive that, for $k \geqq 2$, $E_k(n)$ is divisible by

$$(n - k + 1)(n - k + 2)\ldots(n - 1)[n(n + 1)]^{[3-(-1)^k]/2}$$

3.21. *Procedure (a) is a particular case of procedure (b).*   In fact, if A_{n+1} is implied by A_n alone, it is *a fortiori* (even with stronger reason) implied by $A_1, A_2, \ldots, A_{n-1}$ and A_n together.   That is, if statement (II*a*) happens to be correct, statement (II*b*) must be correct.   Hence, if we accept procedure (*b*), we are obliged to accept procedure (*a*).

Procedure (b) can be reduced to procedure (a).   Define B_n as the simultaneous assertion of the n propositions $A_1, A_2, \ldots, A_{n-1}$ and A_n.   Then

statement (I) means: B_1 *is true.*

statement (II*b*) boils down to: B_n *implies* B_{n+1}.

Hence, the statements (I) and (II*b*) about the sequence $A_1, A_2, A_3, \ldots$ boil down to statements (I) and (II*a*) with B_n substituted for A_n, (for $n = 1, 2, 3, \ldots$, of course).

3.22. Fig. 3.3 can be conceived as representing the case in which Bernie, Charlie, Dick, Roy, and Artie (blocks from northwest to southeast) put up the tent and the other five boys (Ricky, Abe, Al, Alex, and Bill—blocks from northeast to southwest) cook the supper.   Starting from this concrete case, you should be able to see that to each division of the ten boys into two differently labeled teams of five there corresponds a shortest zigzag path from top to bottom in Fig. 3.3 and, conversely, to each zigzag path of this kind there corresponds such a division; the correspondence is one-to-one. Therefore, the desired number of divisions is 252; see Fig. 3.3.

3.23. We are facing here a general situation, a *representative special case* of which (MPR, vol. 1, p. 25, ex. 10) appears in ex. 3.22 and Fig. 3.3.

Number the individuals from 1 to n and let correspond the kth "base" (horizontal row) of the Pascal triangle to the kth individual. An individual belongs to the subset if, and only if, the zigzag path arrives at the corresponding base coming down along a block running *from northwest to southeast*. In this manner, any subset of size r contained in the given set of size n can be visualized as a zigzag path ending in a fixed point, and we count the subsets by counting the zigzag paths. Cf. MPR, vol. 2, pp. 105–106, ex. 31.

3.24. $\dfrac{n(n-1)}{1 \cdot 2}$ straight lines, $\dfrac{n(n-1)(n-2)}{1 \cdot 2 \cdot 3}$ triangles.

3.25. Given n points in space in "general position", there are

$$\frac{n(n-1)(n-2)(n-3)}{1 \cdot 2 \cdot 3 \cdot 4}$$

tetrahedra with vertices chosen among the given n points.

3.26.
$$\binom{n}{2} - n = \frac{n(n-3)}{2}$$

3.27. Two diagonals intersecting *inside* the given convex polygon are the diagonals of a *convex* quadrilateral the four vertices of which are chosen among the n vertices of the given polygon. Therefore the number of the intersections in question is

$$\frac{n(n-1)(n-2)(n-3)}{1 \cdot 2 \cdot 3 \cdot 4}$$

3.28. The red face can be chosen in

$$\binom{6}{1} = 6$$

different ways. From the remaining five faces, the two blue faces can be chosen in

$$\binom{5}{2} = 10$$

different ways. Hence the total number of possibilities for distributing the three colors among the six faces in the required manner is

$$\binom{6}{1}\binom{5}{2} = 6 \times 10 = 60$$

3.29. $$\binom{n}{r}\binom{s+t}{s} = \frac{n!}{r!(n-r)!}\ \frac{(s+t)!}{s!t!} = \frac{n!}{r!s!t!}$$

3.30. A set of n individuals is divided into h nonoverlapping subsets (that is, two different subsets have no member in common); the first subset has r_1 members, the second r_2 members, ... and the last subset has r_h members so that

$$r_1 + r_2 + r_3 + \cdots + r_h = n$$

There are

$$\frac{n!}{r_1!r_2!r_3!\ldots r_h!}$$

different subdivisions of this kind.   The numbering or labeling of the subsets is essential: if some of the numbers $r_1, r_2, \ldots, r_h$ happen to be equal, we must carefully distinguish between differently labeled subsets of equal size.   Thus in ex. 3.22, we distinguish between the five individuals who put up the tent and the other five who cook the supper; or, which boils down to the same, in Fig. 3.3 we distinguish between two zigzag paths that are mirror images of each other with respect to the middle line of the figure (which joins the initial A to the final A).   Or, in ex. 3.29, the r faces have a predetermined color different from that of the s faces, even if the numerical values of r and s happen to coincide.

3.31. This fact is accessible through all four approaches indicated in sect. 3.8 and also through ex. 3.23.

(1) The network of streets is symmetrical with respect to the vertical through the apex of the Pascal triangle.

(2) The same symmetry appears both in the recursion formula and in the boundary condition.

(3) Using the notation for the factorial

$$1 \cdot 2 \cdot 3 \ldots m = m!$$

we have

$$\binom{n}{r} = \frac{n(n-1)\ldots(n-r+1)}{1\ \cdot\ 2\ \ \ldots\ \ r}$$

$$= \frac{n(n-1)\ldots(n-r+1)(n-r)\ldots2\cdot1}{1\ \cdot\ 2\ \ \ldots\ \ r\ \ \ \ (n-r)\ldots2\cdot1}$$

$$= \frac{n!}{r!(n-r)!} = \frac{n!}{(n-r)!r!} = \binom{n}{n-r}$$

(4) Since $(a+b)^n$ remains unchanged when we interchange a and b, its expansion must show the same coefficient for $a^r b^{n-r}$ and $a^{n-r}b^r$.

(5) When, from a set of n individuals, we pick out a subset of r individuals, we leave another subset of $n-r$ individuals.   Therefore, there are as many subsets of one kind as of the other kind.

3.32. $$\binom{n}{0} + \binom{n}{1} + \binom{n}{2} + \cdots + \binom{n}{n} = 2^n.$$

Proof: put $a = b = 1$ in the expansion of $(a+b)^n$. *Another proof*: There are 2^n shortest zigzag paths from the apex of the Pascal triangle to its nth base; this is obvious since, in picking a southward path in Fig. 3.3, you have a choice between two alternatives in passing any street corner (any base). *Still another proof*: In a set of n individuals, there are 2^n subsets, including the empty set and the full set (which are accounted for by $\binom{n}{0}$ and $\binom{n}{n}$, respectively); this is obvious, since in picking a subset you may accept or refuse any one of the n individuals.

3.33.
$$\binom{n}{0} - \binom{n}{1} + \binom{n}{2} - \cdots + (-1)^n \binom{n}{n} = 0$$

for $n \geqq 1$. Put $a = 1$ and $b = -1$ in the expansion of $(a + b)^n$.

Another proof: By boundary condition and recursion formula

$$\binom{n}{0} = \binom{n-1}{0}$$

$$- \binom{n}{1} = -\binom{n-1}{0} - \binom{n-1}{1}$$

$$\binom{n}{2} = \binom{n-1}{1} + \binom{n-1}{2}$$

. . . .

$$(-1)^{n-1} \binom{n}{n-1} = (-1)^{n-1}\binom{n-1}{n-2} + (-1)^{n-1}\binom{n-1}{n-1}$$

$$(-1)^n \binom{n}{n} = (-1)^n \binom{n-1}{n-1}$$

Add!

Still another proof: Each zigzag path attaining the $(n-1)$th base splits into two zigzag paths going to the nth base, of which one goes to a "positive" corner ($r = 0, 2, 4, \ldots$) and the other to a "negative" corner ($r = 1, 3, 5, \ldots$).

3.34. Analogously (fourth avenue)

$$1 + 5 + 15 + 35 = 56$$

generally (rth avenue)

$$\binom{r}{r} + \binom{r+1}{r} + \binom{r+2}{r} + \cdots + \binom{n}{r} = \binom{n+1}{r+1}$$

Proof by mathematical induction: The assertion is true for $n = r$: in fact,

$$\binom{r}{r} = \binom{r+1}{r+1}$$

by virtue of the boundary condition.

Assume now that the assertion holds for a certain value of n. Adding the same quantity to both sides of the assumed equation, we find

$$\binom{r}{r} + \binom{r+1}{r} + \cdots + \binom{n}{r} + \binom{n+1}{r} = \binom{n+1}{r+1} + \binom{n+1}{r} = \binom{n+2}{r+1}$$

by virtue of the recursion formula, and so the truth of the assertion follows for the next value, $n + 1$.

This proves the theorem for $n \geq r$.

Another proof: In Fig. 3.7 (I), A is the apex and L a given point specified by $n + 1$ and $r + 1$; the total number of shortest zigzag paths from A to L is $\binom{n + 1}{r + 1}$. Each of these paths must use some street in going from the rth avenue to the $(r + 1)$th avenue; the number of paths using the successive streets is

$$\binom{r}{r}, \quad \binom{r + 1}{r}, \quad \binom{r + 2}{r}, \quad \ldots, \quad \binom{n}{r}$$

respectively, and so the sum of these numbers is the total number of the paths in question, $\binom{n + 1}{r + 1}$, as it has been asserted.

3.35. Adding the numbers first along the northwest boundary line (0th avenue), then along the first avenue, then along the second,... and finally along the fifth avenue in Fig. 3.5 we obtain

$$6, \quad 21, \quad 56, \quad 126, \quad 252, \quad 462$$

respectively, and the sum of these numbers is 923, which we seek in vain in the neighborhood of the fragment of the Pascal triangle exhibited in Fig. 3.5. Yet we find quite close the next number

$$924 = \binom{12}{6}$$

Observe now that we could have saved the trouble of performing our additions (also the last, the seventh addition) by using ex. 3.34 and a table of binomial coefficients, and you can easily prove, in ascending from our representative example, that generally

$$\sum_{l=0}^{m} \sum_{r=0}^{n} \binom{l + r}{r} = \binom{m + n + 2}{m + 1} - 1$$

3.36. On the left-hand side of the proposed equation, the first factors are taken from the fifth base and the second factors from the fourth base of the Pascal triangle; the right-hand side can be found in the ninth base. In the example $1 \cdot 1 + 5 \cdot 3 + 10 \cdot 3 + 10 \cdot 1 = 56$, the fifth, the third, and the eighth base are analogously involved. The more general situation considered in sect. 3.9 involves the nth, again the nth, and the $2n$th base. These examples suggest the general theorem:

$$\binom{m}{0}\binom{n}{r} + \binom{m}{1}\binom{n}{r - 1} + \binom{m}{2}\binom{n}{r - 2} + \cdots + \binom{m}{r}\binom{n}{0}$$
$$= \binom{m + n}{r}$$

We admit here, in fact, an extension of the meaning of our symbols; a formal statement follows in ex. 3.65 (III).

Both proofs found in sect. 3.9 can be extended to the present more general case. The geometric approach is suggested by the comparison of (II) and (III) in Fig. 3.7. The analytic approach consists in computing in two different ways the coefficient of x^r in the expansion of

$$(1 + x)^m (1 + x)^n = (1 + x)^{m+n}$$

3.37. On the left-hand side of the proposed equation, the first factors are taken from the first avenue and the second factors from the second avenue of the Pascal triangle; the right-hand side can be found on fourth avenue. In the example

$$1 \cdot 10 + 3 \cdot 6 + 6 \cdot 3 + 10 \cdot 1 = 56$$

the second, again the second, and the fifth avenue are analogously involved. We can interpret the general situation dealt with in ex. 3.34 and Fig. 3.7 (I) as involving the 0th, the rth and the $(r + 1)$th avenue in an analogous way. These examples suggest the general theorem:

$$\binom{r}{r}\binom{s+n}{s} + \binom{r+1}{r}\binom{s+n-1}{s}$$

$$+ \binom{r+2}{r}\binom{s+n-2}{s} + \cdots + \binom{r+n}{r}\binom{s}{s} = \binom{r+s+n+1}{r+s+1}$$

Geometric proof (more general than the geometric proof in ex. 3.34, analogous to that in sect. 3.9 and ex. 3.36): In Fig. 3.7(IV), the point L is specified by the numbers $r + 1 + s + n$ (total number of blocks) and $r + 1 + s$ (blocks to the right downward) and so the total number of shortest zigzag paths from the apex A to L is

$$\binom{r+s+n+1}{r+s+1}$$

Each of these paths must use some street in going from the rth avenue to the $(r + 1)$th avenue; according to the street used, we classify the paths on the left-hand side of the asserted formula, and count the paths of each class separately; on the right-hand side, we count all the paths in question together.

It would be desirable to parallel here also the other, analytic proof of sect. 3.9 and ex. 3.36 where the formula is derived from the consideration of the product of two series; yet this seems to be less immediate and there is a gap. It would be desirable too to find some (algebraic?) connection between the two similar formulas, obtained here and in the foregoing ex. 3.36; there is another gap.

3.38. The nth triangular number is

$$1 + 2 + 3 + \cdots + n = \frac{n(n+1)}{2} = \binom{n+1}{2}$$

The triangular numbers 1, 3, 6, 10, ... form the second avenue of the Pascal triangle.

3.39. The nth pyramidal number is

$$\binom{2}{2} + \binom{3}{2} + \binom{4}{2} + \cdots + \binom{n+1}{2} = \binom{n+2}{3} = \frac{n(n+1)(n+2)}{6}$$

We have used ex. 3.34. The pyramidal numbers 1, 4, 10, 20,... form the third avenue of the Pascal triangle.

Remark. The expressions for the triangular and pyramidal numbers were known before the general explicit formula for the binomial coefficients (sect. 3.7) and may have led, by induction, to the discovery of the general formula.

3.40.
$$1^2 + 2^2 + \cdots + n^2 = \frac{n(n+1)(2n+1)}{6}$$

3.41. Shortest zigzag paths joining the apex to the point characterized by the two numbers

$$n = n_1 + n_2 + \cdots + n_h$$

(total number of blocks) and

$$r = r_1 + r_2 + \cdots + r_h$$

(blocks from northwest to southeast) which, however, are subject to the *restriction* that they must pass through $h - 1$ given intermediate points, analogously characterized by the numbers

$$
\begin{aligned}
&n_1 && \text{and} && r_1 \\
&n_1 + n_2 && \text{and} && r_1 + r_2 \\
&\quad \cdot \quad \cdot \quad \cdot \quad \cdot \quad \cdot && && \cdot \quad \cdot \quad \cdot \quad \cdot \quad \cdot \\
&n_1 + n_2 + \cdots + n_{h-1} && \text{and} && r_1 + r_2 + \cdots + r_{h-1}
\end{aligned}
$$

3.42. (*a*) Fig. 3.10 represents *two* paths belonging to the set, but *not* to the subset (1). They have the same initial point A, the same final point C, and pass through the same intermediate point B which lies on the line of symmetry and cuts each path into two arcs, AB and BC. The arcs AB are symmetric to each other with respect to the line of symmetry and neither of them has an *interior* point in common with this line; the arcs BC coincide. Of these two paths, one belongs to the subset (2) and the other to the subset (3). Conversely, any path belonging to these subsets can be matched with another path in the manner presented by Fig. 3.10: look for the second common point B of the path with the line of symmetry (the apex A is the first such common point). Such matching establishes a one-to-one correspondence between the subsets (2) and (3).

(*b*) We could match the paths differently: whereas in Fig. 3.10 the two arcs AB are symmetrical to each other with respect to the straight line through the points A and B, in Fig. 3.11 they are symmetrical with respect to the midpoint of the segment AB.

(*c*) From (*a*) or (*b*) it follows that

$$\binom{n}{r} = N + 2\binom{n-1}{r}$$

Using first one then the other of the two following relations

$$\binom{n}{r} = \binom{n-1}{r-1} + \binom{n-1}{r}, \qquad \binom{n-1}{r} = \frac{n-r}{n}\binom{n}{r}$$

we obtain two different expressions

$$N = \binom{n-1}{r-1} - \binom{n-1}{r} = \frac{2r-n}{n}\binom{n}{r}$$

We have derived this in supposing that $2r > n$. Yet we can easily get rid of this restriction by using the symmetry of the Pascal triangle.

3.43. *With mathematical induction.* Verify the predicted result for $n = 1, 2, 3, (m = 0, 1)$ by inspecting the figure.

From 2m to 2m + 1. By producing a path of length $2m$ which has no point in common with the line of symmetry except the apex, we obtain two paths of length $2m + 1$ of the same nature. Assuming the predicted result for $n = 2m$, we obtain so for $n = 2m + 1$

$$2\binom{2m}{m}$$

as the value of the required number.

From 2m + 1 to 2m + 2. By producing a path of length $2m + 1$ of the specified nature (see above) we obtain in most cases two paths of length $2m + 2$ of the same nature: the paths ending in the two points of the $(2m + 1)$th base nearest to the line of symmetry are exceptional. Visualizing this case, assuming the result for $n = 2m + 1$, and using the appropriate particular case of ex. 3.42, we obtain so, for $n = 2m + 2$, as the value of the required number

$$4\binom{2m}{m} - 2\frac{1}{2m+1}\binom{2m+1}{m+1}$$

which turns out, after suitable transformation, equal to

$$\binom{2m+2}{m+1}$$

Without mathematical induction. Use the first expression obtained for N in the solution of ex. 3.42 under (c), and extend the following sum to values of r restricted by $n/2 < r \leq n$:

$$2\sum\left[\binom{n-1}{r-1} - \binom{n-1}{r}\right]$$

is the value of the required number and yields the predicted result, if we carefully distinguish between the cases $n = 2m$ and $n = 2m + 1$.

3.44.   0, 1,   6, 21,   50,   90, 126, 141, 126...
            0, 1,   7, 28,   77, 161, 266, 357, 393, 357...

1, 393, and 1 are not, the other numbers of the seventh base are, divisible by 7.

3.45. Analogous to ex. 3.1.

3.46. Analogous to ex. 3.31.

3.47. Analogous to ex. 3.32.

3.48. Analogous to ex. 3.33.

3.49. Analogous to sect. 3.9; a wider generalization is analogous to ex. 3.36.

3.50. The lines sloping from northeast to southwest

$$
\begin{array}{cccccc}
1, & 1, & 1, & 1, & 1, & \ldots \\
1, & 2, & 3, & 4, & 5, & \ldots \\
1, & 3, & 6, & 10, & 15, & \ldots
\end{array}
$$

are also "avenues" in the Pascal triangle.

3.51. The symmetry, visible in the first lines, persists; it is enough to write out two bases (the seventh and the eighth) to the middle

$$
\frac{1}{8} \qquad \frac{1}{56} \qquad \frac{1}{168} \qquad \frac{1}{280}
$$

$$
\frac{1}{9} \qquad \frac{1}{72} \qquad \frac{1}{252} \qquad \frac{1}{504} \qquad \frac{1}{630}
$$

3.52. In a given "base" of the harmonic triangle, the denominators are proportional to the binomial coefficients, and the factor of proportionality is visible from the extreme terms. More explicitly, we find in corresponding location in the two triangles the numbers

$$
\binom{n}{r} \qquad \frac{1}{(n+1)\binom{n}{r}}
$$

Pascal   Leibnitz

Proof. For $r = 0$, the boundary condition of the harmonic triangle is verified. To verify its recursion formula, use first the recursion formula, then the explicit form, of the binomial coefficients:

$$
\frac{1}{(n+1)\binom{n}{r-1}} + \frac{1}{(n+1)\binom{n}{r}} = \frac{\binom{n+1}{r}}{(n+1)\binom{n}{r-1}\binom{n}{r}}
$$

$$
= \frac{1}{n+1} \cdot \frac{(n+1)!}{r!(n+1-r)!} \cdot \frac{(r-1)!(n-r+1)!}{n!} \cdot \frac{r!(n-r)!}{n!}
$$

$$
= \frac{(r-1)!(n-r)!}{n!} = \frac{1}{n\binom{n-1}{r-1}}
$$

3.53. On the left-hand side, there is the initial term of an avenue in Fig. 3.13, and on the right-hand side the sum of all terms in the next avenue. For a proof, see the solution of ex. 3.54.

3.54. Use the recursion formula of the Leibnitz triangle:

$$\frac{1}{6} - \frac{1}{12} = \frac{1}{12}$$

$$\frac{1}{12} - \frac{1}{20} = \frac{1}{30}$$

$$\frac{1}{20} - \frac{1}{30} = \frac{1}{60}$$

$$\frac{1}{30} - \frac{1}{42} = \frac{1}{105}$$

$$\cdot \quad \cdot \quad \cdot \quad \cdot \quad \cdot$$

Add ! (A "faraway" term of the second avenue is "negligible.")   From this representative particular case, we easily pass to the general proposition: In the Leibnitz triangle, the sum of all the (infinitely many) terms of the avenue beginning with, and to the southwest from, a certain initial term, is the northwest neighbor of the initial term.   By changing

    "Leibnitz"   "infinitely many"   "southwest"   "northwest"
into
    "Pascal"   "finite number of"   "northeast"   "southeast"

we pass from the present result to that of ex. 3.34, in which we may see a further manifestation of that "analogy by contrast" observed in ex. 3.51.

3.55. In view of the explicit formula for the general term of the Harmonic Triangle (ex. 3.52) the displayed $(r - 1)$th line differs only in a factor from the corresponding line of ex. 3.53, for $r = 2, 3, \ldots$, and its sum is

$$\frac{1}{(r - 1)!(r - 1)}$$

3.57. The product is $= 1$.   The reader acquainted with the theory of infinite series understands the equation

$$1 + x + x^2 + \cdots + x^n + \cdots = (1 - x)^{-1}$$

in a less formal meaning, knows the condition under which it possesses that meaning, and knows also a satisfactory derivation.

3.58. $a_0 + a_1 + a_2 + \cdots + a_n$; ex. 3.57 is a particular case.

3.59. Each series corresponds to an avenue of the Pascal triangle.   For the initial series see ex. 3.57.   By repeated application of ex. 3.58 and ex. 3.34 we find that

$$1 + 2x + 3x^2 + 4x^3 + \cdots = (1 + x + x^2 + \cdots)^2$$
$$1 + 3x + 6x^2 + 10x^3 + \cdots = (1 + x + x^2 + \cdots)^3$$

and, generally,

$$\binom{r}{r} + \binom{r + 1}{r}x + \binom{r + 2}{r}x^2 + \cdots + \binom{r + n}{r}x^n + \cdots$$
$$= (1 + x + x^2 + \cdots)^{r+1} = (1 - x)^{-r-1}$$

For a formal proof, use mathematical induction.

3.60. Compute in two different ways the coefficient of x^n in the product

$$(1 - x)^{-r-1}(1 - x)^{-s-1}$$

This is strictly analogous to the analytic approach in the solution of ex. 3.36, which goes back to sect. 3.9(3).

3.61. 1, 0, 0, 0, respectively, which can be regarded as a confirmation of the conjecture N.

3.62. $\dfrac{2}{3}$, $-\dfrac{1}{9}$, $\dfrac{4}{81}$, $-\dfrac{7}{243}$ respectively, computed by two essentially different

procedures, which can be regarded as another confirmation of the conjecture N.

3.63. $(1 + x)^{1/3}(1 + x)^{2/3}$

$$= \left(1 + \frac{x}{3} - \frac{x^2}{9} + \frac{5x^3}{81} - \frac{10x^4}{243} + \cdots\right)\left(1 + \frac{2x}{3} - \frac{x^2}{9} + \frac{4x^3}{81} - \frac{7x^4}{243} + \cdots\right)$$

$$= 1 + x + 0x^2 + 0x^3 + 0x^4 + \cdots$$

which yields a further confirmation for the conjecture N.

3.64.

$$1 + \frac{-1}{1} x + \frac{(-1)(-2)}{1 \cdot 2} x^2 + \frac{(-1)(-2)(-3)}{1 \cdot 2 \cdot 3} x^3 + \cdots$$

$$= 1 - x + x^2 - x^3 + \cdots$$

$$= [1 - (-x)]^{-1} = (1 + x)^{-1}$$

by virtue of ex. 3.57, which confirms conjecture N from a quite different side. Can the other series of ex. 3.59 also be derived from conjecture N?

3.65. $\dbinom{r - 1 - x}{r} = \dfrac{r - 1 - x}{1} \cdot \dfrac{r - 2 - x}{2} \cdots \dfrac{-x}{r}$

$$= (-1)^r \frac{x}{1} \cdots \frac{x - r + 2}{r - 1} \cdot \frac{x - r + 1}{r}$$

3.66. According to conjecture N, the coefficient of x^n in the expansion of $(1 + x)^{-r-1}$ is $\dbinom{-r - 1}{n} = (-1)^n \dbinom{n + r}{n} = (-1)^n \dbinom{r + n}{r}$; we have first used ex. 3.65 (II), and then we have supposed that *r is a non-negative integer* and used ex. 3.31. Replacing x by $-x$, and so x^n by $(-1)^n x^n$, we obtain the general result of ex. 3.59, which proves the conjecture N in an extensive particular case: for negative integral values of a.

3.67. From

$$\left[\dbinom{a}{0} + \dbinom{a}{1} x + \cdots + \dbinom{a}{r} x^r + \cdots\right]$$

$$\times \left[\dbinom{b}{0} + \cdots + \dbinom{b}{r - 1} x^{r-1} + \dbinom{b}{r} x^r + \cdots\right]$$

$$= \dbinom{a + b}{0} + \dbinom{a + b}{1} x + \cdots + \dbinom{a + b}{r} x^r + \cdots$$

we infer (ex. 3.56) that

$$(*)\quad \binom{a}{0}\binom{b}{r} + \binom{a}{1}\binom{b}{r-1} + \cdots + \binom{a}{r}\binom{b}{0} = \binom{a+b}{r}$$

If we put $a = m$ and $b = n$, this goes over into the result of ex. 3.36, but the range is different: m and n are restricted to be non-negative integers, a and b are unrestricted, arbitrary numbers.

3.68. Relation (*), derived from the conjecture N, is not proved: it is just a conjecture.

The particular case of (*) in which a and b are positive integers has been proved in ex. 3.36. In fact, in view of the solution of ex. 3.66, the particular case of (*) in which a and b are negative integers is equivalent to the result of ex. 3.57, and so it is also proved. (Observe that (*) provides so the desired connection between ex. 3.36 and ex. 3.37; see the remark at the end of the solution of ex. 3.37.)

Could we use ex. 3.36, which is an extensive particular case of the desired (*), as a stepping stone to prove the full statement (*)? (Yes, we can, if we know the relevant algebraic facts: a polynomial in two variables x and y must vanish identically if it vanishes for all positive integral values of x and y.)

Set

$$\binom{a}{0} + \binom{a}{1}x + \binom{a}{2}x^2 + \cdots + \binom{a}{n}x^n + \cdots = f_a(x)$$

The relation (*) is essentially equivalent to the relation

$$f_a(x)f_b(x) = f_{a+b}(x)$$

Now, take (*) for granted; there follows

$$f_a(x)f_a(x)f_a(x) = f_{2a}(x)f_a(x) = f_{3a}(x)$$

and, generally,

$$f_a(x)^n = f_{na}(x)$$

for any positive integer n. Let m be a (positive or negative) integer; since we have verified already conjecture N for positive and negative integral values of a (see ex. 3.1 and ex. 3.66, respectively) we infer that

$$[f_{m/n}(x)]^n = f_m(x) = (1+x)^m$$
$$f_{m/n}(x) = (1+x)^{m/n}$$

and so we have *derived from (*) the conjecture N for all rational values of the exponent a.*

(In fact, the last step is rather risky: in extracting the nth root, we failed to indicate which one of its possible values is meant, and thus we left a gap which we can hardly fill if we remain on the purely formal standpoint of ex. 3.56. Still, we have discovered essential materials for the construction of a full proof. One century and a half after Newton's letter, in 1826, there appeared a memoir of the great Norwegian mathematician Niels Henrik Abel in which he discussed the convergence and the value of the binomial series, also for complex values of x and a, and greatly advanced the general theory of infinite series; see his *Œuvres complètes*, 1881, vol. 1, p. 219–250.)

3.69. We find 1, 2, 6, 20 on the line of symmetry of the Pascal triangle. Explanation: the coefficient of x^n is

$$(-4)^n \binom{-\frac{1}{2}}{n} = 4^n \frac{1 \cdot 3 \cdot 5 \ldots (2n-1)}{2 \cdot 4 \cdot 6 \ldots 2n}$$

$$= \frac{1 \cdot 3 \cdot 5 \ldots (2n-1) \cdot 2 \cdot 4 \cdot 6 \ldots 2n}{n!n!}$$

$$= \binom{2n}{n}$$

3.70.

$$a_0 u_0 = b_0$$
$$a_0{}^2 u_1 = a_0 b_1 - a_1 b_0$$
$$a_0{}^3 u_2 = a_0{}^2 b_2 - a_0 a_1 b_1 + (a_1{}^2 - a_0 a_2) b_0$$
$$a_0{}^4 u_3 = a_0{}^3 b_3 - a_0{}^2 a_1 b_2 + (a_0 a_1{}^2 - a_0{}^2 a_2) b_1 - (a_1{}^3 - 2a_0 a_1 a_2 + a_0{}^2 a_3) b_0$$

3.71. The cases $n = 0, 1, 2, 3$ treated in ex. 3.70 suggest that $a_0{}^{n+1} u_n$ is a polynomial in the a's and b's all terms of which have

(1) the same degree n in the a's
(2) the same degree 1 in the b's
(3) the same weight n in the a's and b's jointly.

Reasons:

(1) If a_n is replaced by $a_n c$ (for $n = 0, 1, 2, \ldots$, with arbitrary c) u_n must be replaced by $u_n c^{-1}$.

(2) If b_n is replaced by $b_n c$, u_n must be replaced by $u_n c$.

(3) If a_n and b_n are replaced by $a_n c^n$ and $b_n c^n$ respectively (as a result of substituting cx for x) also u_n must be replaced by $u_n c^n$.

3.72. $u_n = b_n - b_{n-1}$: this value must result if, having expressed u_n in terms of a's and b's, we set $a_0 = a_1 = a_2 = a_3 = \cdots = 1$. This is a valuable check; carry it through for $n = 0, 1, 2, 3$ (ex. 3.70).

3.73. $u_n = b_0 + b_1 + b_2 + \cdots + b_n$ (see ex. 3.58): this value must result if, having expressed u_n in terms of a's and b's, we set $a_0 = 1$, $a_1 = -1$, $a_2 = a_3 = \cdots = 0$. This is a valuable check; carry it through for $n = 0, 1, 2, 3$ (ex. 3.70.)

3.74.

$$1 - \frac{x}{6} + \frac{x^2}{40} - \frac{x^3}{336} + \cdots + \frac{(-1)^n x^n}{(2 \cdot 4 \cdot 6 \ldots 2n)(2n+1)} + \cdots$$

Cf. MPR, vol. 1, p. 84, ex. 2.

3.75.

$$a_1 u_1 = 1$$
$$-a_1{}^3 u_2 = a_2$$
$$a_1{}^5 u_3 = 2a_2{}^2 - a_1 a_3$$
$$-a_1{}^7 u_4 = 5a_2{}^3 - 5a_1 a_2 a_3 + a_1{}^2 a_4$$
$$a_1{}^9 u_5 = 14a_2{}^4 - 21a_1 a_2{}^2 a_3 + 3a_1{}^2 a_3{}^2 + 6a_1{}^2 a_2 a_4 - a_1{}^3 a_5$$

3.76. The cases treated in ex. 3.75 suggest that $a_1{}^{2n-1}u_n$ is a polynomial in the a's each term of which is

(1) of degree $n - 1$ and
(2) of weight $2n - 2$.

Reasons:

(1) If a_n is replaced by $a_n c$ (as a result of substituting $c^{-1}x$ for x) u_n must be replaced by $u_n c^{-n}$.

(2) If a_n is replaced by $a_n c^n$ (as a result of substituting cy for y) u_n must be replaced by $u_n c^{-1}$.

3.77. $x = \dfrac{y}{1-y}$, $\quad y = \dfrac{x}{1+x}$ and so

$$y = x - x^2 + x^3 - x^4 + \cdots$$

Hence, if we set $a_n = 1$, in ex. 3.75, we must get $u_n = (-1)^{n-1}$. This is a valuable check; carry it through for $n = 1, 2, 3, 4, 5$.

3.78. $1 - 4x = (1 + y)^{-2}$ $\quad$ or

$$y = -1 + (1 - 4x)^{-\frac{1}{2}} = 2x + 6x^2 + \cdots + \binom{2n}{n}x^n + \cdots$$

see ex. 3.69.

3.79. $y = -1 + (1 + 4ax)^{\frac{1}{2}}(2a)^{-1}$
$$= x - ax^2 + 2a^2x^3 - 5a^3x^4 + 14a^4x^5 - \cdots$$

The coefficient of x^n is

$$\frac{(4a)^n}{2a}\binom{\frac{1}{2}}{n} = \frac{(-1)^{n-1}a^{n-1}}{n}\binom{2n-2}{n-1}$$

(computation similar to ex. 3.69) and u_n of ex. 3.75 must reduce to this value if $a_1 = 1, a_2 = a, a_3 = a_4 = \cdots = 0$. Cf. MPR, vol. 1, p. 102, ex. 7, 8, 9.

3.80.

$$y = x - \frac{x^2}{2} + \frac{x^3}{3} - \frac{x^4}{4} + \cdots + \frac{(-1)^{n-1}x^n}{n} + \cdots$$

3.81. $u_0 = u_1 = u_2 = 1$, $\quad u_3 = \dfrac{4}{3}$, $u_4 = \dfrac{7}{6}$.

3.82. Mathematical induction: The assertion is true for $n = 3$. Assume that $n > 3$ and that the assertion has been proved for the coefficients preceding u_n so that

$$u_{n-1} > 1, \quad u_{n-2} > 1, \ldots, u_3 > 1$$

We know that $u_0 = u_1 = u_2 = 1$ and, therefore,

$$nu_n = u_0u_{n-1} + u_1u_{n-2} + \cdots + u_{n-1}u_0 > n$$

3.83. Set

$$y = u_0 + u_1x + u_2x^2 + \cdots + u_nx^n + \cdots$$

$$\frac{d^2y}{dx^2} = 2\cdot1u_2 + \cdots + n(n-1)u_nx^{n-2} + \cdots$$

From the differential equation

$$n(n - 1)u_n = -u_{n-2}$$

From the initial condition

$$u_0 = 1, \qquad u_1 = 0$$

Finally, for $m = 1, 2, 3, \ldots,$

$$u_{2m} = \frac{(-1)^m}{2m!}, \qquad u_{2m-1} = 0$$

$$y = 1 - \frac{x^2}{2!} + \frac{x^4}{4!} - \frac{x^6}{6!} + \cdots$$

3.84. $B_n = B_{n-5} + A_n$
$C_n = C_{n-10} + B_n$
$D_n = D_{n-25} + C_n$
$E_n = E_{n-50} + D_n$

The last equation yields, for $n = 100$,

$$E_{100} = E_{50} + D_{100}$$

and the foregoing equation yields for $n = 20$,

$$D_{20} = C_{20}$$

since $D_{-5} = 0$: any of the introduced quantities with a negative subscript must be properly regarded as equal to 0.   These examples should illustrate the main feature of the system of equations obtained: we can compute any unknown (such as E_{100}) if a certain unknown of the same kind with a lower subscript (as E_{50}) and another unknown of a lower kind (denoted by the foregoing letter of the alphabet, as D_{100}) have been computed previously.   (There are cases in which only one previously computed unknown is needed, see D_{20}.   In other cases we need some of the "boundary values" which we have known before setting up our equations: I mean B_0, C_0, D_0, E_0, and A_n for $n = 0, 1, 2, 3, \ldots$)   In short, we compute the unknowns by going back to lower subscripts or to former letters of the alphabet and, eventually, to the boundary values. (Do not let the difference of notation hide the analogy between this computation and the determination of the binomial coefficients by recursion formula and boundary condition; see sect. 3.6(2).)

The reader should set up a convenient system of computation which he can check against the following values:

$$B_{10} = 3, \qquad C_{25} = 12, \qquad D_{50} = 49, \qquad E_{100} = 292$$

(For more details, and a concrete interpretation of the problem, see HSI, p. 238, ex. 20, and *American Mathematical Monthly*, **63**, 1956, pp. 689–697.)

3.85. Mathematical induction: assuming the proposed form for $y^{(n)}$, we find by differentiating once more

$$y^{(n+1)} = (-1)^{n+1}(n + 1)!x^{-n-2} \log x$$
$$+ (-1)^n x^{-n-2}[n! + (n + 1)c_n]$$

which is of the desired form provided that

$$c_{n+1} = n! + (n + 1)c_n$$

Transforming this into

$$\frac{c_{n+1}}{(n + 1)!} = \frac{c_n}{n!} + \frac{1}{n + 1}$$

and using $c_1 = 1$, we find

$$c_n = n!\left(1 + \frac{1}{2} + \frac{1}{3} + \frac{1}{4} + \cdots + \frac{1}{n}\right)$$

3.86. To find the sum of a geometric series is a closely related problem: both the result and the usual method of derivation are used in the following.

Call S the proposed sum. Then

$$(1 - x)S = 1 + x + x^2 + \cdots + x^{n-1} - nx^n$$

$$= \frac{1 - x^n}{1 - x} - nx^n$$

Hence the required short expression:

$$S = \frac{1 - (n + 1)x^n + nx^{n+1}}{(1 - x)^2}$$

3.87. Use method, notation, and result of ex. 3.86: call T the proposed sum and consider

$$(1 - x)T = 1 + 3x + 5x^2 + 7x^3 + \cdots + (2n - 1)x^{n-1} - n^2x^n$$
$$= 2S - (1 + x + x^2 + \cdots + x^{n-1}) - n^2x^n$$

and so by straightforward algebra

$$T = \frac{1 + x - (n + 1)^2x^n + (2n^2 + 2n - 1)x^{n+1} - n^2x^{n+2}}{(1 - x)^3}$$

3.88. In following ex. 3.86 and 3.87, we could find an expression for

$$1^k + 2^kx + 3^kx^2 + \cdots + n^kx^{n-1}$$

by recursion, by reducing the case k to the cases $k - 1, k - 2, \ldots, 2, 1, 0$.

3.89. Mathematical induction. The assertion is obviously true for $n = 1$, and

$$\frac{a_n(n + \alpha) - a_1(1 + \beta)}{\alpha - \beta} + a_{n+1} = \frac{a_{n+1}(n + 1 + \beta) - a_1(1 + \beta)}{\alpha - \beta} + a_{n+1}$$

3.90. Apply ex. 3.89 with

$$a_1 = \frac{p}{q}, \qquad \alpha = p, \qquad \beta = q - 1$$

The proposed sum turns out to equal

$$\frac{p}{p - q + 1}\left(\frac{p + 1}{q}\frac{p + 2}{q + 1}\cdots\frac{p + n}{q + n - 1} - 1\right)$$

3.91. (1) $8, 4\sqrt{2}, 4\sqrt{3}, 6$.

(2) $C_n = 2n \tan(\pi/n)$, $I_n = 2n \sin(\pi/n)$.

Hence straightforward verification by familiar trigonometric identities.

3.92. For more examples on mathematical induction see p.75 footnote 5. Problems connected with those in Parts II and III can be found in books on Calculus of Probability or Combinatorial Analysis. Problems connected with those in Part IV, or with ex. 3.53 to 3.55, can be found in books on Infinite Series or Complex Variables. Problems closely related to ex. 3.81, 3.82 and 3.83 fill an extensive chapter of the Theory of Differential Equations.

There is an inexhaustible source of themes for further examples. Here is just one instance: *polynomial coefficients* (cf. ex. 3.28, 3.29, 3.30). The coefficients of the expansion of

$$(a + b + c)^n$$

for $n = 0, 1, 2, 3,\ldots$ can be associated with the lattice points in an octant of space analogously to the coefficients of

$$(a + b)^n$$

which are, in the Pascal triangle, associated with the lattice points in one quarter of a plane. What is, in the space arrangement, analogous to the boundary condition, to the recursion formula, to the avenues, streets, and bases of the Pascal triangle, to ex. 3.31–3.39? What is the connection with ex. 3.44–3.50? We have not yet mentioned the number-theoretic properties of binomial, or polynomial, coefficients. And so on.

No solution: **3.56.**

Chapter 4

4.1. Let A denote the vertex of the pyramid opposite its base (the "apex"). Dissect the base of the pyramid into n triangles of area

$$B_1, B_2,\ldots, B_n$$

respectively. Each of these triangles forms the base of a tetrahedron of which A is the opposite vertex and h the height; the pyramid is dissected (by planes passing through A) into these n tetrahedra of volume

$$V_1, V_2,\ldots, V_n$$

respectively. Obviously,

$$B_1 + B_2 + \cdots + B_n = B$$
$$V_1 + V_2 + \cdots + V_n = V$$

Supposing that the desired expression for the volume has been proved for the particular case of the tetrahedra, we have

$$V_1 = \frac{B_1 h}{3}, \quad V_2 = \frac{B_2 h}{3}, \quad \ldots, \quad V_n = \frac{B_n h}{3}$$

Addition (superposition!) of these special relations yields the general relation

$$V = \frac{Bh}{3}$$

4.2. A polynomial of degree k is of the form

$$f(x) = a_0 x^k + a_1 x^{k-1} + \cdots + a_k$$

where $a_0 \neq 0$. Substitute successively $1, 2, 3, \ldots, n$ for x and add: you obtain, in using the notation of ex. 3.11,

$$f(1) + f(2) + \cdots + f(n) = a_0 S_k(n) + a_1 S_{k-1}(n) + \cdots + a_k S_0(n)$$

The right-hand side is, by the result of ex. 3.3, a polynomial of degree $k + 1$ in n.

4.3. We can write the result of ex. 3.34 in the form

$$\binom{0}{r} + \binom{1}{r} + \binom{2}{r} + \cdots + \binom{n}{r} = \binom{n+1}{r+1}$$

see ex. 3.65(III). Assuming ex. 4.4, we write the polynomial considered in the form

$$f(x) = b_0 \binom{x}{k} + b_1 \binom{x}{k-1} + \cdots + b_k \binom{x}{0}$$

where (see the solution of ex. 4.4) $b_0 = k! a_0 \neq 0$. Substitute successively $0, 1, 2, 3, \ldots, n$ for x and add: you obtain, by the result of ex. 3.34 restated above,

$$f(0) + f(1) + f(2) + \cdots + f(n) = b_0 \binom{n+1}{k+1} + b_1 \binom{n+1}{k} + \cdots + b_k \binom{n+1}{1}$$

The right-hand side is a polynomial of degree $k + 1$ in n.

4.4. Comparing the coefficient of x^k (the highest power of x present) on both sides of the proposed identity, we find

$$a_0 = b_0 / k!$$

Hence, it follows from the proposed identity

$$a_0 x^k + a_1 x^{k-1} + \cdots + a_k - k! a_0 \binom{x}{k} = b_1 \binom{x}{k-1} + \cdots + b_k \binom{x}{0}$$

Comparing the coefficient of x^{k-1} on both sides, we express b_1 in terms of a_0 and a_1, and in continuing in this fashion we determine $b_0, b_1, b_2, \ldots, b_k$ successively, by recursion.

4.5. We have to determine four numbers b_0, b_1, b_2 and b_3 so that

$$x^3 = b_0 \binom{x}{3} + b_1 \binom{x}{2} + b_2 \binom{x}{1} + b_3 \binom{x}{0}$$

holds identically in x. This means

$$x^3 = \frac{b_0}{6} (x^3 - 3x^2 + 2x) + \frac{b_1}{2} (x^2 - x) + b_2 x + b_3$$

Comparison of the coefficients of x^3, x^2, x^1, and x^0 yields the equations

$$1 = b_0/6$$
$$0 = -b_0/2 + b_1/2$$
$$0 = b_0/3 - b_1/2 + b_2$$
$$0 = b_3$$

respectively, from which we derive

$$b_0 = 6, \qquad b_1 = 6, \qquad b_2 = 1, \qquad b_3 = 0$$

Hence, by the procedure of ex. 4.3 ($k = 3$)

$$1^3 + 2^3 + \cdots + n^3 = 6\binom{n+1}{4} + 6\binom{n+1}{3} + \binom{n+1}{2}$$
$$= \frac{(n+1)^2 n^2}{4}$$

by straightforward algebra.

4.6. As it has been proved in ex. 4.3 there are five constants c_0, c_1, c_2, c_3, and c_4 such that

$$1^3 + 2^3 + 3^3 + \cdots + n^3 = c_0 n^4 + c_1 n^3 + c_2 n^2 + c_3 n + c_4$$

for all positive integral values of n. In setting successively $n = 1, 2, 3, 4$, and 5, we obtain a system of five equations for the five unknowns c_0, c_1, c_2, c_3, and c_4. By solving these equations we obtain

$$c_0 = 1/4, \qquad c_1 = 1/2, \qquad c_2 = 1/4, \qquad c_3 = 0, \qquad c_4 = 0$$

that is, we obtain the same result as in ex. 4.5, but with more trouble.

4.7. Ex. 4.3 yields a new proof for the result of ex. 3.3, except one point: the coefficient of n^{k+1} in the expression of $S_k(n)$ remains undetermined by the procedure of ex. 4.3. (A little additional remark, however, will yield also this coefficient.)

4.8. Yes: a straight line is represented by an equation of the form

$$y = ax + b$$

the right-hand side of which is a polynomial of degree ≤ 1.

4.9. The straight line coinciding with the x axis appears intuitively as the simplest interpolating curve; it corresponds to the identically vanishing polynomial. Any *other* interpolating polynomial is necessarily of higher degree, namely of degree n at least, since it has n different zeros $x_1, x_2, \ldots, x_n$.

4.10. Lagrange's interpolating polynomial, given by the final formula of sect. 4.3, is of degree $\leq n - 1$; it is the only interpolating polynomial of such a low degree, I say. In fact, if two polynomials, both of degree $\leq n - 1$, take the same values at the n given abscissas, their difference has n different zeros, that is more zeros than the degree would permit, unless this difference vanishes identically. Lagrange's interpolating polynomial, being the only one of degree $\leq n - 1$, *is* of the lowest possible degree.

4.12. (*a*) Obvious, in view of the rule

$$(c_1 y_1 + c_2 y_2)' = c_1 y_1' + c_2 y_2'$$

for constant c_1 and c_2.

(*b*) $y = e^{rx}$ is a solution of the differential equation iff r is a root of the algebraic equation in r

$$r^n + a_1 r^{n-1} + a_2 r^{n-2} + \cdots + a_n = 0$$

(*c*) If the equation in r under (*b*) has d *different* roots $r_1, r_2, \ldots, r_d$ and $c_1, c_2, \ldots, c_d$ are arbitrary constants

$$y = c_1 e^{r_1 x} + c_2 e^{r_2 x} + \cdots + c_d e^{r_d x}$$

is a solution of the differential equation, and its most general solution (as it can be shown) when $d = n$.

4.13. The equation in r is

$$r^2 + 1 = 0$$

and so the general solution of the differential equation is

$$y = c_1 e^{ix} + c_2 e^{-ix}$$

The initial conditions yield the equations

$$c_1 + c_2 = 1, \qquad ic_1 - ic_2 = 0$$

which determine c_1 and c_2 and so the desired particular solution

$$y = (e^{ix} + e^{-ix})/2$$

Observe that also $y = \cos x$ satisfies both the differential equation and the boundary conditions.   See also ex. 3.83.

4.14. (*a*) Obvious.

(*b*) $y_k = r^k$ is a solution of the difference equation iff r is a root of the algebraic equation given in ex. 4.12(*b*).

(*c*) If the equation of ex. 4.12(*b*) has d *different* roots $r_1, r_2, \ldots, r_d$ and $c_1, c_2, \ldots, c_d$ are arbitrary constants

$$y_k = c_1 r_1{}^k + c_2 r_2{}^k + \cdots + c_d r_d{}^k$$

is a solution of the difference equation and its most general solution (as it can be shown) when $d = n$.

4.15. The equation in r is

$$r^2 - r - 1 = 0$$

and so the general solution of the difference equation is

$$y_k = c_1 \left(\frac{1 + \sqrt{5}}{2}\right)^k + c_2 \left(\frac{1 - \sqrt{5}}{2}\right)^k$$

This yields for $k = 0$ and $k = 1$ (initial conditions) the equations

$$c_1 + c_2 = 0, \qquad c_1 \frac{1 + \sqrt{5}}{2} + c_2 \frac{1 - \sqrt{5}}{2} = 1$$

which determine c_1 and c_2 and so the desired expression of the Fibonacci numbers

$$y_k = \frac{1}{\sqrt{5}} \left[\left(\frac{1 + \sqrt{5}}{2} \right)^k - \left(\frac{1 - \sqrt{5}}{2} \right)^k \right]$$

4.16. If the actual motion is obtainable by *superposition* of the three virtual motions, the coordinates of the moving point at the time t are

$$x = x_1 + x_2 + x_3 = tv \cos \alpha,$$
$$y = y_1 + y_2 + y_3 = tv \sin \alpha - \tfrac{1}{2}gt^2$$

Elimination of t yields the trajectory of the projectile

$$y = x \tan \alpha - \frac{gx^2}{2v^2 \cos^2 \alpha}$$

which is a *parabola*.

4.18. There are two unknowns: the base and the height of the tetrahedron. See Fig. 4.5a.

4.19. Let

 V stand for the volume of the tetrahedron,
 B for its base,
 H for its height,
 h for that height of its base that is perpendicular to the edge of given length a.

Then

$$V = \frac{BH}{3}, \qquad B = \frac{ah}{2}$$

and, therefore,

$$V = \frac{ahH}{6}$$

Yet neither h nor H is given.

4.20. The orthogonal projection of our tetrahedron (of ex. 4.17) on a plane, perpendicular to the line of length b and passing through one of its endpoints, is a square. The diagonals of this square are of length a, its area is $a^2/2$, and the square itself is, see Fig. 4.5b, the base of a right prism with height b. This prism, see Fig. 4.5b, is split into five (nonoverlapping) tetrahedra; our tetrahedron of ex. 4.17 is one of them (we call its volume V); the other four are congruent, the base of each is an isosceles right triangle with area $a^2/4$, and the height of each is b. Hence

$$a^2b/2 = V + 4a^2b/12$$
$$V = a^2b/6$$

4.21. The plane that passes through an edge of length a and the midpoint of the opposite edge is a plane of symmetry for our tetrahedron and divides it into two congruent tetrahedra (see Fig. 4.5c) the common base of which (an isosceles triangle) has obviously the area $ab/2$ and the height of which is $a/2$. Hence the desired volume

$$V = \frac{2}{3}\cdot\frac{a}{2}\cdot\frac{ab}{2} = \frac{a^2b}{6}$$

(There are two such planes of symmetry which jointly divide our tetrahedron into four congruent tetrahedra; this yields another, but little different, access to the solution.)

4.22. We can regard our tetrahedron as an extreme (degenerate, limiting) prismoid, with height b and each base shrunk into a line-segment of length a; the midsection is a square with side $a/2$; see Fig. 4.5d. Hence

$$h = b, \qquad L = 0, \qquad M = a^2/4, \qquad N = 0$$

and the prismoidal formula yields $V = a^2b/6$.

4.23. If the expression found for V in ex. 4.19 agrees with the result derived in three different ways in ex. 4.20, 4.21, and 4.22, we should have

$$Hh = ab$$

Yet we can show this relation independently, by computing, in two different ways, the area of the isosceles triangle, in which the tetrahedron is intersected by a plane of symmetry (ex. 4.21, Fig. 4.5e). And so we brought to a successful end a fourth (somewhat tortuous) derivation started in ex. 4.18 and continued through ex. 4.19.

4.24. The route from ex. 4.18 through ex. 4.19 to ex. 4.23 is too long and tortuous. The solution in ex. 4.21 appears the most elegant: it exploits fully the symmetry of the figure—but just for this reason it may be less applicable to nonsymmetric cases. Thus, *prima facie* evidence favors ex. 4.20. Do you see another indication in favor of ex. 4.20?

4.25. $L = M = N$, and so $V = Lh$.

4.26. $N = 0$, $M = L/4$, and so $V = Lh/3$.

4.27. Let L_i, M_i, N_i, and V_i denote the quantities that are so related to P_i as L, M, N, and V are to P respectively, for $i = 1, 2,\ldots, n$. All prismoids have the same height h. Obviously

$$\begin{aligned}
L_1 + L_2 + \cdots + L_n &= L \\
M_1 + M_2 + \cdots + M_n &= M \\
N_1 + N_2 + \cdots + N_n &= N \\
V_1 + V_2 + \cdots + V_n &= V
\end{aligned}$$

By combining these equations we obtain

$$\sum_{i=1}^{n} \left(\frac{L_i + 4M_i + N_i}{6} h - V_i\right) = \frac{L + 4M + N}{6} h - V$$

We regard the right-hand side as *one* term; the left-hand side is a sum of n analogous terms. If n terms out of these $n + 1$ terms linked by our equation vanish, the remaining one term must vanish too.

4.28. The orthogonal projection of our tetrahedron onto the plane we have passed through l is a quadrilateral (a square in the particular case of ex. 4.20, Fig. 4.5*b*, but irregular in general). One diagonal of this quadrilateral is the edge l, the other diagonal is parallel and equal to n. This quadrilateral is the base of a prism with height h; the prism is split into five tetrahedra; one of them is our tetrahedron, the other four are pyramids in the situation described by ex. 4.26, and so the prismoidal formula is valid for them. This formula is also valid for the prism, by ex. 4.25, and so also for our tetrahedron, by ex. 4.27.

4.29. Fig. S4.29 shows a prismoid; $B, C, \ldots, K$ are the vertices of the lower base (in the plane of the paper), and $B', C', \ldots, K'$ are the vertices of the upper base.

(1) Consider the pyramid of which the base is the upper base of the prismoid and the apex (the vertex opposite the base) a point A (freely chosen) in the lower base.

(2) Join the point A to the vertices $B, C, \ldots, K$ of the lower base. Each segment so obtained is associated with a side of the upper base (an edge of the prismoid); the segment and the side form a pair of opposite edges of a tetrahedron. (For instance, the segment AB is associated with the side $B'C'$ and they determine together, as opposite edges, the tetrahedron $ABB'C'$.)

(3) The lines drawn from A to the vertices $B, C, \ldots, K$ dissect the lower base into triangles. Each such triangle is associated with a vertex of the upper base; the triangle forms the base, the associated vertex the apex, of a pyramid (which is, in fact, a tetrahedron; for instance, ABC is associated with the vertex C', and they determine together the pyramid ABC-C').

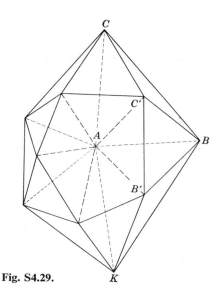

Fig. S4.29.

Our prismoid is dissected into the solids introduced in (1), (2), and (3). (Of the upper base (1) takes the area, (2) the sides, (3) the vertices. Of the lower base (1) takes just one point, (2) dividing segments, (3) the area.) Apply ex. 4.26 to the pyramids (1) and (3) and ex. 4.28 to the tetrahedra (2). Using ex. 4.27, you prove the prismoidal formula for the prismoid $BC...KB'C'...K'$ of Fig. S4.29.

4.30. The solution of ex. 4.28 is incomplete, since it treats only one out of three possible cases. Consider two line segments: l and n, and the orthogonal projection n' of n on the plane parallel to n that passes through l. Consider the two straight lines containing these two segments, respectively, and the point of intersection I of these two lines. There are three possible situations: the point I may belong

>   (0) to none of the two segments l and n',
>   (1) to just one segment, or
>   (2) to both segments.

Ex. 4.20 treats only the case (2). Yet a tetrahedron in situation (1) can be regarded as the difference of two tetrahedra in situation (2), and a tetrahedron in situation (0) can be regarded as the difference of two in situation (1). In view of ex. 4.27, this remark completes the proof of ex. 4.28.

4.31. Fig. S4.29 is subject to two restrictions:
(1) Both bases are convex.
(2) Each vertex of one base is associated (there is a one-to-one correspondence) with a side of the other base: two lateral edges start from the vertex and end in the endpoints of the side. (For instance, vertex B corresponds to the edge $B'C'$, vertex C' to the edge BC.)

Condition (2) is actually less restrictive than it might appear: many shapes which do not fall under it directly are limiting (degenerate) cases, and the proof extends to such shapes (by continuity, or appropriate interpretation).

The proof in ex. 4.35 is free from restrictions (1) and (2), but uses integral calculus.

4.32. $n = 0$; then $L = M = N = 1$, $I = 2$: valid.
$n = 2m - 1$, odd; $-L = N = 1$, $M = I = 0$: valid.
$n = 2m$, even; $L = N = 1$, $M = 0$, $I = 2/(n + 1)$:
valid for $n = 2$, but for no other positive even integer.

4.33. $f(x) = a + bx + cx^2 + dx^3$: superposition of the particular cases $n = 0, 1, 2, 3$ of ex. 4.32.

4.34. The substitution

$$x = a + \frac{h(t + 1)}{2}$$

transforms the interval $a \leq x \leq a + h$ into the interval $-1 \leq t \leq 1$ and any polynomial in x of degree ≤ 3 into another polynomial of the same kind in t.

4.35. We introduce a system of rectangular coordinates x, y, z. We place the prismoid so that its lower base lies in the plane $z = 0$ and its upper base in the plane $z = h$.   The volume of the prismoid is expressed by

(1)
$$V = \int_0^h Q(z)\, dz$$

where $Q(z)$ denotes the area of the intersection of the prismoid with the plane parallel to, and at the distance z from, the lower base.

This intersection is a polygon with n sides if the prismoid has n lateral edges; its area is expressed by

(2)
$$Q(z) = \tfrac{1}{2} \sum_{i=1}^{n} (x_i y_{i+1} - x_{i+1} y_i)$$

if the lateral edge number i is given by the pair of equations

(3)
$$x_i = a_i z + c_i, \qquad y_i = b_i z + d_i$$

a_i, b_i, c_i, d_i are constants specifying the position of the edge; it is understood that edge number $n + 1$ coincides with edge number 1, so that

$$a_{n+1} = a_1, \qquad b_{n+1} = b_1, \ldots, y_{n+1} = y_1$$

Equations (2) and (3) show that $Q(z)$ is a *polynomial in z of degree not exceeding* 2, and so, by ex. 4.34, Simpson's rule, stated in ex. 4.32, is applicable to the integral (1), and this yields the prismoidal formula stated in ex. 4.22 since, obviously,

$$Q(0) = L, \qquad Q(h/2) = M, \qquad Q(h) = N$$

represent the areas of the lower base, the midsection and the upper base, respectively.

No solution: **4.11, 4.17, 4.36.**

Chapter 5

5.1. The unknown is the number V.

The data are the numbers a and h.

The condition is that V measures the volume of a right prism, the height of which is h and the base of which is a square with sides of length a.

5.2. There are two unknowns, the real numbers x and y.   Or there is just one bipartite unknown, with components x and y, which we may interpret geometrically as a point in a plane with rectangular coordinates x and y.

The condition is fully expressed by the proposed equation.

We need not talk about data.   (If we modified the problem by taking r^2 as the right-hand side of the proposed equation instead of 1, r would be a datum.)

A solution is $x = 1$, $y = 0$; another solution is $x = 3/5$, $y = -4/5$; and so on.   In the geometrical interpretation, the full set of solutions consists of the points on the periphery of a circle with radius 1 and center at the origin.

5.3. There is no solution: the set of solutions is the empty set.

5.4. There are eight solutions:

$(2, 3)$ $(3, 2)$ $(-2, 3)$ $(-3, 2)$ $(2, -3)$ $(3, -2)$ $(-2, -3)$ $(-3, -2)$

The set consists of the lattice-points on the periphery of the circle with radius $\sqrt{13}$ and center at the origin. (A point of which both rectangular coordinates are integers is called a *lattice-point*. The configuration of lattice-points is important in number theory, crystallography, etc.)

5.5. We interpret the tripartite unknown (x, y, z) as a point in space with rectangular coordinates x, y, and z.

(1) The set of solutions consists of the points *in the interior* of an octahedron of which the center is at the origin and the six vertices are at the points

$(1, 0, 0)$ $(-1, 0, 0)$ $(0, 1, 0)$ $(0, -1, 0)$ $(0, 0, 1)$ $(0, 0, -1)$

(2) The set of solutions consists of the points in the interior and *on the surface* of the octahedron.

5.6. The following statement renders conspicuous the required principal parts:

If a, b, and c are the lengths of the sides of a right triangle, and a is the length of the side opposite the right angle

then
$$a^2 = b^2 + c^2$$

5.7. We have to restate the theorem as the simultaneous assertion of two propositions in the usual "if-then" form where hypothesis and conclusion are conspicuous:

"If n is a square
then $d(n)$ is odd."
"If $d(n)$ is odd
then n is a square."

Here is a condensed statement, which uses "iff" (for "if and only if"):
"Iff n is a square then $d(n)$ is odd."

5.16. Begin by considering the case of a convex polygon and postpone whatever modifications may be needed to treat the general case.

(1) Given the lengths of $n - 1$ line-segments joining a chosen vertex of the polygon to the other $n - 1$ vertices, and the $n - 2$ angles included by pairs of consecutive line-segments.

(2) Divide the polygon by $n - 3$ diagonals into $n - 2$ triangles which are all determined (each by three sides) if the lengths of the dividing diagonals and of the n sides of the polygon are given.

(3) Take the particular case of the division into triangles considered under (2) supplied by the line-segments considered under (1). Number the triangles so that each triangle (except the first) has one side in common with the preceding triangle. Give any three independent data for the first triangle and, for each of the following $n - 3$ triangles, 2 data independent of each other and of the side that belongs also to the preceding triangle.

(4) Giving 2 rectangular coordinates for each of the n vertices, $2n$ data in

all, we would determine not only the polygon, but also a coordinate system the position of which is unessential.   The position of the coordinate system depends on 3 parameters and so only $2n - 3$ data are essential.

5.17. To determine the base, $2n - 3$ data are required, see ex. 5.16.   To determine the apex (the vertex opposite to the base) give the 3 rectangular coordinates of the apex with respect to a coordinate system of which a coordinate plane is the base, the origin a chosen vertex of the base, and a coordinate axis contains a side of the base starting from the chosen vertex.   Hence, $2n$ data are required.

5.18. As in ex. 5.17, $2n$.

5.19. The polynomial is of the form

$$f_0 x_v^n + f_1 x_v^{n-1} + \cdots + f_{n-1} x_v + f_n$$

where f_j is a polynomial of degree j in $v - 1$ variables.   Using ex. 3.34, we prove by mathematical induction that the number of data needed (the number of coefficients in the expansion in powers of $x_1, x_2, \ldots, x_v$) is

$$\binom{n + v}{v} = \binom{0 + v - 1}{v - 1} + \binom{1 + v - 1}{v - 1} + \cdots + \binom{n + v - 1}{v - 1}$$

No solution: **5.8, 5.9, 5.10, 5.11, 5.12, 5.13, 5.14, 5.15.**

Chapter 6

6.1. (1) 9 of shape $r(x) = 0$
(2) 9 of shape $r(x) = 0$
(3) 36 of shape $r(x, y) = 0$
(4) 7 of shape $r(x, y, z, w) = 0$
take into account obvious cancellations.
There are altogether 61 clauses.

6.2. Regard x as having the n components $x_1, x_2, \ldots, x_n$.

6.3. Let $x_1 = y_1$, $x_4 = y_3$, regard y_2 as having the two components x_2 and x_3, and regard the combination (the simultaneous imposition) of clauses r_2 and r_3 as one clause; then you have (with suitable notation) the "recursive" system

$$s_1(y_1) = 0$$
$$s_2(y_1, y_2) = 0$$
$$s_3(y_1, y_2, y_3) = 0$$

6.4. Regard y_1 as having the components x_1, x_2, x_3, and y_2 as having the components x_4, x_5, x_6, set $y_3 = x_7$, combine the first three clauses r_1, r_2, r_3 into one s_1, and the next three clauses r_4, r_5, r_6 into s_2: then you obtain the same final system as in ex. 6.3.

6.5. Essentially the same as the plan developed in sect. 6.4(2).

6.6. Particular case of the system in sect. 6.4(1).   See ex. 3.21.

6.7. Two loci for a straight line; cf. sect. 6.2(5).   In fact, all chords of given length in a given circle are tangent to an (easily constructible) circle concentric with the given circle.

6.8. Construct the point A on a, and the point B on b, each at the distance $l/2$ from the intersection of the lines a and b: draw the circle that touches a at A and b at B; since A may have two positions and the same is true of B, there are four such circles.   One of these circles is an escribed circle of the desired triangle: the desired line x must be tangent to one of four circular arcs.   We have here two loci for the straight line x; cf. sect. 6.2(5) and ex. 6.7.

6.9. HEARSAY.

6.11. (1) Begin with the "constant" c of the magic square.   The sum of all nine unknowns, x_{ik} is, on the one hand,

$$= 1 + 2 + 3 + \cdots + 9 = 45$$

and it is, on the other hand, the sum of three rows and, therefore,

$$= 3c$$

From $45 = 3c$ follows $c = 15$.

(2) Add the three rows and the two diagonals; their sum is $5c$.   Subtract hence those rows and columns that do *not* contain the central element x_{22}; their number is 4, their sum is $4c$.   You obtain so

$$3x_{22} = 5c - 4c = 15$$

whence $x_{22} = 5$.

(3) In order to fill those rows and columns that do *not* contain the central x_{22}, list all different ways of representing 15 as the sum of three different numbers chosen among the following eight: 1, 2, 3, 4, 6, 7, 8, 9.   A systematic survey yields:

$$
\begin{aligned}
15 &= \mathbf{1} + 6 + 8 \\
&= 2 + 6 + \mathbf{7} \\
&= 2 + 4 + \mathbf{9} \\
&= \mathbf{3} + 4 + 8
\end{aligned}
$$

(4) Those numbers that arise in *just one* of the above four representations of 15 are distinguished by heavy print; they must be placed in the middle of a row or column.   The others (in ordinary print) arise *just twice*: they must be placed in a corner of the magic square.

(5) Start with any number in heavy print (for instance **1**) and set it $= x_{12}$. One of the numbers in ordinary print in the same representation of 15 (6 or 8 in our example) must be $= x_{11}$.   The first time, you choose between 4 alternatives, the second time between 2, and you have no further choice: in using the four representations of 15 collected under (3), you can proceed further in just one way.   You may obtain, for example, the magic square

6	1	8
7	5	3
2	9	4

and the $4 \times 2 = 8$ squares you can obtain are, in a sense, "congruent": you may derive all of them from any one of them by rotations and reflections. The square displayed shows the number 61 which arises in the solution of ex. 6.1.

6.12. (1) If the first digit of a number is ≥ 2, multiplication by 9 increases the number of digits. Therefore, the required number is of the form $1abc$.

(2) Moreover $1abc \times 9 = 9\ldots$, and so the number must be of the form $1ab9$.

(3) Therefore

$$(10^3 + 10^2a + 10b + 9)9 = 9\cdot 10^3 + 10^2b + 10a + 1$$
$$89a + 8 = b$$

Hence, $a = 0$, $b = 8$: the required number is $1089 = 33^2$.

6.13. (1) Since $ab \times ba$ yields a three-digit number, $a\cdot b < 10$. Assume $a < b$; then there are only ten possible cases:

$$a = 1, \quad 2 \leqq b \leqq 9; \qquad a = 2, \quad b = 3 \text{ or } 4$$

(2) $(10a + b)(10b + a) = 100c + 10d + c$

$$10(a^2 + b^2 - d) = 101(c - ab)$$

Hence $a^2 + b^2 - d$ is divisible by 101; but

$$-9 < a^2 + b^2 - d \leqq 82$$

Therefore, $a^2 + b^2 - d = 0$.

(3) $a^2 + b^2 = d \leqq 9$. Hence $b < 3$, and so $a = 1$, $b = 2$; hence $c = 2$, $d = 5$.

6.14. (Mathematical Log, vol. II, no. 2.) Let us try to find such a paradoxical pair of triangles.

(1) Among those five parts there can not be three sides: otherwise the triangles would be congruent and all six parts would be identical.

(2) And so the triangles agree in two sides and three angles. Yet, if they have the same angles, they are similar.

(3) Let a, b, c be the sides of the first triangle, and b, c, d the sides of the second triangle: if they are, in this order, corresponding sides in our two similar triangles we must have

$$\frac{a}{b} = \frac{b}{c} = \frac{c}{d}$$

That is, the sides form a *geometric progression*. This can be done; here is an example:

$$a, \quad b, \quad c, \quad d$$

are equal to

$$8, \quad 12, \quad 18, \quad 27$$

respectively. Observe that $8 + 12 > 18$ and that the triangles, with sides 8, 12, 18, and 12, 18, 27 respectively, are similar, since their sides are proportional, and so they have the same angles.

6.15. (Mathematical Log, vol. III, no. 2 and 3.)

(*a*) Find three integers x, y, and z such that
$$x + y + z = 9, \qquad 1 \leqq x < y < z$$

A systematic survey yields just three solutions (just three ways of splitting 9 dollars);

$$9 = 1 + 2 + 6$$
$$= 1 + 3 + 5$$
$$= 2 + 3 + 4$$

(*b*) Arrange these three rows in a square so that also each column has the same sum 9.

Essentially (that is, except for permutations of the rows and the columns) there is just one such arrangement (which we present in a neat symmetric form):

$$
\begin{array}{ccc}
6 & 2 & 1 \\
2 & 4 & 3 \\
1 & 3 & 5
\end{array}
$$

(*c*) Now bring into play the remaining "minor" clauses of the condition: Since 6 is the greatest number in the square, the first row is for Al and the first column for ice cream.   The only number in the square that equals twice the number in the intersection of the same row with the first column is 4; hence the second row is for Bill and the second column for sandwiches.   And so Chris spent for soda pop the number in the intersection of the last row with the last column, 5 dollars.

6.16. (Mathematical Log, vol. III, no. 2 and 3)

(*a*) The wife buys x presents for x cents each, and the husband y presents for y cents each: The problem requires
$$x^2 - y^2 = 75$$

Now $75 = 3 \times 5 \times 5$ has just six divisors:
$$(x - y)(x + y) = 1 \times 75 = 3 \times 25 = 5 \times 15$$

and so there are just three alternatives:

$x - y = 1$	$x - y = 3$	$x - y = 5$
or	or	
$x + y = 75$	$x + y = 25$	$x + y = 15$

which yield the table:

wife	husband
38	37
14	11
10	5

(*b*) Now bring into play the remaining "minor" clauses of the condition. They show *unambiguously*

Ann 38	37	Bill Brown
	14	11 Joe Jones
Betty 10	5	

and so Mary's last name *must* be Jones.

6.17. (Cf. Archimedes, vol. 12, 1960, p. 91.)   Obviously, the number of cases is restricted from the start (4! = 24).   Yet, if you are smart, you need not examine all these cases.

(*a*) Let

$$b, \qquad\qquad g, \qquad\qquad w, \qquad\qquad s,$$

stand for the number of bottles consumed by the wife of

Brown,   Green,   White,   Smith,

respectively.   Then

$$b + g + w + s = 14$$
$$b + 2g + 3w + 4s = 30$$

and so

$$g + 2w + 3s = 16$$

(*b*) As the last equation shows, either g and s are both odd or they are both even.   Hence, there are only 4 cases that need to be examined:

g	s	$w = 8 - (g + 3s)/2$
3	5	-1
5	3	1
2	4	1
4	2	3

Only the last case is admissible.   Therefore

$$s = 2, \qquad w = 3, \qquad g = 4, \qquad b = 5$$

and the ladies are

Ann Smith,   Betty White,   Carol Green,   Dorothy Brown.

6.18. The division of the condition into clauses is often useful in solving puzzles.   The reader may find appropriate examples in collections of mathematical puzzles, for instance in H. E. Dudeney, *Amusements in Mathematics* (Dover).   The *Otto Dunkel Memorial Problem Book* published as supplement to the *American Mathematical Monthly* **64** (1957) contains some suitable material; no. E776 on p. 61 deserves to be quoted as an exceptionally neat example of its type.

6.20. (*b*) sect. 6.2(4), $l = 5$.   (*c*) ex. 6.6, $n = 4$.   (*d*) sect. 6.4(1), $n = 4$. (*e*) sect. 2.5(3), sect. 6.4(2).

6.22. In this system of three equations each of the three unknowns x, y, and z plays exactly the same role: A cyclical permutation of x, y, and z interchanges the 3 equations, but leaves their system unchanged.   Therefore, *if* the unknowns are uniquely determined, we must have $x = y = z$; supposing this, we have immediately $6x = 30$, $x = y = z = 5$.

There remains to prove that the unknowns are uniquely determined; this can be shown by some usual procedure to solve a system of linear equations.

6.23.
$$y + z = a$$
$$x \quad\ \ + z = b$$
$$x + y \quad\ \ = c$$

Any permutation of x, y, and z leaves the system of the left-hand sides unchanged. Set $x + y + z = s$ (which also remains unchanged by a permutation of x, y, and z); adding the three equations we easily find

$$s = (a + b + c)/2$$

and the system reduces to three equations, each containing just one unknown

$$s - x = a, \qquad s - y = b, \qquad s - z = c$$

The whole system (not only the left-hand sides) is symmetric with respect to the pairs (x, a), (y, b), (z, c).

6.24. (Stanford 1958.) Set

$$x + y + u + v = s$$

and see ex. 6.23.

No solution: **6.10, 6.19, 6.21, 6.25.**

APPENDIX

Hints to Teachers, and to Teachers of Teachers

Teachers who wish to make use of this book in their profession should not neglect the hints addressed to all readers, but they should pay attention also to the following remarks.

1. As explained in the Preface, this book is designed to give opportunity for *creative work on an appropriate level* to prospective high school mathematics teachers (also to teachers already in service). Such opportunity is desirable, I think: a teacher who has had no personal experience of some sort of creative work can scarcely expect to be able to inspire, to lead, to help, or even to recognize the creative activity of his students.

The average teacher cannot be expected to do research on some very advanced subject. Yet the solution of a nonroutine mathematical problem is genuine creative work. The problems proposed in this book (which are not marked with a dagger) do not require much knowledge beyond the high school level, but they do require some degree, and sometimes a high degree, of concentration and judgment. The solution of problems of this kind is, in my opinion, the kind of creative work that ought to be introduced into the high school mathematics teachers' curriculum. In fact, in solving this kind of problems, the prospective teacher has an opportunity to acquire *thorough knowledge of high school mathematics*—real knowledge, ready to use, not acquired by mere memorizing but by applying it to interesting problems. Then, which is even more important, he may acquire some *know-how*, some skill in handling high school mathematics, some insight into the essentials of problem-solving. All this will enable him to lead, and to judge, his students' work more efficiently.

2. The present volume contains only the first half of a full course. Especially, teaching methods are only implicitly suggested by this volume; they will be explicitly discussed in a chapter of the second volume.

Yet this first volume contains much problem material that could be used in certain (especially, in more advanced) high school courses. I propose as a useful exercise to teachers to *reflect on possible classroom use* of the problems they are doing.

The best time for such reflection may be when the solution has been obtained and well digested. Then you *look back* at your problem and ask yourself: "Where could I use this problem? How much previous knowledge is needed? Which other problems should be treated first to prepare the class for this one? How could I present this problem? How could I present it (be specific) to such and such a class—or how could I present it to Jimmy Jones?" All these questions are good questions and there are many other good questions—but the best question is the one that comes spontaneously to your mind.

3. Although this first volume does not present a full course, it contains enough material to serve as textbook for a *Seminar in Problem Solving*. I conducted such seminars in various Institutes for teachers; several colleagues interested in starting such a seminar asked me for my materials; I know of a few colleges where such seminars, or similar classes, have been actually offered lately; and it is, I think, highly desirable that many more colleges should start experimenting with such seminars. It is in view of this situation that I have decided to publish this first volume before the second, in spite of the obvious risks of such an incomplete publication.

4. After some trials, I worked out a procedure for my seminar, a description of which at this place may be useful.[1]

Typical problems, which indicate a useful pattern, are solved in class discussion led by the instructor; the text of the first four chapters reproduces (as closely as it can be done in print) such class discussions. Then the discussion leads to recognizing and formulating the pattern involved—the text of the chapters quoted shows also how this is done.

The homework of the students consists of problems (such as the problems printed here following each chapter) which offer an opportunity to apply, to clarify, and to amplify the pattern obtained (and also the methodical remarks made) in class.

5. I used my seminar (and this is an essential feature of it) to give the participants some practice in explaining problems and guiding their solu-

---

[1] Some of the following will be, and a few foregoing sentences have been, extracted from an article in *The Journal of Education, Vancouver and Victoria*; see the Bibliography.

tion, in fact, some opportunity for *practice teaching*, for which in most of the usual curricula there is not enough opportunity.

When the homework is returned, this or that point (a more original solution, a more touchy problem) is presented to the class on the blackboard by one of the participants who did that point particularly well, or particularly badly. Later, when the class has become more familiar with the style of the performance, a participant takes for a while the instructor's place in leading the class discussion. Yet the best practice is offered in *group work*. This is done in three steps.

First, at the beginning of a certain practice session, each participant receives a different problem (each just one problem) which he is supposed to solve in that session; he is not supposed to communicate with his comrades, but he may receive some help from the instructor.

Then, between this session and the next, each participant should check, complete, review, and, if possible, simplify his solution, look out for some other approach to the result and, by these means and any other means, master the problem as fully as he can. He should also do some planning for presenting his problem and its solution to a class. He is given opportunity to consult the instructor about any of the above points.

Finally, in the next practice session, the participants form *discussion groups*: each group consists of four members (there may be one odd group); the participants form these groups by mutual consent, without intervention of the instructor. One member of the group takes the role of the teacher, the other members act as students. The "teacher" presents his problem to the "students," tries to challenge their initiative and tries to guide them to the solution, in the same style as the instructor does it in class discussions. When the solution has been obtained, a short friendly criticism of the presentation follows. Then another member takes the role of the teacher and presents his problem, and the procedure is repeated until each member of the group has had his turn. Then the participants partially regroup (each of two neighboring groups may send a member as "teacher" to the other group) so that each participant has occasion to polish his performance in presenting his problem several times. Some particularly interesting problems or particularly good presentations are shown to, and afterwards discussed by, the whole class. Congenial groups may spontaneously undertake the discussion of problems which are new to all participants; this should be encouraged, of course.

Such problem solving by discussion groups became quickly very popular in my classes, and I have the impression that the seminars as a whole were a success. Many of the participants were experienced teachers and several of them felt that their participation suggested to them useful ideas about conducting their own classes.

6. This volume may help the college instructor who conducts a Seminar in Problem Solving (especially when he conducts it for the first time). He may follow the procedure just described (in sect. 4 and 5). In class discussions, he may use the text of any one of the four first chapters. The problems printed at the end of the chapter are suitable for homework: serious work may be needed to expand the sketch of a solution given at the end of the volume into a fully presented solution. (Yet the instructor cannot choose at random: he should have a good look at the problem, at its solution, and also at the problems surrounding it, before assigning it.) For examinations and term papers, the instructor may wish to avoid problems printed in this volume; then he may consult appropriate textbooks (and also ex. 1.50, 2.78, and 3.92). For group work (see sect. 5) the problems should be harder, but they need not be closely connected with the chapters treated; suitable problems may be selected from this book, also from later chapters.

Chapters 5 and 6 may be also discussed, or assigned for reading. The role of these chapters, however, can be better explained in the second volume.

Of course, after having gained some experience, the instructor may adopt the spirit of this book without following its details too closely.

CHAPTER 7

GEOMETRIC REPRESENTATION OF THE PROGRESS OF THE SOLUTION

It is very helpful to represent these things in this fashion since nothing enters the mind more readily than geometric figures.
DESCARTES: *Œuvres,* vol. X, p. 413; Rules for the Direction of the Mind, Rule XII.

7.1. Metaphors.

It happened about fifty years ago when I was a student; I had to explain an elementary problem of solid geometry to a boy whom I was preparing for an examination, but I lost the thread and got stuck. I could have kicked myself that I failed in such a simple task, and sat down the next evening to work through the solution so thoroughly that I shall never again forget it. Trying to see intuitively the natural progress of the solution and the concatenation of the essential ideas involved, I arrived eventually at a geometric representation of the problem-solving process. This was my first discovery, and the beginning of my lifelong interest, in problem solving.

I was guided to the geometric image that finally emerged by a group of usual metaphorical expressions. That the language is full of metaphors (dead, half-dead, and live) has been observed often enough. I do not know whether it has been also observed that many of these metaphors are interdependent: they are connected, they are associated, they form clusters, more or less loose, more or less overlapping families. At any rate, there is an extensive family of metaphorical expressions which have two things in common: they are all concerned with that basic human activity of problem solving and they are all suggesting the same geometric configurations.

1

Discovering the solution is finding a *connection* between formerly separated things or ideas (the things we have and the things we want, the data and the unknown, the hypothesis and the conclusion). The farther apart the things connected stood originally, the more credit is due to the discoverer for connecting them. We sometimes see the connection under the guise of a *bridge:* a great discovery strikes us as bridging over a deep chasm between two widely separated ideas. We often see the connection realized by a *chain:* a proof appears as a *concatenation* of arguments, as a chain, perhaps a long chain, of conclusions. The chain is not stronger than its weakest *link,* and there is no valid proof, no uninterrupted chain of reasoning, if even one link is missing. More often still we use a *thread* for mental connecting. We have all listened to the professor who lost the thread, or got himself entangled in the threads, of his argument. He was obliged to peek in his notes before resuming the thread of his lecture, and we were all tired when he gathered up the threads to a final conclusion. A slender thread becomes a mere geometric line, the things connected become mere geometric points, and there emerges, almost unavoidably, a diagrammatic, graphic image of the concatenation of mathematical conclusions.

Let us look now at some geometric figures instead of merely listening to figures of speech.

7.2. What is the problem?

We need an example, and I choose a very simple problem of solid geometry:[1]

Find the volume F of the frustum of a right pyramid with square base. Given the altitude h of the frustum, the length a of a side of its upper base, and the length b of a side of its lower base.

(A pyramid with square base is a *right pyramid* if its height meets the base at the center. The *frustum* of a pyramid is the portion of the pyramid between its base and a plane parallel to its base; this parallel plane contains a face of the frustum which we call the *upper base* of the frustum; its *lower base* is the base of the original full pyramid, its *altitude* is the perpendicular distance between its bases.)

The first step in solving the problem is to concentrate on its aim. *What do you want?* we ask ourselves and we represent as sharply as we can the shape of which we wish to find the volume *F* (see the left side of Fig. 7.1). The mental situation is appropriately represented by a single point, labeled *F*, on which our whole attention should be focused (see the right-hand side of Fig. 7.1).

---

[1] Very similar to, but even simpler than, the problem originally considered. See the Bibliography, papers 1 and 3 of the author.

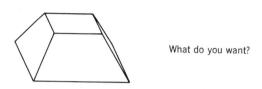

F
o

What do you want?

Fig. 7.1. Concentrate on one point: the aim.

Yet we cannot find the unknown F if nothing is given. *What are the data?* we ask ourselves, or *What have you?* and our attention emphasizes those lines of the figure of which the length is given, a, b, and h, see the left-hand side of Fig. 7.2. (The square with side a is above, the square with side b below, the solid considered.) To portray the changed mental situation, three new points emerge, labeled a, h, and b, representing the data and separated from the unknown F by a gap, an *open space* in Fig. 7.2. This open space symbolizes the open question; our problem aims at connecting the unknown F with the data a, h, and b, we have to bridge the gap.

7.3. That's an idea!

We start working at our problem by visualizing its aim, its unknown, its data. This initial phase of our work is fittingly pictured by Figs.7.1 and 7.2. Yet how should we proceed from here, which course should we adopt?

If you cannot solve the proposed problem, look around for an appropriate related problem.

In our case, we do not have to look far. In fact, *what is the unknown?* The volume of the frustum of a pyramid. And what is such a frustum? How is it *defined?* As a portion of a full pyramid. Which portion? The portion between—No, enough of that; let us *restate it differently:*

F o

What have you?

o   o   o
a h b

Fig. 7.2. The open question: a gap to bridge over.

The frustum is the portion that remains when we cut off from the full pyramid a smaller pyramid by a plane parallel to the base. In our case, see Fig. 7.3, the base of the big (full) pyramid is a square of area b^2. *If* we knew the volumes of these two pyramids, say B and A, respectively, we could find the volume of the frustum:

$$F = B - A$$

Let us find these volumes B and A. That's an idea!

Thus we have reduced our original problem, to find F, to two appropriately related auxiliary problems, to find A and B. To express this reduction graphically, we introduce two new points, labeled A and B, into the space between the data a, h, b and the unknown F. We join both A and B to F with slanting lines and thereby indicate the essential relation between these three quantities: starting from A and B we can reach F; the solution of the problem about F can be based on the solution of the two problems about A and B.

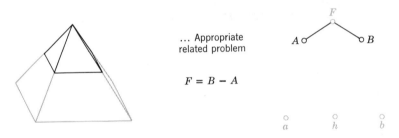

... Appropriate related problem

$F = B - A$

Fig. 7.3. If you cannot solve the proposed problem, look around for an . . . .

Our work is not yet finished; we have two new unknowns to find, A and B; there are two pendent points separated from the data by a gap in Fig. 7.3. The situation looks hopeful, however; the pyramid is a more familiar figure than the frustum, and although we have two unknowns, A and B, to find instead of one, F, the two are of similar nature and similarly related to the data a and b, respectively. Correspondingly, the graphical representation of the mental situation in Fig. 7.3 is symmetrical. The line FA is inclined toward the given a, FB toward the given b. We have started bridging the open space between the original unknown and the data; the remaining gap is narrower.

7.4. Developing the idea

Where are we now? *What do you want?* We want to find the unknowns A and B. *What is the unknown A?* The volume of a pyramid.

How can you get this kind of thing? How can you find this kind of unknown? From what data can you derive this kind of unknown? The volume of the pyramid can be computed if we have two data, the area of the base and the height of the pyramid; the volume is, in fact, the product of these two quantities divided by 3. The height is not given, but we can still consider it. Let us call it x. Then

$$A = \frac{a^2x}{3}$$

On the left-hand side of Fig. 7.4 the small pyramid above the frustum appears in more detail; its height x is emphasized. The present stage of our work is graphically represented on the right-hand side of Fig. 7.4; a new point x appears above the data, and slanting lines join A to x and a, indicating that A can be reached from x and a, that A can be expressed in terms of x and a. Although there are still two unknowns to find (there are still loose ends dangling in mid air in Fig. 7.4) progress

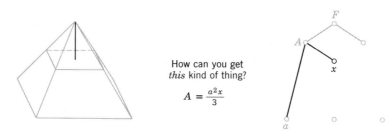

How can you get
this kind of thing?

$$A = \frac{a^2x}{3}$$

Fig. 7.4. A first connection with the data, but loose ends dangling in midair.

has been made; we have succeeded in connecting the unknown F at least with one of the data, with a.

At any rate the next step is obvious. The unknowns A and B are of similar nature (they are symmetrically represented in Fig. 7.3); we have expressed the volume A in terms of base and height, and we can express the volume B analogously:

$$B = \frac{b^2(x + h)}{3}$$

In the left-hand portion of Fig. 7.5, the big pyramid containing the frustum appears in more detail, its height $x + h$ is emphasized. In the right-hand portion of Fig. 7.5 three new slanting lines appear, joining B to b, h, and x. These lines indicate that B can be attained from b, h, and x, that B can be expressed in terms of b, h, and x. And so just one

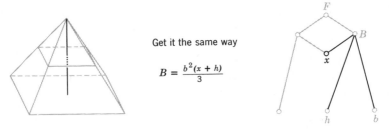

Fig. 7.5. Just one question remains pending.

pending point—the point x—remains unconnected with the data. The gap is narrowed down; it is now between x and the data.

What is the remaining unknown? It is x, the length of a line. *How can you find this kind of unknown?* How can you get this kind of thing?

The most usual thing is to obtain the length of a line from a triangle—from a right triangle, if possible—or from a pair of similar triangles. Yet there is no usable triangle in the figure, and there should be one with a side x. Such a triangle would lie in a plane passing through the altitude of the small pyramid with volume A; this plane would also pass through the altitude of the big pyramid with volume B, which is similar to the small one. Yes, similar triangles in a plane passing through the altitude—and *parallel to a given side* of the base of one of these pyramids! That's it! That finishes it!

A pair of similar triangles appears in Fig. 7.6 from which x can be conveniently computed by the proportion

$$\frac{x}{x+h} = \frac{a}{b}$$

But details are not important at this stage; what is all important now is that x can be expressed in terms of the three data a, h, and b. The three new slanting lines emerging in the right-hand portion of Fig. 7.6 indicate just that by joining x to a, h, and b.

Fig. 7.6. We have succeeded in bridging the gap.

Done! We have succeeded in bridging the gap, in establishing an uninterrupted connection between the unknown F and the data a, h, and b through the intermediaries (auxiliary unknowns) A, B, and x.

7.5. Carrying it out

Is the problem solved? Not yet—not quite. We are required to express the volume F of the frustum in terms of the data, a, h, and b, and this is not yet done. Yet the more important and more exciting first part of our work is behind us; the remaining task is a much calmer affair, much more straightforward.

There was an element of adventure in the first part of our work. At each stage, we *hoped* that the next step would bring us nearer to our goal, to bridging the gap. Yes, we hoped, but we were not certain; at each stage we had to *invent,* and risk, the next step. But now no more invention or risk is needed; we foresee that we can safely reach the unknown F from the data a, h, and b just by following the threads of the uninterrupted connection represented in Fig. 7.6.

We begin the second part of our work where we have ended the first part. We tackle first the auxiliary unknown x that we have introduced last; from the last equation of sect. 7.4 we obtain

$$x = \frac{ah}{b - a}$$

Then we substitute this value for x in the two foregoing equations of sect. 7.4, obtaining

$$A = \frac{a^3 h}{3(b - a)}, \qquad B = \frac{b^3 h}{3(b - a)}$$

(The analogy between these two results is comforting.) We use last the equation that we obtained first, in sect. 7.3:

$$F = B - A = \frac{b^3 - a^3}{b - a} \frac{h}{3}$$

$$F = \frac{a^2 + ab + b^2}{3} h$$

This is the desired expression.

The work of the present section is aptly symbolized by Fig. 7.7 where each connecting line carries an arrow indicating the direction in which we have just used the connection. We started from the data a, h, and b, and proceeded hence through the intermediate auxiliary unknowns x, A, and B to our original, principal unknown F, expressing these quantities one after the other in terms of the data.

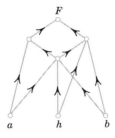

Fig. 7.7. Working from the data to the unknown.

7.6. A slow-motion picture

Figures 7.1 to 7.7 show successive stages of the solution; we assemble these seven figures into one composite picture, Fig. 7.8. (Figure 7.7 is merged into Fig. 7.6. Fig. 7.8 is reproduced in black and red on the endpapers immediately inside the cover at the beginning and the end of the book; the details on which the attention is focused are emphasized by red—red meaning, of course, "burning interest.") View the sequence in Fig. 7.8 from left to right. Viewing it in fast succession, we have a kind of cinematographic representation of the problem-solvers progress, of the march of discovery. Or, viewing Fig. 7.8 at low speed, we have a sort of slow-motion picture of the evolving solution which leaves us time for careful observation of significant details.

In Fig. 7.8, each stage of the solution (each mental situation of the problem-solver) is represented at several levels; the items pertaining to the same situation are vertically arranged, one under the other. Thus the solution progresses along four parallel lines, on four different levels.

On the uppermost level, the *image level*, we see the evolution of the investigated geometric figure in the problem solver's mind. At each stage, the problem solver has a mental picture of the geometric figure he explores, but this picture changes in transition to the next stage; some details may recede into the background, other details come to our attention, new details are added.

Going down to the next layer, we attain the *relational* level. In graphical representation, the objects considered (unknown, data, auxiliary unknowns) are symbolized as points, and the relations connecting the objects are indicated by lines connecting the symbolizing points.

Immediately below the relational level we find the *mathematical* level, which consists of formulas and which contrasts with the relational level. On the relational level, the system of all relations obtained up to the moment in question is exhibited; the last obtained relation is emphasized by color (it is in the focus of attention) but it is not exhibited with more details than the preceding. On the mathematical

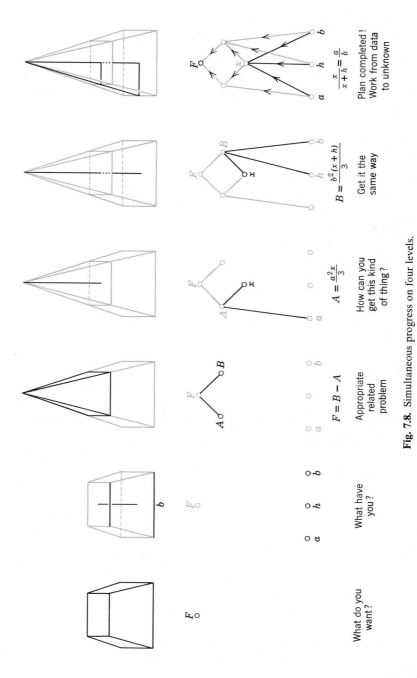

Fig. 7.8. Simultaneous progress on four levels.

9

level, the last obtained relation is fully shown and the preceding relations are not shown at all.

Going down still further we attain what is for us the basic layer, the *heuristic* level. At each stage, it exhibits a simple, natural (generally applicable!) question or suggestion which has enabled us to reach that stage. Our basic concern is to study the nature of such questions and suggestions.

7.7. A preview

We wish to view and review Fig. 7.8; we wish to compare it and correlate it with our past experience. Thus behind the story this figure tells, we wish to perceive points of general interest which deserve further study. In the graphical narrative of the solution of a single problem, we shall find useful preindications about the general questions treated in the following chapters. We shall survey these chapters one by one, in due order.

(Each of the following sections of the present chapter previews a following chapter and has the same title and a corresponding number as the chapter that is previews.)

And now let us try to dig down through the particular story to the underlying general ideas.

7.8. Plans and programs

In viewing the stages of Fig. 7.8 in sequence, we see how the problem solver's attention ranges over the geometric figure he explores, how he takes possession of more and more details of this figure, and how he builds up step by step the system of connections which constitutes the plan of his solution. In viewing the unfolding of the solution attentively, we may discern in it several phases and activities.

We have already observed (in sect. 7.5) the contrast between the two portions of the solution: in the first portion (illustrated by Figs. 7.1 to 7.6) we work downward, from the unknown to the data, in the second portion (illustrated by the single Fig. 7.7) we work upward, from the data to the unknown.

Yet even in the first portion we may distinguish two phases. In the initial phase, Figs. 7.1 and 7.2, the problem solver's main effort is aimed at understanding his problem. In the latter phase, Figs. 7.3 to 7.6, he develops the system of logical connections, he constructs a *plan* of the solution.

This latter phase, the devising of a plan, seems to be the most essential part of the problem solver's work; we shall examine it more closely in chapter 8.

7.9. Problems within problems

In surveying Fig. 7.8, we may observe that the problem solver, in working at his original (proposed, main) problem, encountered several *auxiliary problems* ("helping" problems, subproblems). In seeking the volume of the frustum, he was led to seek the volume of a full pyramid, then that of another full pyramid, then the length of a line. To arrive at his original unknown F, he had to pass through the auxiliary unknowns A, B, and x. Just a little experience in solving mathematical problems may be enough to convince us that such breaking up of the proposed problem into several subproblems is typical (see sect. 2.5(3), for instance).

We shall thoroughly examine the role of auxiliary problems and distinguish various kinds of auxiliary problems.

7.10. The coming of the idea

Which of the various steps of the solution illustrated by Fig. 7.8 is the most important? The popping up of the full pyramid (Fig. 7.3) I think—and I think too that most people who have had some experience in these matters and have devoted some thought to their experience will agree. Introducing the full pyramid and conceiving the frustum as the difference of two full pyramids is the *decisive idea* of the solution; for most problem solvers the rest of the solution will be easier, more obvious, more routine—it may be almost completely routine for the more experienced problem solvers.

The emergence of the decisive idea is not very impressive in the present case, but we should not forget that the problem we are discussing is very simple. Conceiving the decisive idea, seeing a sudden light after a long period of tension and hesitation may be very impressive; it may be a great experience which the reader should not miss.

7.11. The working of the mind

In the graphical representation of Fig. 7.8, the most conspicuous sign of progress is the introduction of more and more detail. As the problem solver advances, more and more lines appear both in the geometric figure and in the relational diagram. Behind the increasing complexity of the figure we should perceive a growing structure in the problem solver's mind. At each significant step he brings in some relevant knowledge; he recognizes some familiar configuration, he applies some known theorem. Thus the work of the problem solver's mind appears as recalling relevant elements of his experience and connecting them with the problem at hand, a work of *mobilization* and *organization*.

We have already had the opportunity in the foregoing to make such remarks (especially in ex. 2.74), and we shall have the opportunity to explore this point as well as other aspects of the working of the mind.

7.12. The discipline of the mind

Figure 7.8, which represents the progress of the solution on four different levels, gives some indication about the problem solver's work. We certainly want to know how he is working, but we are even more anxious to know how he ought to be working. Can Fig. 7.8 give some indication of this?

On the lowest level of Fig. 7.8 there is a sequence of questions and suggestions motivating the successive steps of the problem solver's work. These questions and suggestions are simple, natural, and very general; they have guided the problem solver to the solution of the simple problem we have chosen as example, and they may guide us in innumerable other cases. If there is a discipline of the mind (some system of guidelines, some body of maxims or rules in the direction of the universal method sought by Descartes and Leibnitz) there is a good chance that the questions and suggestions aligned in the basic layer of Fig. 7.8 somehow belong to it. We must thoroughly examine this point.

Examples and Comments on Chapter 7

7.1. *Another approach* to the problem stated in sect. 7.2. The base of the frustum is in a horizontal plane (on the desk). By four vertical planes passing through the four sides of the top of the frustum (see Fig. 7.9) we divide it into 9 polyhedrons:

> a prism with square base of volume Q,
> four prisms with triangular base of volume T each,
> four pyramids with square base of volume P each.

Compute F, following the approach indicated by Fig. 7.10.

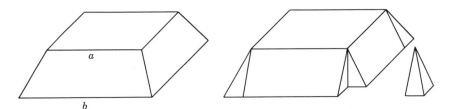

Fig. 7.9. By four vertical planes.

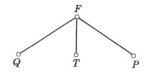

Fig. 7.10. Another approach.

7.2. Express the two solutions of the same problem given in sect. 2.5(3) and 2.5(4) diagrammatically (the quantities involved by points, the connecting relations by connecting lines).

7.3. *The search for a proof.* Proposition 4 in the Eleventh Book of Euclid's Elements can be stated as follows:

If a straight line passes through the point of intersection of two given straight lines and is perpendicular to both, then it is also perpendicular to any third straight line that lies in the plane of the two given lines and passes through their point of intersection.

We wish to analyze a proof of this proposition, visualize its structure, and understand the motives of its discovery in using the geometric representation of the problem solver's progress developed in this chapter. In relying on the analogy with the discussion illustrated by Figs. 7.1 to 7.8, we shall deal with the present case somewhat more concisely.

We are here principally concerned with the formation of the proof in the problem solver's mind. But the proposition to be proved is interesting too; it states a basic fact of solid geometry. Even the logical form of the proposition is interesting. The teacher who said "Two bad boys spoil the whole class" was possibly wrong, but his statement has the same form as Euclid's proposition we are about to prove.

(1) *Working backward.* We draw Fig. 7.11, introduce suitable notation, and then we restate the proposition we intend to prove in standard form, split into hypothesis and conclusion:

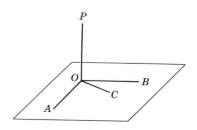

Fig. 7.11. Two bad boys spoil the whole class.

Hypothesis: The three lines OA, OB, and OC pass through the same point O and are in the same plane, but they are different; and

$$PO \perp OA \qquad PO \perp OB$$

Conclusion: Then also

$$PO \perp OC$$

"What is the conclusion?"

That the line PO is perpendicular to OC. Or that $\angle POC$ is a right angle. "What is a right angle? *How is it defined?"*

A right angle is one that is equal to its supplement. There could be some advantage in restating the conclusion in this sense. Produce the line PO to P' (so that P, O, and P' are collinear, and O and the plane containing O lie between P and P'). Then (see Fig. 7.12*a*) the desired conclusion is

$$\angle POC = \angle P'OC$$

"Why do you think that this form of the conclusion is more advantageous?"

We often prove the equality of angles from congruent triangles. In the present case we could prove the desired conclusion if we knew that

$$\triangle POC \cong \triangle P'OC$$

(see Fig. 7.12*b*). To prove this, however, we should, can, and shall suppose P' so constructed that

$$PO = P'O$$

In fact, what do we need to prove that those two triangles are congruent? We know two pairs of equal sides, $PO = P'O$ by construction, and also

$$OC = OC$$

obviously. To finish the proof it would be enough to know (see Fig. 7.12*c*) that

$$PC = P'C$$

In the foregoing, we have worked starting from the desired conclusion toward the given hypothesis; we have worked *backward*. We went a good distance working backward from the conclusion, although the continuation of the path that should lead us to the hypothesis is still in the clouds. Our work is symbolized by Fig. 7.13, which shows graphically which statements we can conclude from which other statements—as Figs. 7.1 to 7.8 show which quantities we can compute from which other quantities. In Fig. 7.13 each of the foregoing equations (congruences) is represented by its left-hand side:

$$\angle POC = \angle P'OC \qquad \text{by } \angle POC$$
$$\triangle POC \cong \triangle P'OC \qquad \text{by } \triangle POC$$
$$OC = OC \qquad \text{by } OC$$

and so on. It is, in fact, enough to write the left-hand side, because we can derive from it the right-hand side by substituting P' for P, that is, by passing from the space above the plane through A, B, C, and O to the space below it.

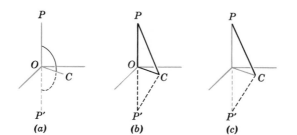

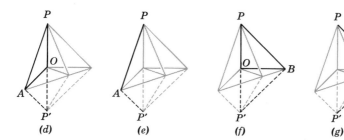

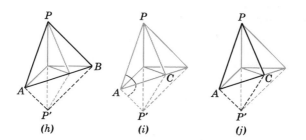

Fig. 7.12. The changing aspect of the geometric figure.

(2) *Restating the problem.* Having worked some time at the conclusion, we should now turn our attention to the hypothesis of the proposition we wish to prove.

"What is the hypothesis?"

We should restate the hypothesis so that it harmonizes with the restated conclusion; we should bring hypothesis and conclusion nearer to each other and not pull them apart. We have to prove (the restated conclusion) that

$$\angle POC = \angle P'OC$$

in supposing (let us restate analogously the hypothesis) that

$$\angle POA = \angle P'OA \quad \text{and} \quad \angle POB = \angle P'OB$$

The whole proposition sounds quite good; it appears homogeneous. Yet we have still to add to the hypothesis an essential clause which says that the three differ-

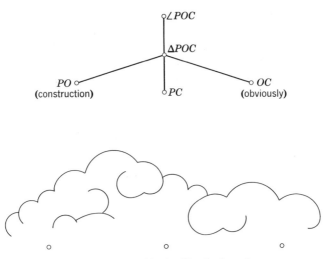

Fig. 7.13. Working backward.

ent lines OA, OB, and OC are in the same plane. Also this clause should be some-how related to the conclusion. But how?

In fact, an essential idea is needed here to observe that we are allowed to place the points A, B, and C along a straight line and that it may be advantageous to place them so. (Any straight line that does not pass through O and is not parallel to any one of the three given lines OA, OB, and OC may be chosen.) Thus we arrive at a reformulation of the proposition we wish to prove.

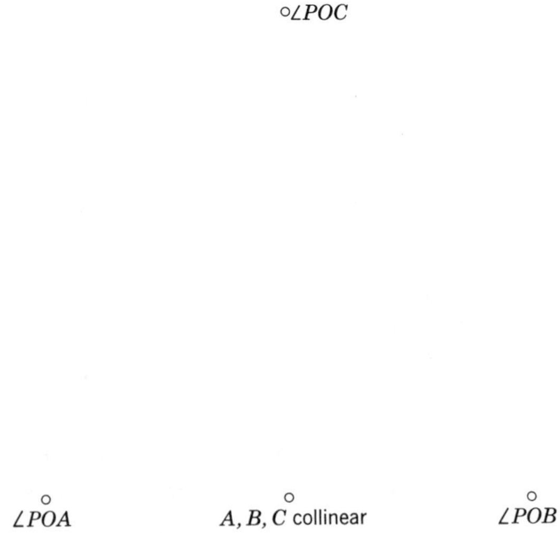

○$\angle POC$

○   ○   ○
$\angle POA$ A, B, C collinear $\angle POB$

Fig. 7.14. The gap between the hypothesis and the conclusion.

Hypothesis: The points A, B, and C are collinear, and the line passing through them does not contain the point O. Moreover

$$\angle POA = \angle P'OA \qquad \angle POB = \angle P'OB$$

Conclusion: Then

$$\angle POC = \angle P'OC$$

Figure 7.14 symbolizes this statement.

(3) *Working forward.* In working at the hypothesis we consider the same kind of relations as we have considered in working at the conclusion, but we consider them in reverse order.

Since

$$\begin{aligned} \angle POA &= \angle P'OA && \text{by hypothesis} \\ PO &= P'O && \text{by construction} \\ OA &= OA && \text{obviously} \end{aligned}$$

we conclude that

$$\triangle POA \cong \triangle P'OA$$

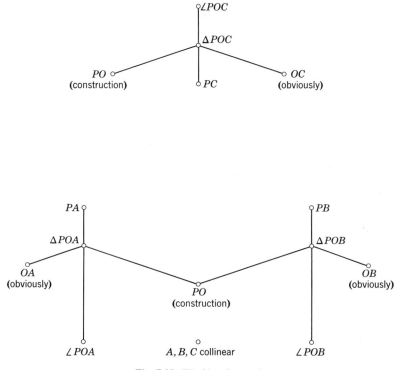

Fig. 7.15. Working forward.

(see Fig. 7.12*d*) and hence that

$$PA = P'A$$

(see Fig. 7.12*e*). An exactly parallel reasoning shows that

$$PB = P'B$$

(see Figs. 7.12*f* and 7.12*g*).

In the foregoing we have worked *forward;* that, is in the direction from the given hypothesis toward the desired conclusion.

Figure 7.15 exhibits the mental situation resulting from the work just done and from the work recorded by Fig. 7.13. As Fig. 7.15 shows, it remains to prove that $PC = P'C$ from the already proved similar statements $PA = P'A$ and $PB = P'B$ and the heretofore unused hypothesis that A, B, and C are collinear. Comparing the present situation with that represented by Fig. 7.14 we have some reason to be hopeful; the gap that we have to bridge became definitely less wide.

(4) *Working from both sides.* The remaining part of the proof may occur so fast to the problem solver (or to the reader) that the final conclusion may seem instantaneous. Yet let us record the details.

The desired relation

$$PC = P'C$$

(Fig. 7.12*c*) could be derived from congruent triangles. (This is working backward.) In fact, we can easily derive from the already established two relations

$$PA = P'A \qquad PB = P'B$$

and the self-evident

$$AB = AB$$

the congruence (Fig. 7.12*h*)

$$\triangle PAB = \triangle P'AB$$

(We have worked forward.) These are, however, not the triangles we need. To derive $PC = P'C$ (which would finish the proof) we should know, for example, that

$$\triangle PAC = \triangle P'AC$$

We could conclude this (now, we are working backward) from the already established

$$PA = P'A$$

and the self-evident

$$AC = AC$$

if we only knew that

$$\angle PAC = \angle P'AC$$

Now, in fact, we know that

$$\angle PAB = \angle P'AB$$

(see Fig. 7.12*i*) from the already established congruence of the triangles $\triangle PAB$ and $\triangle P'AB$. (Figure 7.16 expresses the mental situation at this instant.) Yet as A, B, and C are collinear, by hypothesis,

$$\angle PAB = \angle PAC \qquad \text{and} \qquad \angle P'AB = \angle P'AC$$

And with this remark we close the last gap (see Fig. 7.17; review the whole Fig. 7.12).

The last step, the transition from Fig. 7.16 to Fig. 7.17 deserves particular attention; only this last step uses the vitally important clause of the hypothesis that A, B, and C are collinear.

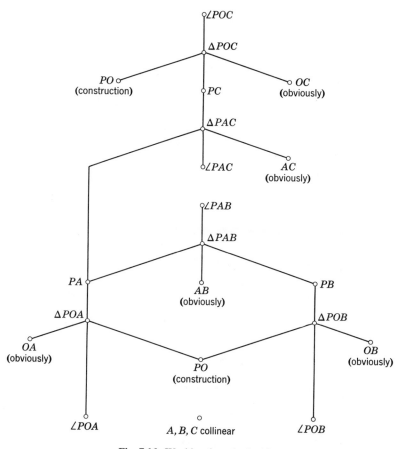

Fig. 7.16. Working from both sides.

7.4. *Elementary diagrams.* In sect. 7.2 to 7.6 we have studied a problem to find, in the foregoing ex. 7.3, a problem to prove. In both cases we have used diagrams consisting of points and connecting lines to illustrate the progress and the structure of the solution. Comparing the two cases, we wish to clarify the meaning of these diagrams.

Let us consider an "elementary diagram," as in Fig. 7.18. This diagram consists of $n + 1$ points of which one, A, is on a higher level than the n others, B, C, D, ..., and L. The higher placed point A is joined by a line going downward to each of the other n points. Such elementary diagrams are the "bricks" of which the diagrams encountered in Fig. 7.3 to 7.8 and Fig. 7.13 to 7.17 are built up. What does such an elementary diagram express?

When the elementary diagram belongs to a problem to find such as the one considered in Fig. 7.3 to 7.8, the points A, B, C, D, ..., and L represent *quantities*

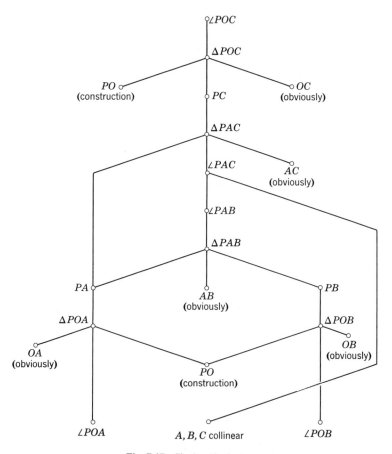

Fig. 7.17. Closing the last gap.

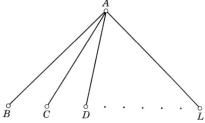

Fig. 7.18. We could have *A* if we had *B* and *C* and *D* . . . and *L*.

(lengths, volumes, . . .) and they represent *statements* when the diagram belongs to a problem to prove as in Fig. 7.13 to 7.17. In the former case, Fig. 7.18 expresses that *we can compute the quantity A if we are given the quantities B, C, D, . . . , and L.* In the latter case, Fig. 7.18 expresses that *we can conclude the statement A from the statements B, C, D, . . . , and L.* In other words, the elementary diagram expresses in the first case that the quantity *A* is a *known function* of the quantities *B, C, D, . . . ,* and *L*, and in the second case that the statement *A* is a *consequence* of the statements *B, C, D, . . . ,* and *L*. We can also say that the elementary diagram Fig. 7.18 answers a question, in one case the question: "From what *data* can we compute *A*?" and in the other the question: "From what *premisses* can we conclude *A*?"

And so we can foresee how we could use such diagrams to illustrate the solution of any kind of problem. In a practical problem the points *A, B, C, D,* and *L* may represent things we have or we wish to acquire. Figure 7.18 shows that we could acquire *A* if we had *B* and *C* and *D* . . . and *L*, or that the means *B, C, D, . . . ,* and *L* suffice jointly to attain the end *A*. The diagram answers the question: "What should I get first if I wish to get *A*?"

7.5. *More problems.* Although we can try to represent diagrammatically the solution of all sorts of problems (see ex. 7.4), the representation can become constrained, unnatural. Seek problems with solutions easily represented and instructively clarified by the diagram.

CHAPTER 8

PLANS AND PROGRAMS

From desire ariseth the thought of some means we have
seen produce the like of that which we aim at; and from
the thought of that, the thought of means to that mean;
and so continually, till we come to some beginning within
our own power.

THOMAS HOBBES: *Leviathan,* Chapter III.

8.1. A pattern of planning

Hobbes' words prefixed to this chapter describe a basically important pattern—the pattern of a procedure of problem solving—with admirable concision and precision. Let us read between the lines, let us try to see the full scope of the procedure, the variety of cases to which it is applicable.

We have a problem. That is we have an aim A which we cannot immediately attain and we are searching for some appropriate action to attain it. This aim A may be practical or theoretical, perhaps mathematical—a mathematical object (number, triangle, ...) which we wish to find (compute, construct, ...) or a proposition which we wish to prove. At any rate, we desire to attain our aim A.

"From desire ariseth the thought of some means"—this is well-observed mental behavior. The end suggests the means; a wish is usually quickly followed by the thought of some action that could lead to its fulfillment. I think of some object I would like to have and soon after I remember a shop where I could buy it.

Yet let us return to Hobbes' text: "From desire ariseth the thought of some means" B that could produce the desired A. This thought probably originates from some previous experience: "We have seen B produce the like of A that we aim at." At any rate, we think that we could obtain A if we had B. "And from the thought of B the thought arises of some means, say C, to that mean B"; we could obtain B if we had C. "And so on continually"—we could obtain C if we had D—"till we

22

come to some beginning within our own power"; we could obtain D if we had E—but we do have E! This E ends our train of thought; E is in our possession, "within our own power," is given, is known.

Our train of thought ran through several ifs—"this if that," we could obtain this if we had that. In fact, we have considered

$$A \text{ if } B, B \text{ if } C, C \text{ if } D, D \text{ if } E$$

and we have stopped at E since we have E unconditionally, without any further if.

(It is almost unnecessary to mention that the number of ifs, or steps, is unessential; the four steps and five "targets" or "objectives" arising in our example must be considered as representing n steps and $n + 1$ objectives.)

The foregoing was *planning*. It should be followed, of course, by the *execution* of the plan. Starting from E which is a "beginning within our own power" we should obtain D; having obtained D, we should proceed to C, from C to B, and finally from B to our desired aim A.

Observe that planning and execution proceed in opposite directions. In planning, we started from A (the aim, the unknown, the conclusion) and we ended by reaching E (the things we have, the data, the hypothesis.) Yet in carrying out the plan we worked from E to A so that A, the aim, is the first thing we thought of and the last thing we laid hands on. If we regard motion toward the goal as progress, we must regard the direction in which our planning has moved as *regressive*. Thus the important pattern of problem solving described by Hobbes can be appropriately called *regressive planning*, or *working backwards;* the Greek geometers called it *analysis* which means "backward solution." The complementary work of execution which proceeds from the things in our possession to the aim (from E to A in our case) is called, in contradistinction, *progressive*, or working forward, or *synthesis* which means "putting together."[1]

The reader should visualize this working backward in planning and working forward in executing the plan on some obvious example. "I could get that desirable object A in that shop if I paid for it that sum B; I could get that sum B if . . . ." I hope that the reader will succeed in devising a neat plan, and will not meet with disappointments in carrying it through.

8.2. A more general pattern

Let us confront the pattern developed in the foregoing section with the example carefully analyzed in chapter 7 and illustrated by Fig. 7.8.

---

[1] *Cf.* HSI, pp. 141–148, Pappus, and pp. 225–232, Working backwards.

The example unmistakably shows the general trend of the pattern: working backward from the unknown to the data in the planning phase, but working forward from the data to the unknown in executing the plan. Yet the details of the example do not fall under the pattern.

Let us look at the very first step. In the pattern of sect. 8.1, A is reduced to B, the primary aim is reduced to a secondary aim, attaining A depends on attaining B. In the example of Fig. 7.8, however, the computation of the unknown (the volume of the frustum) is reduced to the computation of *two* new unknowns (two volumes); there is not only one secondary aim, but there are two such aims.

If, however, we reconsider our example represented by Fig. 7.8 and various remarks made in chapter 7 on this graphical representation (see especially ex. 7.4), it should not be difficult for us to conceive a generalization of the pattern of sect. 8.1 that encompasses the case of Fig. 7.8 and along with it an unlimited variety of worthwhile cases.

Our aim is A. We cannot attain A immediately, but we notice that we could attain it if we had several things, B', B'', B''', ... Well, we do not have these things, but we start thinking how we could get them; that is, we set up B', B'', B''', ... as our secondary aims. We may perceive after some reflection that we could attain all our secondary aims B', B'', B''', ... if we had several other things C', C'', C''', ... In fact, we do not have these things (C', C'', C''', ...), but we can try to get them; we set them up as our tertiary aims, and so on. We are spinning the web of our plan. We may be obliged to say several times "We could have this, if we had that and that and that" till we finally reach solid ground, things we really have. The web of our plan consists of accessory aims, all subordinate to our primary aim A, and of their interconnections. There may be many subordinate aims, and the details of the web may be too complex for words, but they can be appropriately represented by the points and lines of a diagram of the kind we developed in chapter 7. (For instance, in sect. 2.5(3) our primary aim was D, our secondary aims were a, b, and c, and our tertiary aims p, q, and r. See also ex. 7.2.)

I think that the foregoing indicates clearly enough a general pattern which contains the pattern described in sect. 8.1 as a particular case, and which we shall call the *pattern of working backwards*. It is a pattern of planning; the planning starts from the aim (the thing we want, the unknown, the conclusion) and works backward toward the things "within our power" (the things we have, the data, the hypothesis.) It is part of the plan that when we shall have reached those things "within our power" we shall use them as a "beginning," and by retracing our steps we shall work forward toward the aim. *Cf.* ex. 9.2(3).

8.3. A program

Are the two numbers $\sqrt{3} + \sqrt{11}$ *and* $\sqrt{5} + \sqrt{8}$ *equal?* *If they are not, which one is greater?* (It is understood that all arising square roots are taken with the positive sign.)

With a little experience in algebraic manipulation, we may readily conceive a plan to handle this question; we may even conceive a plan so clear and determinate that it deserves a special term and may be called a *program.*

The two proposed numbers are either equal, or the first is greater, or the second is greater. There are three eligible relations between the two numbers, expressed by the signs $=$, $>$, and $<$, but just one of these three relations is actually valid—we do not know for the moment which one, although we hope that we shall know it soon. Let us denote the one among the three relations that is actually valid by ? and write

$$\sqrt{3} + \sqrt{11} \quad ? \quad \sqrt{5} + \sqrt{8}$$

Whichever of the three relations may be actually valid, we can perform certain algebraic operations equally applicable to all three. To begin with, we can square both sides and then the same relation will hold between the squares:

$$3 + 2\sqrt{33} + 11 \quad ? \quad 5 + 2\sqrt{40} + 8$$

By this operation we have diminished the number of square roots; at the outset we had four, and now we have only two. By subsequent operations we shall get rid of the remaining square roots by and by, and then we shall see which one of the three possible relations is represented by the sign ?.

The reader need not foresee the operations planned in every detail, but he should realize that they can be carried out without hesitation and are bound to lead to the desired decision. Then he may agree that the situation deserves a special term and that such a definite plan should be called a program, (see sect. 8.5).

With these remarks we have, in fact, attained the aim of the present section and there is no need to carry out the programmed steps. Still, let us carry them out:

$$1 + 2\sqrt{33} \quad ? \quad 2\sqrt{40}$$
$$1 + 4\sqrt{33} + 132 \quad ? \quad 160$$
$$4\sqrt{33} \quad ? \quad 27$$
$$528 \quad ? \quad 729$$

Yet there is no question here, we know which side is larger, and retracing our steps we find that

$$\sqrt{3} + \sqrt{11} < \sqrt{5} + \sqrt{8}$$

8.4. A choice between several plans

On each side of a given (arbitrary) *triangle describe an equilateral triangle exterior to the given triangle, and join the centers of the three equilateral triangles. Show that the triangle so obtained is* EQUILATERAL.

In Fig. 8.1, $\triangle ABC$ is the given triangle, and A', B', and C' are the centers of the equilateral triangles described on BC, CA, and AB, respectively. We are required to show that $\triangle A'B'C'$ is equilateral, but it seems odd, almost unbelievable, that this triangle is always equilateral, that the final shape produced by the construction is independent of the arbitrary initial shape. We suspect that the proof cannot be easy.

At any rate we do not like the points A', B', and C' appearing so isolated from the rest of Fig. 8.1. This defect, however, is not serious. As it is readily seen, $\triangle BA'C$ is isosceles, $A'B = A'C$, and $\angle BA'C = 120°$. We introduce this triangle and two analogous triangles into the figure and thus obtain the "more coherent" Fig. 8.2.

Yet we still do not know how to approach our goal. *How can you prove such a conclusion?* In Euclid's manner? By analytic geometry? By trigonometry?

(1) How can we prove in Euclid's manner that $A'B' = A'C'$? By showing that $A'B'$ and $A'C'$ are corresponding sides in congruent triangles. Yet there are no usable triangles in the figure, and we do not see how we could introduce usable triangles. This is discouraging—let us try another approach.

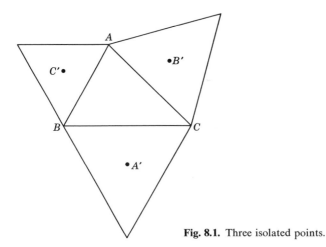

Fig. 8.1. Three isolated points.

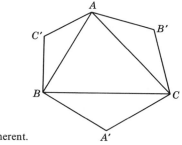

Fig. 8.2. More coherent.

(2) How can we prove by analytic geometry that $A'B' = A'C'$? We should regard the coordinates of the points A, B, and C as given quantities and the coordinates of the points A', B', and C' as unknowns. Having expressed these unknowns in terms of the data, we can express also the distances in question in terms of the data and examine whether they are equal or not. It is a pretty clear plan, but we should handle six unknowns and six data—no, it is not too inviting—let us try the third approach.

(3) How can we prove by trigonometry that $A'B' = A'C'$? We should regard the sides a, b, and c of $\triangle ABC$ as given quantities, and the three distances

$$B'C' = x, \qquad C'A' = y, \qquad A'B' = z$$

as unknowns. Having computed the unknowns, we should examine whether actually $x = y = z$. This looks better than (2); we have only three unknowns and three data.

(4) In fact, we need not compute three unknowns, two are enough: if $y = z$, *any* two sides are equal—and that is enough.

(5) In fact, we need not even compute two unknowns, one is enough if we proceed a little more subtly; it is enough to express just x in terms of a, b, and c provided that we manage to obtain an expression *symmetric* in a, b, and c. (An expression is symmetric in a, b, and c if it remains unchanged when we interchange a, b, and c. If such an expression is valid for x it must be valid just as well for y and z.)

This plan, although it depends on the ingenuity of the problem solver, on the coming of some little novel idea, appears rather attractive—the reader should try to carry it out (see Fig. 8.3, ex. 8.3).

(6) Has our story a moral? I think, it has.

If you see several plans, none of them too sure, if there are several roads diverging from the point where you are, explore a bit of each road before you venture too far along any one—any one could lead you to a dead end.

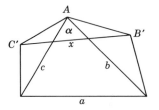

Fig. 8.3. Concentrate on one side.

8.5. Plans and programs

We may regard our plan as a road along which we intend to travel. There are plans and plans, however. We would like to have a plan of some action that leads us right to the goal, but, unfortunately we do not always succeed in devising such a complete plan, and also less far-reaching actions need some planning. We may see just a little bit of the road, we may see a long stretch of it, or we may see the full extent of our road right to the goal. Then, we may see our road hazily or clearly. Again, we may expect, and prepare for, various eventualities along the portion of the road that we do not see well or do not see at all. A desirable eventuality, the hope for which never leaves us entirely, is to have a bright idea which will immediately clear up everything.

Quite often, we do not have yet a complete plan; there are gaps in the plan, some vital ideas are still missing. Yet we go ahead, we start executing our plan in relying on some bright, or at least novel, idea that will come to us and fill the gaps.

How much we rely on such an idea may be regarded as the most important distinction between plans and plans. If we do not depend at all on novel ideas, but are confident that the steps already pondered and foreseen by the plan are sufficient to reach our goal, we have a plan clear and determined enough to be called a *program*. We may spend much of our time in working at various imperfect plans till we succeed in developing one of them into a program.

Compare sect. 8.3 with sect. 8.4, for instance.

8.6. Patterns and plans

Under appropriate circumstances, each pattern we have studied in the foregoing suggests a plan—but does not immediately yield a definite plan, a program.

For instance, we have a problem of geometric construction. We may try to solve it following the pattern of two loci. This is a plan, indeed, but it takes additional ideas to find a suitable point to the construction of which the problem can be reduced, and to split the condition suitably so that we obtain two loci for that point.

Or, we decide to solve a geometric problem by reducing it to equations, following the Cartesian pattern. This is a plan, indeed, but it takes additional ideas to set up as many equations as unknowns and further ideas to solve the system of equations.

Working backward is a very general and useful pattern of planning but obviously we need some ideas from the subject matter to work across the gap from the unknown to the data. When the regressive planning has succeeded and the logical network spanning the gap has been perfected, the situation is very different. Then we have a program for working foreward from the data to the unknown.

Examples and Comments on Chapter 8

8.1. *Backward or forward? Regressive or progressive? Analysis or synthesis?* In our terminology (see sect. 8.2) the term "working backward" stands for a certain strategy of problem solving, a typical way of planning the solution. Is it the only possible strategy? Is it the best strategy?

(1) Take "our example," the example we made a graphic study of in chapter 7. The plan of the solution at which we have finally arrived is represented by Fig. 7.7; it is a web of points and lines, of intermediate unknowns and interconnections, stretching across the originally open space between the unknown and the data. We started spinning this web from the unknown, working toward the data; the successive stages of the work are displayed in Fig. 7.8. We called this direction of work "regressive" or "backward" (in Fig. 7.8 it is downward).

The final plan, the full system of interconnections (see Fig. 7.7—the web may be more complex in another case) does not show in which direction it has been built up. Another problem solver could have built it up starting from the data and working in the direction of the arrows in Fig. 7.7 (which we have followed in executing the plan). Discovering the plan in this direction would be *progressive* work, *working forward*.

Still another problem solver (having, perhaps, a more complicated problem) could devise such a plan without working all the time in the same direction. Starting from either end, he could work sometimes from the unknown toward the data, sometimes from the data toward the unknown; he may work alternately from both ends; he may even establish some promising connection in the middle between things which are not yet connected with either end. Thus, devising a plan in working backward is by no means the only possibility. For a concrete case, see ex. 7.3.

(2) In our example summarized by Fig. 7.8 we have devised the plan of the solution working backward. Let us compare our work with the work of a problem solver who happened somehow to devise the same plan working forward.

In fact, we have started from the unknown, we have asked ourselves questions emphasizing the unknown: *What do you want? What is the unknown? How can you get this kind of thing? How can you find this kind of unknown? From what data can you derive this kind of unknown?* And thus we have found two "data," the volumes A and B, from which the unknown can be derived, in terms of which the unknown F can be expressed: $F = B - A$. This stage of our work is indicated by Fig. 8.4 (which is a part of Fig. 7.3).

Yet the other problem solver started differently: from the data. He asked himself questions emphasizing the data: *What have you? What are the data? What are such things good for? How can you use such data? What can you derive from such data?* And somehow he observed that he can derive from the data the length (altitude) x, express x in terms of a, h, and b (from a proportion as we have observed too later in the game, see Fig. 7.6). This stage of his work is indicated by Fig. 8.5.

Let us return to our work, to the stage represented by Fig. 8.4. Having expressed the unknown F in terms of A and B, we are left with two new unknowns, A and B, two new (auxiliary) problems:

> Compute A in terms of the data a, h, and b.
> Compute B in terms of the data a, h, and b.

These are clear-cut mathematical problems, of the same kind as our original problem. Working regressively, we ask again: *How can you find such unknowns? From what data can you derive such unknowns?*

Let us now return to the other problem solver; he has reached the stage represented by Fig. 8.5. Having expressed x in terms of the data a, h, and b, he can regard x as given, and so he has more data which may give him a better chance to find the original unknown. Progressing in the same way, however, he has no clear-cut auxiliary problem, but he has to ask himself the less definite question: How can I use x? *What are such things good for? What can I derive from such data?*

Yet the most striking difference between us and the other problem solver, between the two situations represented by Fig. 8.4 and 8.5, is in the outlook. What happens if we succeed in solving our auxiliary problems? What happens if he succeeds in answering his question?

If we succeed in expressing our auxiliary unknowns in terms of the data (A and B in terms of a, h, and b), we can express so also our original unknown

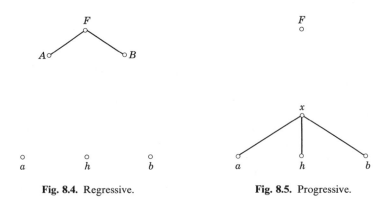

Fig. 8.4. Regressive.   **Fig. 8.5.** Progressive.

($F = B - A$) and our problem is solved. If the progressive worker succeeds in expressing some quantity, say y, in terms of the data, he still faces an indefinite question: How can he use y? Except for one case, of course: if he is lucky, he can express F itself, and then he also has solved his problem.

(3) Both ways of planning, the regressive and the progressive, can succeed, and both can fail. Working backward, we may arrive at an auxiliary problem which we cannot solve. Working foreward, we may derive more and more quantities from the data, but these quantities may serve no useful purpose; we may not be able to derive from them our original unknown.

Both ways of planning require a combination of various activities. Yet, when working backward we may expect to spend most of our time in doing clear cut problems. When working foreward, we may expect to spend much of our time in hesitating between problems we might do, or in doing problems which are no help.

On the whole, planning regressively, working backward, "analysis" (in the terminology of the Greek geometers) is preferable. There can be no hard and fast rule, but the wise thing may be to look first at the unknown (the conclusion, the thing you want) then at the data (the condition, the hypothesis, the things you have). Start working backward from the unknown unless you have some special reason not to do so—of course, if a bright idea urges you to start from the data, go ahead!

(4) Let us make just a few more short remarks, although there are still many things to say.[2]

In some cases there is little choice. In many a practical problem the thing we want to find (to construct, to acquire, . . .) is quite definite, but the things we could possibly use for attaining our aim we scarcely know, and they are impossible to survey, there are so many of them. We can hardly have a good reason to begin with any one item in that unmanageable heap of data, and so we are almost obliged to plan regressively.

After arriving at our plan regressively, we execute it progressively (recall sect. 7.5); but it is execution and not planning for we have conceived all the ideas before, and now we just carry them out. This may even raise the suspicion that those people who start planning progressively may have conceived some idea before—conceived it, I mean, implicitly, perhaps subconsciously.

A lady student explained it thus: synthesis by itself (without foregoing analysis) would be difficult—somewhat like trying to make a cake with the ingredients in front of you and no recipe to follow.

And, of course, you should not be pedantic or fussy. Having started working backward, from the unknown, you may see a good opportunity to take a step forward from the data; then take that step by all means.

8.2. *A wise man begins in the end.* A friend of mine, a good mathematician and a good philosopher, once told me that when he tries to find a proof for a theorem he often begins by writing down Q.E.D. ("quod erat demonstrandum," what was required to prove), and this act of writing down the traditional phrase which comes at the very end of a proof puts him in the right mood.

There is a proverb: "A wise man begins in the end, a fool ends in the beginning."

---

[2] *Cf.* HSI, pp. 141–148, Pappus, and pp. 225–232, Working backwards.

8.3. Carry out the plan indicated in sect. 8.4(5).

8.4. *A choice between three plans.* Let a denote the radius of the base and h the altitude of a right circular cylinder. A plane intersects the base along a diameter, is tangent to (has just one point in common with) the perimeter of the top, and divides the volume of the cylinder into two unequal portions. Compute the smaller portion (the volume of a "hoof") which is between the base and the intersecting plan.

This problem and its first solution are due to Archimedes.[3]

We use solid analytic geometry. The axis of the cylinder is the z-axis, and its base lies in the x, y-plane of a rectangular coordinate system. The plane dividing the volume of the cylinder intersects the x, y-plane in the y-axis. Therefore, the equation of the perimeter of the base is

$$x^2 + y^2 = a^2$$

and the equation of the dividing plane is

$$\frac{z}{h} = \frac{x}{a}$$

We may use either integral calculus or Cavalieri's principle to compute the desired volume. With both methods we have to consider a family of parallel cross sections of the "hoof." There are three obvious plans: we may choose cross sections perpendicular to the x-axis, perpendicular to the y-axis, or perpendicular to the z-axis.

Which plan do you prefer? Carry it out.

8.5. *A choice between two plans.*

(1) In doing a crossword puzzle we hesitate between two words. One has 4 letters, of which 1 is known and 3 are unknown, the other has 8 letters of which 3 are known and 5 are unknown. Which one of these two words should we try to find first? *On the basis of the numerical data* offered, is it possible to make a rational choice between the two?

It is hardly possible, I think, but there is a challenge.

†(2) The question raised must be restated in a more general and (as far as possible) more precise form.

A word consists of $k + l$ letters of which k are known and l are unknown. We undertake to find the word, and we are considering the *degree of difficulty* of this undertaking.

Let us first suppose that we know those k letters *completely:* the nature and position of each of them are specified (as in the example

$$\text{IN}\,_-{}_-\,\text{R}\,_-{}_-{}_-$$

where $k = 3, l = 5$). In this case we may define the degree of difficulty of finding the word as the number N of those English words in present-day usage that consist of $k + l$ letters of which k letters are of the specified nature and are in the specified position. (Of course, any well-defined increasing function of N, such as $\log N$, could be just as well taken for the degree of difficulty.)

---

[3] See Proposition 11 of his *Method,* pp. 36–38 of the Supplement to his *Works* edited by T. L. Heath (Dover.).

Theoretically, this definition may appear reasonable; the greater the number N of the admissible words, the greater is the difficulty to choose one of them. Practically, there are several awkward points. How should we decide whether an English word is or is not "in present-day usage"? Is the definition satisfactory from the standpoint of the crossword fan? At any rate, the practical determination of the number N appears prohibitively tiresome and unprofitable.

†(3) Thus we are led to a different and higher aim; we wish to determine the degree of difficulty *insofar as* it depends on k and l alone—the difficulty "other things being equal"—perhaps the "average difficulty." We wish to lump together all the cases with the same k and l, and take only the numerical data k and l into account. If we could attain this higher aim, the degree of difficulty would be a numerical function $f(k, l)$ of k and l. It seems obvious that $f(k, l)$ should be a decreasing function of k and an increasing function of l. Yet, for instance, we do not know yet which one is greater, $f(1, 3)$ or $f(3, 5)$.

†(4) If the letters in English words were *independent* of each other, the number N of English words with k given and l freely eligible letters could be simply expressed as follows:

$$N = 26^l$$

Number N is used in the meaning explained under (2). Hence the degree of difficulty could be defined, for instance as

$$f(k, l) = \frac{\log N}{\log 26} = l$$

This choice of $f(k, l)$ is consistent in itself, but it sidesteps the crucial question: How much do the k known letters restrict the choice of the l still eligible (but, in fact, not quite freely eligible) letters?

It is very doubtful that a somewhat realistic formula for $f(k, l)$ can be devised. At any rate, we expect that such a formula would differ from the one just proposed at least in two respects: it should be a strictly decreasing function of k, and it should be adaptable, if not to all, at least to a few languages.

†(5) Here is a crude and completely speculative trial proposal:

$$f(k, l) = \frac{\log [26 - \alpha k] [26 - \alpha(k + 1)] \cdots [26 - \alpha(k + l - 1)]}{\log 26}$$

Suitable choice of the positive parameter α should adapt the formula to a language written with the 26 Roman letters. The formula should be used only for words of length

$$k + l < \frac{25}{\alpha} + 1$$

(6) The foregoing considerations may shed some light on the scope of heuristic and on the precision that it could possibly attain, and herein lies their justification.

8.6. *A plan, indeed.* "I will just sit here, look at the figure, and wait for a good idea." This, indeed, is a plan. Perhaps a little too primitive. Perhaps a little too optimistic: you seem to rely just a little too much on your ability to produce good ideas. Still, such a plan may work, occasionally.

8.7. Looking back at problems you have solved in the past, reconsider some problems you have solved, or *could* have solved, in working backward.

8.8. *Do not commit yourself.* We consider a half-concrete example. We are required to prove a proposition of elementary geometry the conclusion of which states, "... then the angles $\angle ABC$ and $\angle EFG$ are equal." We have to derive this conclusion from a certain hypothesis the details of which, however, are irrelevant to the purpose of this comment, and so we ignore those details.

At a certain (probably early) stage of the problem solving process we concentrate on the conclusion: *What is the conclusion?*

We have to prove that

$$\angle ABC = \angle EFG$$

How can you prove such a conclusion? From what hypothesis can you derive such a conclusion?

We succeed in remembering several pertinent facts learned in the past; several possibilities of deriving such a conclusion as we have to prove. Two angles are equal,

(1) if they are corresponding angles in congruent triangles; or

(2) if they are corresponding angles in similar triangles; or

(3) if they are corresponding angles formed by two parallels with a transversal; or

(4) if they have the same third angle as complement; or

(5) if they are inscribed in the same circle and intercept the same arc.

We have here five different known theorems, each of which could be applicable to our case, five different hypotheses from each of which we could derive the de-

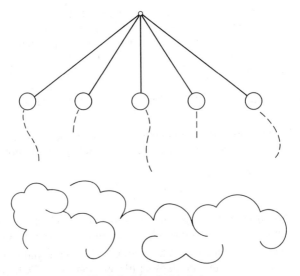

Fig. 8.6. A doubtful choice.

sired conclusion. We may start working with any one. For instance, we could try (1); we could introduce two likely triangles, say $\triangle ABC$ and $\triangle EFG$, and then try to prove that they are congruent. If we succeeded, the desired conclusion would immediately follow! And how could we prove that $\triangle ABC \cong \triangle EFG$? With this question we embark upon regressive planning. Yet we could begin such regressive planning just as well with any one of the other theorems quoted— has any one of them some chance of success? Which one has the best chances of success? If we cannot answer these questions, if the response they evoke is just some dim uncertain feeling, the choice we are facing is doubtful indeed. We are at the crossroads. We have to choose between several paths; the beginning of each is clear enough, but the continuation is uncertain and the end is hidden by clouds. Figure 8.6 attempts to express the situation graphically.

The aim of this example was just to make the reader realize the perplexing situation, the uncertainty of the choice between several plans. The advice I would give in such a situation is: *Do not commit yourself too early, do not commit yourself to one course more than necessary.* Do one thing, but do not forget the others.

The good problem solver plans like a good general; he realizes that the attack planned may fail and does not neglect the line of retreat. A good plan must have some built-in *flexibility,* some adaptability to unforseeable difficulties.[4]

[4] See MPR, Vol. 2, pp. 148–152.

CHAPTER 9

PROBLEMS WITHIN PROBLEMS

When, in either construction or demonstration, we assume anything which has not been proved but requires argument, then we regard what has been assumed as doubtful in itself and worthy of investigation, and call it a lemma.

PROCLUS: *Commentary on Euclid,* On Proposition 1 of Book I.

When a problem arises, we should be able to see soon whether it will be profitable to examine some other problems first, and which others, and in which order.

DESCARTES: *Œuvres,* vol. X, p. 381; Rules for the Direction of the Mind, Rule VI.

What is the best you can do for this problem? Leave it alone and invent another problem.

THE TRADITIONAL MATHEMATICS PROFESSOR.

9.1. Auxiliary problems: means to an end

Some observations of Wolfgang Köhler on anthropoid apes have great interest for us. Let us describe, schematically, one of his experiments.[1]

Inside the cage there is a chimpanzee and he is hungry. Outside the cage there lies a banana on the ground. The chimpanzee can pass his arms through the bars of his cage, but the banana is beyond his reach. The chimpanzee has tried hard to reach the banana, but without

[1] Wolfgang Köhler, *The mentality of apes* (Harcourt, Brace & Co.) pp. 32–34.

success, and now he just sits there. Also on the ground outside his cage, and within his reach, there lies a stick, but he seems to pay no attention to it. Suddenly he stirs, grabs the stick, clumsily pushes the banana with it until he can reach the banana, and then he grabs it and eats it.

This chimpanzee has solved two problems:

A. To grab the banana.
B. To grab the stick.

Problem A arose first. Originally, the chimpanzee did not show the least interest in the stick, which he cannot eat; yet he solved B first. The solution of problem B paved the way to the solution of his original problem A. The chimpanzee had a direct interest in A and only an indirect interest in B; A was end, B only means to him; A was his main, or original, problem, B just an *auxiliary problem* ("helping" problem, subproblem).

Let us outline generally the meaning of this important term: *An auxiliary problem is a problem on which we spend attention or work not for its own sake, but because we hope that such attention or work may help us to solve another problem, our original problem.* An auxiliary problem is means to an end, it should yield access to the goal; the original problem is the end and the goal.[2]

Gaining access to the solution of an apparently inaccessible problem by devising and solving first an appropriate auxiliary problem is the most characteristic kind of intelligent action. We can hardly refuse to regard the performance of the chimp as an act of intelligence.

We are going to classify auxiliary problems, starting from mathematical examples.

9.2. Equivalent problems: bilateral reduction

We begin with an example. Our task is to solve the following system of three equations with three unknowns:

$$(A) \begin{cases} x - y & = -4 \\ x + y + z = 5 \\ x + y - z = 31 \end{cases}$$

From system (A) we derive another system (B):

(1) we leave the first equation of (A) unchanged;
(2) we add the second and third equations of (A);
(3) we subtract from the second equation of (A) its third equation

---

[2] HSI, pp. 50–51; Auxiliary problem.

and so we obtain the three equations of a new system:

$$(B) \begin{cases} x - y = -4 \\ 2(x + y) = 36 \\ 2z = -26 \end{cases}$$

Our derivation of (B) shows that such numbers x, y, z as satisfy (A) must, of necessity, satisfy (B). The converse is also true: numbers x, y, z satisfying (B) must satisfy (A). This seems plausible, but we can also prove it in various ways, for instance as follows. By dividing the last two equations of (B) by 2, we obtain

$$(C) \begin{cases} x - y = -4 \\ x + y = 18 \\ z = -13 \end{cases}$$

and from (C) we can return to (A) by leaving the first equation of (C) unchanged and first adding, then subtracting, the last two. In brief, if three numbers x, y, and z satisfy one of the two systems (A) and (B), they must of necessity satisfy the other.

The systems (A) and (B) are not identical; they do not consist of the same equations. Therefore, we can not properly say that the two corresponding problems, the one to solve (A), the other to solve (B), are identical. Yet we can properly say that these two problems are *equivalent*. Here is the general definition of this use of the term: *Two problems are equivalent for us if we know that the solution of each involves the solution of the other.*[3]

The transition from a problem to an equivalent problem is called a *bilateral* (or reversible, or convertible, or equivalent) *reduction*. For instance, the transition from our original problem which was to solve (A) to the problem to solve (B) is a bilateral reduction. It is a useful reduction too; the system (B) is closer to the solution than the system (A). In fact, (B) is closer to (C) than (A), and (C) is almost at the end of our task; there is already the value of z, and little remains to do to get also the values of x and y.

9.3. Chains of equivalent problems

Let us return to system (C) in the sect. 9.2; we derive from it, by addition and subtraction, the system

$$(D) \begin{cases} 2x = 14 \\ 2y = 22 \\ z = -13 \end{cases}$$

---

[3] HSI, p. 53; Auxiliary problem 6.

and hence

$$(E) \begin{cases} x = 7 \\ y = 11 \\ z = -13 \end{cases}$$

We have here a sequence of five systems (each one is a system of three equations)

$$(A), \quad (B), \quad (C), \quad (D), \quad (E)$$

There is a problem associated with each: to find the values of x, y, and z satisfying the system. [The "problem" is fully solved, and so the term "problem" is not used in its proper, but in an extended, meaning in the case of the system (E).] Each of these problems is equivalent to the foregoing (and also to the following) problem as each link of a chain is joined to the next link; we have here a *chain of equivalent problems*.

In our chain, (A) is the beginning and (E) is the end; (A) is the originally proposed system of equations and (E) exhibits the solution. We have here an ideally perfect way to arrive at the solution. Starting from the proposed problem, we devise a sequence of problems; each problem is equivalent to, and nearer to the solution than, the foregoing problem; proceeding thus from problem to problem, we attain, with a last step, the solution itself.

Yet, even in mathematics, in the quest of the unknown and in the struggle for the proof, we often have to settle for something less than perfection. And so we turn to survey further kinds of auxiliary problems.

9.4. More, or less, ambitious auxiliary problems: unilateral reduction.

We begin with the consideration of a schematically stated problem:

A. Find the volume of a pyramid, being given . . . .

We suppose here that the data are sufficient to determine the pyramid, but the base and the altitude are not among the data, neither of these two quantities is given. This much is important, but what the data otherwise are is unimportant for our present discussion and so we suppress their list.[4]

We know that the volume of a pyramid can be computed if its base and altitude are given—but, as we have just said, neither of these two quantities is given. As they are not given, we shall try to compute them, and so we turn to another problem:

---

[4] For a concrete problem of the form A see ex. 4.17.

B. Find the base and the altitude of a pyramid, being given . . . .

Problem A has one unknown, problem B has two unknowns, and both problems have the same (unlisted) data. There is a one-sided asymmetric relation between these two problems. If we succeed in solving B, we have the base and the altitude of the pyramid, hence we can compute its volume, and so we can solve A. If, however, we succeed in solving A, it is by no means certain that we can also solve B: although the result of A yields a simple relation between the two unknowns of B, there may remain some serious difficulty to find either of them. Thus, we achieve less in solving A than in solving B. We may call A the *less ambitious* and B the *more ambitious* of the two problems.[5]

Let us repeat the foregoing in general terms. There are two problems, A and B, both unsolved, and our state of knowledge is as follows: *we do know how we could derive from the solution of B the solution of A, but we do not know how we should derive from the solution of A the solution of B.* Under such circumstances we say that A is less ambitious than B and that (which means the same) B is more ambitious than A.

The transition from an original problem to an auxiliary problem which is more ambitious, or less ambitious, than the original problem (in any case, not equivalent to the original problem) is called a *unilateral* (or irreversible) *reduction.* In our example, the original problem A is less ambitious than the auxiliary problem B, and so the reduction of A to B is unilateral. The experienced reader may recall several examples similar to the one given here in which such a unilateral reduction was profitable.

Unilateral reduction of the opposite kind, where the auxiliary problem is less ambitious than the original problem, is also often profitable. Here is a schematic example:

A. Compute the unknowns $x_1, x_2, \ldots x_{n-1}$, and x_n, being given . . . .

B. Compute the unknown x_1, being given . . . .

We suppose that the condition and the data determine the unknowns and that they are the same in both problems, in A and in B, but what they are is unimportant here and so we suppress them. Trivially, the solution of A involves the solution of B, but, in general, the solution of B cannot involve the solution of A: according to our definition, A is more ambitious than B. Yet, very often when we are required to solve A, we may introduce with advantage B as auxiliary problem; we did so many times in chapter 3 when we solved A by recursion and, by choosing x_1 as our initial unknown, we began our work with the auxiliary problem B as a stepping stone to A.

[5] HSI, p. 56; Auxiliary problem 8.

9.5. Remoter auxiliary problems

We begin with an example. Let us consider the following problem:

A. Given the length of an edge of a regular tetrahedron, find the radius of the sphere circumscribed about the tetrahedron.

If we do not see some other access to problem A we may try to approach it by considering the following problem:

B. Given the length of a side of an equilateral triangle, find the radius of the circle circumscribed about the triangle.

The transition from A to B is neither a bilateral nor a unilateral reduction in the sense of sections 2 and 4. In fact, we can scarcely see *a priori* how the solution of B would involve that of A or the solution of A that of B—problems A and B do not appear equivalent, and neither appears more ambitious than the other in the sense of our definitions.

Yet problems A and B are not unrelated. Problem B is analogous to the problem A; we have here a little example of that great analogy between plane geometry and solid geometry. Moreover, to most of us the problem B will appear easier than the problem A; we may have even seen already, and could recall with little trouble, the solution of B. In this situation, it is natural to ask: Is it worthwhile to consider problem B? Is there a chance that the consideration of B will facilitate the solution of A?

It may happen that the consideration of B does not contribute appreciably to the solution of problem A; this may even happen if we clearly see the analogy between A and B and if we possess a complete solution of B. Yet it may also happen that B will help, although it is apparently sterile. The comparison of A with the analogous B may render the proposed problem A more interesting, and in such a case B *is* useful. Yet the contribution of B to the solution of A may be even more distinct; there is a chance that the analogy between A and B will suggest some useful remark. For instance, in the "plane" problem B, the desired radius is a simple fraction ($\frac{2}{3}$) of the altitude of the equilateral triangle. This may suggest the question: How about the analogous "solid" problem A? Is that desired radius some simple fraction of the altitude of the regular tetrahedron? This question, or a similar question, can introduce some usable element and pave the way to the solution of A. Or, in solving A, we may need the altitude of one of the faces of a tetrahedron; then, if we know the ratio of the altitude to the radius of the circumcircle for an equilateral triangle (which we have just mentioned) the answer to problem B may contribute a link to the chain that we have to forge to obtain the answer to A.

In general, we may expect that the consideration of a problem B will contribute in some way or other to the solution of the proposed problem A, even if B is neither equivalent to A, nor more ambitious, nor less ambitious, than A. Such a problem B is called a *remoter* auxiliary problem to A.

9.6. Material help, methodological help, stimulating influence, guidance, practice

An auxiliary problem may help to solve the original problem in an inexhaustible variety of ways.

An equivalent auxiliary problem yields, if solved, the *whole solution* of the original problem, and an auxiliary problem more ambitious than the original problem does the same. (The difference between these two kinds of auxiliary problems shows up when we are unable to solve them. If an equivalent auxiliary problem is definitely beyond our reach, so is our original problem; if, however, a more ambitious auxiliary problem turns out inaccessible, the prospects for our original problem need not be so dark.)

Other sorts of auxiliary problems, even if solved, do not guarantee the whole solution of the original problem, but they may offer *material help*. A part of the solution of the auxiliary problem (or even all of it) may become a part of the solution of the original problem to which it may contribute a conclusion, a construction, or a fact on which such a conclusion or construction is based, and so on.

Yet even when no such material help is forthcoming, the auxiliary problem may give *methodological help;* it may suggest the method of solution, an outline of the solution, or the direction in which we should start working, and so on. An auxiliary problem analogous to, but easier than, the original problem is in a good position to offer such methodological help.

We might not be able to point out any part or feature in the final solution of the original problem that was taken over from, or suggested by, a certain auxiliary problem. Still, it is quite possible that the *stimulating influence* of that auxiliary problem contributed a good deal to the discovery of the solution of the original problem. Perhaps that auxiliary problem has rendered, by analogy or contrast, the original problem more understandable or more interesting; or perhaps it has stirred our memory—started a train of thought from which, eventually, some essential relevant fact emerged.

Auxiliary problems may help in still another rather subtle manner. Working on a problem involves decisions. We could continue our work in two directions; there are two paths open for us, one to the right and one to the left. Which one should we choose? Which one is more likely to lead us to the solution? It is important to assess our prospects reasonably and auxiliary problems may give us welcome guidance in this respect. Our attention and work spent on, and our experience gained with, an auxiliary problem have a good chance to influence our judgment in the right direction.

Sometimes we may take on an auxiliary problem just for *practice*. It may happen that our original problem involves concepts with which we are not used to dealing. In such a situation, it may be advisable to try some easier problem involving the same concepts that would thus become a (rather remote) auxiliary problem to our original problem.

Although there are so many different things to gain, it quite often happens that we gain little and lose a lot of time and trouble in working on an auxiliary problem. Therefore, before we get too deeply involved in such a problem, we should try to weigh the prospects and estimate the chances.

Examples and Comments on Chapter 9

9.1. *Reliable sources of auxiliary problems?* An auxiliary problem may "spontaneously grow out" of the proposed problem. Yet it may also happen that when we would like to have an attractive auxiliary problem none comes to mind. In such a case we may wish we had a list of reliable sources from which useful auxiliary problems could be drawn. There are, in fact, various usual modes of forming auxiliary problems, and we shall consider the most obvious ones in what follows; they will lead you to some auxiliary problems in most cases—but it cannot be guaranteed that they will lead you to useful auxiliary problems.

Auxiliary problems may arise at any stage of the problem solving process. Let us assume, however, that we are a little beyond the very first phase. We have already considered and well understood the principal parts of our problem—the unknown, the data and the condition, or the hypothesis and the conclusion—and also the most obvious subdivisions (clauses, etc.) of these principal parts. Yet we see no promising plan, and we wish we had some more accessible or more attractive goal. It is good to know that searching examination of the principal parts of our problem may present us with such a goal, with a usable auxiliary problem. We shall survey the most notable cases in what follows.

9.2. *Respice finem.* The desire to attain the aim is a productive desire, it produces thoughts of actions which possibly could attain the desired aim. The end suggests the means. Therefore, look at the end, do not lose sight of your aim; it guides your thoughts.

Respice finem means "Look at the end" and was a current phrase when Latin was more used.[6] Hobbes expands it: "Look often upon what you would have, as the thing that directs all your thoughts in the way to attain it."[7]

[6] From a medieval hexameter: *Quidquid agis prudenter agas et respice finem.*
[7] *Leviathan,* chapter III.

Looking at the end, we are waiting until the thought of some means emerges. To shorten the waiting time, we should keenly realize the end: *What do you want? What kind of thing do you want? What is the unknown? What is the conclusion?* We should also make a resolute effort to conjure up some appropriate means: *How can you get this kind of thing? Where can you get this kind of thing? In which shop can you buy this kind of thing? How can you find this kind of unknown? How can you derive such a conclusion?*

The last two questions specifically refer to mathematical problems, one to a problem to find, the other to a problem to prove. Let us consider these two cases separately.

(1) *Problems to find.* We consider, as we have done in sect. 9.4, a half-concrete problem: "Find the volume of a pyramid, being given . . . ." The unknown, the volume of a pyramid, is specified, whereas the condition and the data remain unspecified. *How can you find this kind of unknown?* How can we compute the volume of a pyramid? *From which data* can you get this kind of unknown? The proposed problem has data, of course, but the trouble is that at least for the moment we can not derive the unknown from the proposed data. What we really want are *more manageable data;* in fact, we want another, more accessible, *problem with the same unknown.*

If we find such a problem, we may face different situations.

(2) *Problem with the same unknown, formerly solved.* If we are lucky enough to recall such a problem we may proceed to choose *its data as target for auxiliary problems.* This procedure is very often useful. Let us illustrate it by our (previously mentioned, semiconcrete) example.

The unknown of our problem is the volume V of a pyramid. In the most familiar problem with this unknown, the data are B, the area of the base, and h, the length of the altitude. We know the solution of the familiar problem ($V = Bh/3$), and we have recalled it. How can we use this solution? The most natural thing is to try to compute B and h from the data of the proposed (unsolved) problem. In trying this, we choose B and h as our targets; we introduce two auxiliary problems, the unknown of one is B, the unknown of the other is h, and the data in both are the data of our present problem. (For a concrete case, see ex. 4.17, 4.18.)

(3) The foregoing procedure is often applicable, and in many cases it should be *repeatedly* applied.

Let x denote our primary unknown, the unknown of the proposed problem. We are looking for manageable data and we notice that we could find x if we had $y', y'', y''', \ldots$ (by using the solution of a formerly solved problem.) We choose $y', y'', y''', \ldots$ as our new targets, as secondary unknowns. Now, we could find $y', y'', y''', \ldots$ if we had $z', z'', z''', \ldots$ (by using the solutions of several formerly solved problems), and so we set up $z', z'', z''', \ldots$ as targets, as our tertiary unknowns. And so on. We are *working backward,* (see sect. 8.2).

To be well prepared for such work, we should have a store of (simple, often usable) solved problems, and it should be a well-stocked and well-organized store (ex. 12.3).

(4) *Problem with the same unknown, not yet solved.* We may regard such a problem as a stepping stone to the proposed problem, introduce it as an auxiliary

problem, and try to solve it—and such a procedure may be profitable. Yet the prospects are, other things being equal, less favorable than in case (2). In fact, to profit by such a problem in the most obvious way, we should first solve it and then, in addition, we should be able to use it in the manner described under (2).

(5) If we do not see at all how we could find this kind of unknown with which our present problem is blessed, if we cannot recall any formerly solved problem, nor imagine a manageable new one, with the same kind of unknown, we may look for a problem with a *similar kind of unknown*. For instance, if we have to find the volume of a pyramid and we see no other way we may try to recall how we find the area of a triangle and examine various approaches, looking for suggestive analogies.

(6) *Problems to prove.* We could repeat here with little change what we have said about problems to find, but an accelerated survey will be enough.

Also here, it is good to start from a half-concrete example. We have to prove a theorem of the form: "If . . . , then the angle is a right angle." The conclusion of this proposition is specified: "the angle is a right angle" but its hypothesis remains unspecified. *How can you prove such a conclusion? From which hypothesis* can you derive such a conclusion? These questions prompt us to look for a *theorem with the same conclusion* where the statement "the angle is a right angle" is inferred from some other, more manageable hypothesis.

If we are lucky enough to recall a *formerly proved theorem with the same conclusion* we can choose its hypothesis as *target:* we may try to prove the hypothesis of the theorem we have recalled from the hypothesis of the theorem we are trying to establish.

This procedure is often applicable. In many cases we can apply it *repeatedly* and discover the proof for the desired conclusion by *working backward.*

If we come across a theorem with the same conclusion as the proposed theorem but which is equally unproved, we may try to prove it. Such a trial may be profitable, but the prospects should be carefully weighed.

If we cannot recall any formerly proved theorem, nor imagine any manageable new one, with the same conclusion, we may look for a theorem with a *similar conclusion.*

(7) Whatever our problem is, we can be certain in advance that we shall use some formerly acquired knowledge in solving it. Yet, especially if our problem is difficult, we cannot so confidently foretell *which* pieces of knowledge we shall be able to use. Any formerly solved problem, any formerly proved theorem, could be usable, especially if it has some point of contact with our present problem— but we have no time to examine all of them. The foregoing discussion directs our attention to the most likely point of contact. If we have a problem to find, *formerly solved problems with the same kind of unknown*—and if we have a problem to prove, *formerly proved theorems with the same conclusion*—are more likely to be usable. Therefore, we should give high priority to the questions: *How can you find this kind of unknown? How can you prove such a conclusion?*

9.3. *Removing, or adding, a clause.* When our work is progressing slowly or not at all, we become impatient with it and wish we had another problem. Now, it is good to be acquainted with modifications of the problem leading to related problems, the consideration of which has some chance to be useful. Here is a list of the most obvious modifications of this kind.

Problems to find:

(1) Removing a clause from the condition.
(2) Adding a clause to the condition.

Change (1) renders the condition *wider*, (2) renders it *narrower*.

Problems to prove:

(1) Removing an assumption from the hypothesis.
(2) Adding an assumption to the hypothesis.
(3) Removing an assertion from the conclusion.
(4) Adding an assertion to the conclusion.

Both (1) and (4) render the theorem *stronger*, both (2) and (3) render the theorem *weaker*.

The effects of these changes are discussed in exs. 9.4 and 9.5.

9.4. *Widening, or narrowing, the condition.* We consider two conditions, $A(x)$ and $B(x)$, regarding objects x of the same category. We say that $A(x)$ is narrower than $B(x)$, or (which amounts to the same) that $B(x)$ is wider than $A(x)$ iff any object satisfying $A(x)$ necessarily also satisfies $B(x)$. (That is, we use these terms "inclusively;" the case in which $A(x)$ and $B(x)$ have the same scope is included in both terms.)

(1) *Widening the condition* means passing from the proposed problem to another problem which has a wider condition than the proposed. The reader should realize that we have very often performed this operation in the foregoing chapters (of course, without describing it in just these terms.) Thus, in a problem of geometric construction appropriately formulated (or reformulated) the condition refers to a point; we obtain a locus for the point by *keeping only a part of the condition* and removing the other part, that is, by widening the condition. Again, in setting up one equation of a system of equations for several unknowns, we take only one part (requirement, clause, proviso, . . .) of the full condition into account and so, in fact, we widen the condition.

Widening the condition is certainly useful if we can accomplish two things. First, find (describe, list, . . .) the set of all objects satisfying the wider condition. Second, remove from this set those objects that do not satisfy the original condition. I think the reader is aware how these two objectives are attained by the pattern of two loci; he should also review sect. 6.3(3) and some examples and comments dealing with puzzles (see also ex. 6.21).

Yet widening the condition can be useful also in a different manner as the reader familiar with the Cartesian pattern can easily see.

(2) *Narrowing the condition* means passing from the proposed problem to another problem which has a narrower condition than the proposed. We do not have much opportunity to employ this procedure on the level with which we are principally concerned, but here is an example.

We have to solve the equation

$$x^n + a_1 x^{n-1} + a_2 x^{n-2} + \cdots + a_n = 0$$

of higher degree n the coefficients of which, $a_1, a_2, \ldots a_n$, are integers. It is advisable to begin the work by looking for integral roots. In fact, by imposing on x the additional requirement that it should be an integer we narrow the condition. Yet the search for integral roots (they must be divisors of the last coefficient a_n) is comparatively easy, and, if we succeed in finding such a root, we can lower the degree of the proposed equation and so facilitate the search for the remaining roots. (For a concrete case, see ex. 2.31.)

Narrowing the condition is often useful on a more advanced level; see ex. 9.11.

9.5. *Examining a stronger, or a weaker, theorem.* We consider two clearly stated propositions, A and B. If we know that A follows from B (if we can derive A in supposing B true) we say that A is weaker than B and (what means the same) that B is stronger than A. This relation between A and B is particularly interesting when we can neither prove nor disprove A, nor prove or disprove B.

(1) *Examining a possible ground.* We want to prove that two given quantities are unequal. For instance, we want to prove theorem A, which asserts that

$$e < \pi$$

We are lucky enough to observe a third quantity with which both given quantities can be conveniently compared. In our example both e and π are conveniently comparable to the simple number 3. Therefore, to establish A, we consider the theorem B that asserts that

$$e < 3 \qquad \text{and} \qquad 3 < \pi$$

Of course, A immediately follows from B. The newly introduced proposition B asserts more and so it is stronger than the proposition A, to prove which was our original problem.

Observe that if we have to prove an inequality between two irrational numbers we are almost obliged to proceed as we have in our example; we should discover a rational number which separates the two irrational numbers. In so doing we reduce the original proposition to a stronger proposition, as in our example; the discovery of the separating rational number makes the new proposition stronger.

Such things happen in more advanced research at every turn: to prove a proposed theorem A we have to imagine a stronger theorem B from which A follows but which for some reason is more manageable than A. In proving B, we exhibit a "ground" why A is true. Of course, when we are discovering a theorem B from which A follows, we do not know yet whether we shall be able to prove B, we do not know even whether B is true or not. Thus, for the moment, such a B is not yet a "ground" for the proposed A, just a "possible ground." Still, it may be advisable to examine B, this possible ground for A.

(2) *Examining a consequence.* We want to prove that two quantities are equal. For instance, letting S denote the surface area of a sphere with radius r, we want to prove theorem A that asserts that

$$S = 4\pi r^2$$

It may be advisable to try to prove less to begin with, namely theorem B, which asserts that

$$S \leqq 4\pi r^2$$

(We could possibly prove B in approximating the sphere by circumscribed polyhedra.) At any rate, B obviously follows from A, B is a consequence of A, the theorem B is weaker than the theorem A.

The proof of the weaker theorem B, however, could eventually lead us to the proof of the original theorem A. In fact, the considerations used in proving B could suggest a proof for another weaker theorem, for the opposite inequality

$$S \geqq 4\pi r^2$$

(perhaps by a transition from circumscribed polyhedra to inscribed polyhedra). From the combination of the two weaker theorems, however, the original theorem A would follow.

Such things happen in more advanced research at every turn.

If we are unable to prove a proposed theorem A, we imagine a weaker theorem B which we can prove. Then we may manage to use the weaker theorem B as a spring-board, and with the impetus gained from B we attain A. This can happen even with quite elementary theorems. For instance, we may attain the theorem A which deals with the general case by proving first a weaker theorem B which deals with a particular case, and then use B as a spring-board.

Do you know an example?

9.6. Let m and n denote given positive integers, $m > n$. Compare the following problems:

A. Find the common divisors of m and n.
B. Find the common divisors of n and $m - n$.

What is the logical relation between A and B?
If you are required to do A, do you see some advantage in passing from A to B?
Use the hint to find the common divisors of 437 and 323.

†**9.7.** Compare the following problems:

A. Find the maximum of the function $f(x)$.
B. Find the abscissas x where $f'(x)$, the derivative of $f(x)$, vanishes.

What is the logical relation between A and B?
Do you see some advantage in passing from A to B?

9.8. We consider a triangle and let stand:
O for the center of its circumscribed circle
G for its centroid (center of gravity)
E for the point on the line through O and G (Euler line) for which $2OG = GE$
(G is between O and E.)
Consider the two theorems:

A. The three altitudes of the triangle meet in a point.
B. The three altitudes of the triangle pass through the point E.

What is the logical relation between A and B?
Do you see some advantage in passing from A to B?
Prove B.

†**9.9.** Compare the two following problems (the square roots are taken with the positive sign):

A. Prove that

$$\lim_{x \to +\infty} (\sqrt{x + 1} - \sqrt{x}) = 0$$

B. Given the positive number ϵ, find those positive values of x for which

$$\sqrt{x + 1} - \sqrt{x} < \epsilon$$

What is the logical relation between A and B?
Do you see some advantage in passing from A to B?
Solve B.

9.10. Compare the two following problems (n denotes a positive integer):

A. Prove (or disprove) the proposition: If $2^n - 1$ is a prime number, n must be a prime number.
B. Prove (or disprove) the proposition: If n is a composite number, $2^n - 1$ must be a composite number.

What is the logical relation between A and B?
Do you see some advantage in passing from A to B?
Solve B.

9.11. *The search for a counterexample.* A *counterexample* explodes a statement purporting to apply to all objects of a certain category: The counterexample is an object of the proper category to which the allegedly general statement does *not* apply. The search for counterexamples has several interesting features which we should discuss, although we are obliged to go a little beyond our usual level for a sufficiently instructive illustration.

† (1) *A problem to prove.* Prove, or disprove, the following statement:

If the infinite series with real terms $a_1 + a_2 + a_3 + \cdots$ is convergent, the infinite series $a_1{}^3 + a_2{}^3 + a_3{}^3 + \cdots$ is also convergent.

After more or less work we may surmise that the proposed statement is false and we try to explode it by a counterexample.

† (2) *A problem to find as auxiliary problem to a problem to prove.* We seek a counterexample, that is, an infinite sequence that satisfies the hypothesis, but does *not* satisfy the conclusion, of the statement proposed under (1). We have so, in fact, a problem to find. Let us look at its principal parts.

What is the unknown? An infinite sequence $a_1, a_2, a_3, \ldots$, of real numbers.
What is the condition? It consists of two clauses:
(I) the series $a_1 + a_2 + a_3 + \cdots$ is convergent
(II) the series $a_1{}^3 + a_2{}^3 + a_3{}^3 + \cdots$ is divergent
We should note that this problem to find arises as an auxiliary problem to a problem to prove.

(3) *Wanted* ONE (anyone) *object fulfilling the condition.* On the elementary level, we are usually required to find all solutions, all the objects satisfying the condition of the problem. Yet in the present case it is enough to find one solu-

tion, one such object; one counterexample is enough to upset the allegedly general statement.

This situation, different from the usual, may demand a different strategy. Leibnitz[8] has some advice to offer: "All solutions may be required or only some solutions. If just any one solution is required, we should invent additional conditions compatible with the original conditions which often demands great skill."

†(4) *Narrowing the condition.* We survey convergent series, satisfying part (I) of the condition, hoping to meet with one that also satisfies part (II). It is natural to begin our search with the simplest and more familiar cases.

Thus we may think first of convergent series with positive terms a_n. Yet, in such a series, $a_n < 1$ for large n, therefore $a_n^3 < a_n$, and so the series with the general term a_n^3 is also convergent; requirement (II) is not fulfilled. We must examine convergent series with positive and negative terms.

The most familiar case here is that of an alternating series in which the signs of the terms form the pattern

$$+ \ - \ + \ - \ + \ - \ + \ - \ + \ - \ \cdots$$

If the terms a_n of such a series steadily decrease to 0 in absolute value, the series is convergent—yet then the terms a_n^3 behave in the same way, form a convergent series and, again, part (II) is not fulfilled. And so we must proceed to less familiar regions.

As we are reluctant to venture too far away from the familiar we may hit on the idea of imposing a restriction:

(III) the signs of the terms a_n should form the pattern

$$+ \ - \ - \ + \ - \ - \ + \ - \ - \ + \ - \ - \ \cdots$$

Even after adding (III) to (I) and (II), we still retain a wide margin of uncertain and arbitrary choice. And so we may hit on the idea of imposing one more (in fact, not quite definite) restriction:

(IV) the series $a_1^3 + a_2^3 + a_3^3 + \cdots$ should diverge in the manner of the familiar series $1 + \frac{1}{2} + \frac{1}{3} + \cdots$.

The self-imposed additional requirements (III) and (IV) essentially *narrow the condition* (see ex. 9.4). They may guide the search for a counterexample, but they may also restrain it. I think that they are more help than hindrance, yet the reader should try to find a counterexample by himself and form his own opinion.

(5) *An alternating procedure.* This may be a good occasion to mention a procedure with which everybody who wants to acquire some ability to do problems to prove should be familiar. (On the high school level there is usually not much opportunity to acquire or practice such ability.)

A problem to prove is concerned with a clearly stated assertion A of which we do not know whether it is true or false: we are in a state of doubt. The aim of the problem is to remove this doubt, to prove A or to disprove it.

Sometimes we are able to devise an approach which could work both ways, which brings us nearer to proof or disproof whatever is in the cards, and so nearer to the solution in any case. Yet such approaches are rare. If we are not lucky

---

[8] *Opuscules et fragments inédits,* p. 166.

enough to find one we face a decision: should we try to prove the assertion A or should we try to disprove it? We have here a choice between two different directions. To prove A we should look for some propositions from which, or for some strategy by which, we could derive A. To disprove A we should look for a counterexample.

A good scheme is to work alternately, now in one direction, then in the other. When the hope to attain the end in one direction fades, or we get tired of working in that direction, we turn to the other direction, prepared to come back if need be, and so, by learning from our work in both directions, we may eventually succeed.

(6) There is a more sophisticated modification of this alternating procedure which may be needed in more difficult cases and may attain a higher aim.

If we cannot prove the proposed assertion A we try to prove instead a weaker proposition (which we have more chances to prove.) And, if we cannot disprove the proposed assertion we try to disprove instead a stronger proposition (which we have more chances to disprove.) If we *succeed in proving* a proposition P we try next to *disprove* a proposition (appropriately chosen) *stronger* than P. Yet if we *succeed in disproving* a proposition P we try next to *prove* a proposition (appropriately chosen) *weaker* than P. Working toward the proposed A from both sides, we may finally prove A. Or we may pass beyond A, and either prove a proposition stronger than A, or we may disprove A yet still save some part of it in proving a proposition weaker than A.

In this way, by working alternately on proofs and counterexamples, we may attain a fuller knowledge of the facts. We may discover a theorem of which we know not only that it is true (we have proved it), but also that it cannot be too easily improved (we have disproved sharper theorems). We catch here a glimpse into the role of proofs in building up science. (*Cf.* G. Pólya and G. Szegö, *Aufgaben and Lehrsätze aus der Analysis,* vol. 1, p. VII. Also MPR, vol. 1, p. 119, ex. 14.)

(7) For further sophistications, striking historical examples, and philosophical overtones of the alternating procedure discussed see the work of I. Lakatos quoted in the Bibliography.

9.12. *Specialization and generalization* are important sources of useful auxiliary problems.

Let us take as example a problem from the theory of numbers. We wish to investigate the *number of divisors* of the positive integer n for which we introduce the symbol $\tau(n)$. For example (we are *specializing*), 12 has the six divisors 1, 2, 3, 4, 6, and 12 and, therefore, $\tau(12) = 6$; we have counted here the "trivial divisors" 1 and 12 of 12, and we intend to proceed the same way for any n.

One way to *specialize* is to consider individual numbers, observing for instance that $\tau(30) = 8$. Or we may systematically list the values of $\tau(n)$ for $n = 1, 2, 3, \ldots$ in constructing a table which starts as follows.

$$\tau(1) = 1 \qquad \tau(6) = 4$$
$$\tau(2) = 2 \qquad \tau(7) = 2$$
$$\tau(3) = 2 \qquad \tau(8) = 4$$
$$\tau(4) = 3 \qquad \tau(9) = 3$$
$$\tau(5) = 2 \qquad \tau(10) = 4$$

Another way to specialize the problem is to consider particular *classes* of numbers. If p is a prime number,

$$\tau(p) = 2, \qquad \tau(p^2) = 3, \qquad \tau(p^3) = 4$$

and now we may find in *generalizing* the answer for any power of p:

$$\tau(p^a) = a + 1$$

If p and q are two different prime numbers, pq has just four divisors $1, p, q,$ and pq and so

$$\tau(pq) = 4$$

Then we may consider the product of three different primes, and so on. By generalizing, we may attempt to find $\tau(n)$ when $n = p_1 p_2 \ldots p_l$ is the product of l different primes. And so on, sometimes specializing and then again generalizing, we may discover a general expression for $\tau(n)$. (Find it!)

Such are the ways of discovery not only in the theory of numbers (of positive integers) but also in other branches of mathematics and in science in general. In specializing, we try to carve out a more tangible, more accessible part of the problem; by generalizing, we try to extend what we have succeeded in observing in a restricted domain.[9]

9.13. *Analogy* is another fertile source of discovery. In simple cases, we may almost copy the solution of an obviously similar problem. In more delicate cases, a more subtle analogy may not give us immediate material help, but it may indicate the direction in which we should work.

The uses of analogy are of inexhaustible variety; they are illustrated by many examples in the foregoing (and in the following) chapters. Let us quote just one [sect. 1.6(3)]. The problem is to construct an angle in a spherical triangle of which the three sides are given. The construction uses the analogous problem of plane geometry as auxiliary problem: to construct an angle in an ordinary triangle of which the three sides are given.

Remember a few more pairs of analogous problems.

There are, as we have hinted, many other ways to use analogy.[10]

9.14. *And if we fail?* The hope with which we undertake the investigation of an auxiliary problem may be disappointed, our undertaking may fail. Still, the time and effort spent on the auxiliary problem need not be lost; we may learn from failure.

We wish to prove theorem A. We notice a stronger theorem B from which A follows. We undertake to investigate B—if we succeed in proving B, A will also be proved. Yet B turns out to be false. This is disappointing—but our experience with B may lead us to a better evaluation of the prospects of A.

We wish to prove the theorem A. We notice the theorem B, a consequence of A which seems to be more manageable than A. We undertake to investigate B—if we succeed in proving B, we may use it as a stepping stone to prove A.

---

[9] See HSI, Generalization pp. 108–110, Specialization pp. 190–197; and MPR, chapter II and *passim*.

[10] See HSI, Analogy pp. 37–46; and MPR, chapter II and *passim*.

In fact, we manage to prove B, but all our trials to use it as a stepping stone to A fail. This is disappointing—but our experience with B may lead us to a better evaluation of the prospects of A.[11]

9.15. *More problems.* Having observed that an auxiliary problem was helpful in solving some problem, try to understand why it was helpful and where it comes from.

Why? Clarify the relation between the problem and the auxiliary problem; see ex. 9.6–9.10.

Wherefrom? Was the auxiliary problem suggested, or could it have been suggested by regressive work (working backward), generalization, specialization, or analogy? Or was there some other (less usual) source?

---

[11] *Cf.* MPR, especially vol. 2, pp. 18–20.

CHAPTER 10

THE COMING
OF THE IDEA

*My mind was struck by a flash of lightning in which its desire
was fulfilled.*

DANTE: *Paradiso,* Canto XXXIII.

10.1. Seeing the light

The solution of a problem may occur to us quite abruptly. After
brooding over the problem for a long time without apparent progress,
we suddenly conceive a bright idea, we see daylight, we have a flash
of inspiration. It is like going into an unfamiliar hotel room late at
night without knowing even where to switch on the light. You stumble
around in a dark room, perceive confused black masses, feel one or the
other piece of furniture as you are groping for the switch. Then, having
found it, you turn on the light and everything becomes clear. The
confused masses become distinct, take familiar shapes, and appear
well arranged, well adapted to their obvious purpose.

Such may be the experience of solving a problem; a sudden clarifica-
tion that brings light, order, connection, and purpose to details which
before appeared obscure, confused, scattered, and elusive.

In these matters, however, one grain of experience is worth more
than pounds of description. To come closer to personal experience
we should get down to a concrete example. Very elementary mathematical examples may be the best to bring us the work, the suspense,
and the pleasure of discovery and to "accustom our eyes to see the
truth clearly and distinctly." (The last phrase is borrowed from
Descartes.)

10.2. Example

I take the liberty of trying a little experiment with the reader. I shall
state a simple but not too commonplace theorem of geometry, and then

54

I shall try to reconstruct the sequence of ideas that led to its proof. I shall proceed slowly, very slowly, revealing one clue after the other, and revealing each clue gradually. I think that before I have finished the whole story, the reader will seize the main idea (unless there is some special hampering circumstance). But this main idea is rather unexpected, and so the reader may experience the pleasure of a little discovery.

A. *If three circles having the same radius pass through a point, the circle through their other three points of intersection also has the same radius.*

This is the theorem that we have to prove. The statement is short and clear, but does not show the details distinctly enough. If we *draw a figure* (Fig. 10.1) and *introduce suitable notation,* we arrive at the following more explicit restatement:

B. *Three circles k, l, m have the same radius r and pass through the same point O. Moreover, l and m interesect in the point A, m and k in B, k and l in C. Then the circle e through A, B, C has also the radius r.*

Figure 10.1 exhibits the four circles *k, l, m,* and *e* and their four points of intersection *A, B, C,* and *O.* The figure is apt to be unsatisfactory, however, for it is not simple, and it is still incomplete; something seems to be missing; we failed to take into account something essential, it seems.

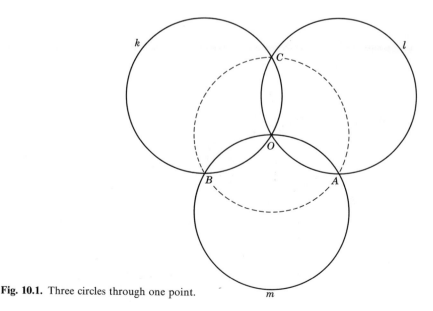

Fig. 10.1. Three circles through one point.

We are dealing with circles. What is a circle? A circle is determined by center and radius; all its points have the same distance, measured by the length of the radius, from the center. We failed to introduce the common radius *r*, and so we failed to *take into account an essential part of the hypothesis.* Let us, therefore, introduce the centers, *K* of *k*, *L* of *l*, and *M* of *m*. Where should we exhibit the radius *r*? There seems to be no reason to treat any one of the three given circles *k*, *l*, and *m* or any one of the three points of intersection *A*, *B*, and *C* better than the others. We are prompted to connect all three centers with all the points of intersection of the respective circle: *K* with *B*, *C*, and *O*, and so forth.

The resulting figure (Fig. 10.2) is disconcertingly crowded. There are so many lines, straight and circular, that we have much trouble in "seeing" the figure satisfactorily; it "will not stand still." It resembles certain drawings in old-fashioned magazines. The drawing is ambiguous on purpose; it presents a certain figure if you look at it in the usual way, but if you turn it to a certain position and look at it in a certain peculiar way, suddenly another figure flashes on you, suggesting some more or less witty comment on the first. Can you recognize in our puzzling figure, overladen with straight lines and circles, a second figure that makes sense?

. . . .

We may hit in a flash on the right figure hidden in our overladen drawing, or we may recognize it gradually. We may be led to it by the

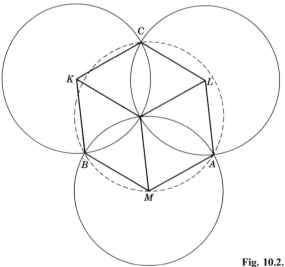

Fig. 10.2. Too crowded.

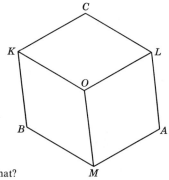

Fig. 10.3. It reminds you—of what?

effort to solve the proposed problem, or by some secondary, unessential circumstance. For instance, when we are about to redraw our unsatisfactory figure, we may observe that the *whole* figure is determined by its *rectilinear part* (Fig. 10.3).

This observation seems to be significant. It certainly simplifies the geometric picture, and it possibly improves the logical situation. It leads us to restate our theorem in the following form.

C. *If the nine segments*

$$K\,O, \qquad K\,C, \qquad K\,B,$$
$$L\,C, \qquad L\,O, \qquad L\,A,$$
$$M\,B, \qquad M\,A, \qquad M\,O,$$

are all equal to r, there exists a point E such that the three segments

$$E\,A, \qquad E\,B, \qquad E\,C$$

are also equal to r.

This statement directs our attention to Fig. 10.3. This figure is attractive; it reminds us of something familiar. (Of what?)

Of course, certain quadrilaterals in Fig. 10.3, such as *OLAM* have, by hypothesis, four equal sides, they are rhombi. A rhombus is a familiar object; having recognized it, we can "see" the figure better. (Of what does the *whole* figure remind us?)

Opposite sides of a rhombus are parallel. Insisting on this remark, we realize that the 9 segments of Fig. 10.3 are of three kinds; segments of the same kind, such as *AL*, *MO*, and *BK*, are parallel to each other. (Of what does the figure remind us *now?*)

We should not forget the conclusion that we are required to attain. Let us assume that the conclusion is true. Introducing into the figure the center *E* of the circle *e*, and its three radii ending in *A*, *B*, and *C*, we

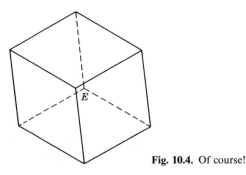

Fig. 10.4. Of course!

obtain (supposedly) still more rhombi, still more parallel segments;
see Fig. 10.4. (Of what does the whole figure remind us *now?*)

. . . .

Of course, Fig. 10.4 is the projection of the 12 edges of a parallelepiped
having the particularity that the projections of all edges are of equal
length.

Figure 10.3 is the projection of a "nontransparent" parallelepiped;
we see only 3 faces, 7 vertices, and 9 edges; 3 faces, 1 vertex, and 3 edges
are invisible in this figure. Figure 10.3 is just a part of Fig. 10.4, but
this part defines the whole figure. If the parallelepiped and the direction of projection are so chosen that the projections of the 9 edges
represented in Fig. 10.3 are all equal to r (as they should be, by hypothesis), the projections of the 3 remaining edges must be equal to r.
These 3 lines of length r are issued from the projection of the 8th, the
invisible vertex, and this projection E is the center of a circle passing
through the points A, B, and C, the radius of which is r.

Our theorem is proved, and proved by a surprising, artistic conception of a plane figure as the projection of a solid.

(The proof uses notions of solid geometry. I hope that this is not
a great wrong, but if so it is easily redressed. Now that we can characterize the situation of the center E so simply, it is easy to examine
the lengths EA, EB, and EC independently of any solid geometry. Yet
we shall not insist on this point here.)

10.3. The nature of the helpful idea

The foregoing exemplifies various points about the nature of the
helpful idea. Its coming was presented as extremely slow; instead of
being uttered triumphantly, it was stuttered. (This was done to give
an opportunity to the reader to participate in the discovery.) Also in
other respects, our example is somewhat one-sided, as any example is
bound to be, because of the immense variety of the phenomenon. Still,
if the reader considers it with sympathetic understanding, in the proper

light, in the proper setting, against the background of his own experience, our example may serve as a useful illustration of various features which occur frequently.

Very often the helpful idea arises *spontaneously*. It brings in some *conspicuous new element,* and it *changes our mode of conception.* A strong *conviction* that the end is within reach follows in its wake.

The *spontaneity* is a very characteristic feature but rather hard to describe. If it happened to the reader that, from the entanglement of the lines and letters in Fig. 10.2, the image of the parallelepiped "jumped" out at him unexpectedly, he will understand better what is meant. He will also understand, perhaps, to a certain extent, what is meant by inspiration, how it is possible to interpret the sudden appearance of an impressive idea as the whispering of an inner voice, or a warning given by a supernatural being.

The *conspicuous new element* arising in our case is the idea of the parallelepiped. It is rather strange that the appearance of a solid figure brings the decisive step in the solution of a problem of plane geometry. It is much more usual that the decisive new element is found within the domain to which the problem belongs. If the problem is one of plane geometry we would expect that the new element will be a new line added to the figure, or a relevant theorem suddenly remembered, or something of this kind.

The *change in the mode of conception* is very spectacular in the present case. The circles recede and vanish in the background, the straight lines come to the foreground. But we cease to conceive them as radii, we now relate everything to the parallelepiped. The former radii, their endpoints, the included quadrilaterals obtain a new significance, they represent now edges, corners, faces of the solid. The change in the mode of conceiving the elements of the problem is spectacular but typical. The same kind of revolutionary restructuring of our whole conception is brought about by the decisive idea in the solution of almost any other problem. As the idea emerges, the elements assume new roles, obtain new meanings. In the solution of geometric problems the elements are reshuffled and regrouped; they are assembled into triangles, or into pairs of triangles with corresponding sides, or into rhombi, or into any other familiar configuration serving the purpose of the search. A line that before the coming of the helpful idea was just a line, obtains some significance: it becomes the side of a triangle whose congruence with some other triangle is essential for the solution; or it becomes a transversal cutting two parallels; or it fits in some other way into a comprehensive picture. After the coming of the idea we *see more*—more meaning, more purpose, and more relations. The coming of the idea is similar to switching on the light in a dark room.

The helpful idea arises with the *conviction* that the end is within reach. A suddenly arising idea, displaying a spectacular new element amid dramatic rearrangement, has an impressive air of importance and carries strong conviction. This conviction is expressed by such exclamations as "Now I have it!" "I have got it!" "That's the trick!" In our present case, if you just see the parallelepiped without seeing that it leads to the solution, you do not yet have the decisive idea. You need more. You need not see in detail *how* the parallelepiped leads to the solution, but you should have a strong feeling that it *will* lead to the solution.

10.4. Ideas depend on chance

Have you got an idea? You say "Yes"? Then you are lucky. You cannot compel helpful ideas to appear. I take my problem earnestly. I put it to myself. I set it to myself. I realize it keenly. I become absorbed in my problem. I am waiting for a helpful idea; will it come? Perhaps at once, perhaps after some time, perhaps not at all.

We need helpful ideas, we naturally desire to have helpful ideas at our service. But, in fact, they are our masters and they are capricious and self-willed. They may flash upon us unexpectedly, but more often they are long in coming, and sometimes they just keep us waiting and do not turn up at all.

Ideas come when they want to come, not when we want them to come. Waiting for ideas in gambling.

If ideas came quite at random, the solution of problems would depend mainly on chance. Many people believe that this is so. Samuel Butler expressed this opinion in four witty lines:

All the inventions that the world contains,
Were not by reason first found out, nor brains;
But pass for theirs who had the luck to light
Upon them by mistake or oversight.

It is difficult to believe that such a widespread opinion should be entirely devoid of foundation, completely wrong. But is it completely right? Must we depend on chance alone when we have a problem to solve? I hope that, after all the preceding chapters, the reader has an opinion.

Examples and Comments on Chapter 10

10.1. *The spontaneity of ideas. A quotation and a comment.*
(1) We quote from *The Age of Reason,* First Part, by Thomas Paine.

Any person who has made observations on the state and progress of the human mind, by observing his own, cannot but have observed that there are two distinct classes of what are called thoughts: those that we produce in ourselves by reflection and the act of thinking, and those that bolt into the mind of their own accord. I have always made it a rule to treat those voluntary visitors with civility, taking care to examine, as well as I was able, if they were worth entertaining; and it is from them I have acquired almost all the knowledge that I have.

(2) Lichtenberg observed that we should not say "I am thinking" but rather "it is thinking" as we say "it is raining" or "it thunders." Lichtenberg suggests that there are spontaneous actions of the mind which we cannot command any more than the great forces of nature.

We could also say that our mind sometimes behaves like a sort of horse or mule, some strange animal whom we have to humor and occasionally to cudgel in order to get the desired service from it—which we quite often do not get.

(Georg Christoph Lichtenberg, 1742–1799, German physicist and writer; the *Aphorismen* may be his best remembered work.)

10.2. *Two experiments.* Some time (but not too much time) spent on crossword puzzles may be quite rewarding; we can learn something about problem solving, how we think, and how we ought to think.

(1) In a crossword puzzle you find the following clue: "A very common kind of heart (10 letters)." You may have no inkling at the outset what the word is or what the clue means. Yet a crossing word that you have succeeded in finding in the meantime yields some information: a letter in the middle. Another crossing word supplies a second letter; then you find a third letter, or a fourth—and suddenly the desired word "bolts into your mind."

Provide yourself with a sheet of paper and turn to the solution of this problem on p. 172. First cover the whole solution with your sheet. Then, sliding it downward, uncover just one line—do you know now the desired word? If not, uncover a second line, then the next, and so on; you may experience the idea "bolting into your mind."

†(2) If you know just a little calculus (not too much) you may try a similar experiment in finding the value of an indefinite integral. Take a sheet and turn to p. 172.

CHAPTER 11

THE WORKING
OF THE MIND

*Mariotte says that the human mind is like a bag: when you are
thinking you are shaking the bag until something falls out of it. Hence
there is no doubt that the result of thinking depends to some extent
on chance. I would add that the human mind is more like a sieve:
when you are thinking you are shaking the sieve until some minute
things pass it. When they pass, the spying attention catches whatever
seems relevant. Again, it is something like this: to catch a thief, the
commander of a city orders the whole population to pass a certain gate
where the man who was robbed is watching. Yet, to save time and
trouble, some method of exclusion may be used. If the man robbed
says that the thief was a man, not a woman, and an adult, not a
youngster or a child, those not concerned are excused from passing
the gate.*

LEIBNITZ: *Opuscules et fragments,* p. 170.

11.1. How we think

A problem solver must know his mind and an athlete must know
his body in about the same way as a jockey knows his horses. I imagine
that a jockey studies horses not for the sake of pure science but to make
them perform better, and that he studies more the habits and whims
of individual horses than horse physiology or horse psychology in
general.

What you start reading now is not a chapter in a textbook of
psychology; it is not exactly a conversation between problem solvers
who talk about the habits of their minds as jockeys may talk about the
habits of their horses; it is, however, more like a conversation than a
formal presentation.

62

11.2. Having a problem

An essential ingredient of the problem is the desire, the will, and the resolution to solve it. A problem that you are supposed to do and which you have quite well understood, is not yet your problem. It becomes your problem, you really have it, when you decide to do it, when you desire to solve it.

You may be involved more or less deeply in your problem—your desire to solve it may be more or less strong. Unless you have a very strong desire, your chances to solve a really hard problem are negligible.

The desire to solve your problem is a *productive* desire: it may eventually produce the solution, it certainly produces a change in your mental behavior.

11.3. Relevancy

You may have a problem so badly that the problem has you; you cannot get rid of your problem, it follows you everywhere.

A man with a problem may be obsessed by his problem. He appears absentminded; he does not notice things which appear obvious to his neighbors, and he forgets things which none of his neighbors would forget. Newton, working intensely on his problems, often forgot to eat his meals.

Yes, the problem solver's attention is *selective:* it refuses to dwell on things which appear irrelevant to his problem and espies the most minute things that appear relevant. It is a "spying" attention as Leibnitz put it.

11.4. Proximity

A student takes a written examination in mathematics. He is not required to do all the proposed problems, but he should do as many as possible. In this situation his best strategy may be to start by looking through all the problems at an appropriate pace and choose those he is most likely to master.

Observe that this supposes that the problem solver is able to assess to some extent the difficulty of his problems, that he can estimate to a degree his "psychological distance" from his problem's solution. In fact, anybody seriously concerned with his problem has a vivid feeling for the proximity of the solution and for the pace of his progress toward the solution. He may not use words but he feels keenly: "It goes well, the solution may be just around the corner," or "It goes so slowly and the solution is still far off," or "I got stuck, there is no progress at all," or "I am drifting away from the solution."

11.5. Prevision

As soon as we are seriously concerned with our problem, we try to foresee, we try to guess; we expect something, we anticipate an outline of the shape of the solution. This outline may be more or less definite—and of course it may be more or less wrong, although I would say not often very wrong.

All problem solvers guess, but the sophisticated and the unsophisticated guess somewhat differently.

A primitive person just sits there with his problem, scratching his head or chewing his pencil, waiting for a bright idea, and doing little or nothing to bring that bright idea nearer. And when the desired idea eventually appears and brings a plausible guess, he simply accepts that guess, regarding it as the solution with little or no criticism.

A more sophisticated problem solver takes his guesses more skeptically. His first guess may be: "There are 25" or "I should tell him this and that." Yet then he checks his guess and may change it: "No, not 25. Yet let me try 30" or "No. It is no use to tell him that, because he could answer thus and so. Yet I could tell him that . . . ." And eventually, by "trial and error," by successive approximations, the problem solver may arrive at the right answer, at an appropriate plan.[1]

A still more sophisticated and more experienced problem solver, when he does not succeed in guessing the whole answer, tries to guess some part of the answer, some feature of the solution, some approach to the solution, or some feature of an approach to the solution. Then he seeks to expand his guess, but also seeks opportunities to check his guess, and so he seeks to adapt his guess to the best information he can get at the moment.

Of course, both the sophisticated and the unsophisticated would like to have a really good guess, a bright idea.

And everybody would like to know what chances his guess has to come true. Such chances cannot be precisely evaluated (this is not the place to discuss remote possibilities of evaluation). Many times, however, the problem solver has a definite feeling about the prospects of his guess. Primitive people who do not even know what a proof is may have the strongest feelings about their guesses; sophisticated people may distinguish fine shades of feeling; but anybody who has conceived a guess has some feeling about the likely fate of his guess. And so we notice still another sort of feeling, besides the feelings of relevancy and proximity, in the problem solver's mind.

Is this point relevant? How far off is the solution? How good is this guess? Such questions accompany each move of the problem

[1] Sect. 2.2(1) and 2.2(5).

solver; they are more felt than formulated and the answers, too, are more felt than formulated. Do such feelings guide the problem solver or do they merely accompany his decisions? Are they causes or symptoms? I don't know, but I do know that if you do not have such feelings, you are not really concerned with your problem.

11.6. Region of search

I seldom part with my wrist watch, but when I do I usually have some trouble to find it. When I miss my watch, I habitually start looking for it at some well-defined place: on my desk, or on a certain shelf where I am used to store little belongings, or at any third place if I happen to remember that I took off my wrist watch just there.

Such behavior is typical. As soon as we are seriously concerned with our problem, we anticipate an outline of its solution. This outline may be vague, it may be hardly conscious, but it manifests itself in our behavior. We may try various solutions, but they are all alike; they are all within that preconceived, but perhaps not consciously preconceived, outline. When none of the solutions tried fits the problem, we feel lost, nothing else comes to mind; we cannot step outside that preconceived outline. We do not look for just any kind of solution, but for a certain kind, a kind within a limited outline. We do not look for a solution just anywhere in the world, but for a solution within a certain limited *region of search*.[2]

To begin our search within a likely limited region may be reasonable. When I am trying to find my missing wrist watch, it is quite reasonable not to look for it anywhere in the universe, or anywhere in the city, or anywhere in the house, but just on my desk where I found it several times in the past. It is quite reasonable to begin by seeking the unknown within that limited region, but it is unreasonable to persevere in seeking it there even when it becomes more and more clear that it is not there.

11.7. Decisions

Problem solving may be contemplative; with primitive people, it may be inarticulate brooding. Or it may be a long, strenuous, winding road to the solution, each turning of which is marked by a decision. Such decisions are prompted (or perhaps merely accompanied) by feelings of relevancy and proximity, by swelling or fading hope. Decisions and prompting feelings are seldom expressed in words, but may be occasionally:

[2] Karl Duncker, On Problem Solving, *Psychological Monographs,* vol. 58, No. 5 (1945). See p. 75.

"Now, let me look at this."

"No, there is not much to see here. Let me look at that."

"There is not much to see here either, but there is something in the air. Let me look at it a little longer."

An important type of decision is to enlarge the region of search, to discard a limitation the narrowness of which starts giving us an oppressive feeling.

11.8. Mobilization and organization

The problem solver's mental activity is very imperfectly known and its complexity may be unfathomable. Yet one result of this activity is perfectly obvious: as the problem solver advances, he collects more and more material.

Let us compare the problem solver's conception of a mathematical problem at the beginning and at the end of his work. When the problem arises, there is a simple picture: the problem solver sees his problem as an undivided whole without details, or with very few details; for instance, he may see just the principal parts, unknown, data, and condition, or hypothesis and conclusion. Yet the final picture is very different: it is complex, full of added details and materials the relevancy of which the problem solver could hardly have suspected at the outset. There are auxiliary lines in the originally almost empty geometric figure, there are auxiliary unknowns, there are materials from the formerly acquired knowledge of the problem solver, especially theorems applied to the problem. That just these theorems will be applicable, the problem solver did not foresee at all at the beginning.

Where do all these materials, auxiliary elements, theorems, etc. come from? The problem solver has collected them; he had to extract them from his memory and purposefully connect them with his problem. We call such collecting *mobilization* and such connecting *organization*.[3]

Solving a problem is similar to building a house. We must collect the right material, but collecting the material is not enough; a heap of stones is not yet a house. To construct the house or the solution, we must put together the parts and organize them to a purposeful whole.

Mobilization and organization cannot actually be separated; they are complementary aspects of the same complex process—of our work aimed at the solution. Such work, when intensive, brings into play all our psychical resources, requires the whole gamut of our mental activities, and presents an inexhaustible variety of aspects. We may be tempted to distinguish some of the manifold mental operations involved and describe them by such terms as isolation and combination, recognizing and remembering, regrouping and supplementing.

---

[3] *Cf.* ex. 2.74.

The following lines attempt to describe these activities. Of course, the reader should not expect, and could not reasonably expect, hard and fast distinctions or rigid and exhaustive definitions.

11.9. Recognizing and remembering

In examining our problem we cheer up when we *recognize* some familiar feature. Thus, in examining a geometric figure, we may recognize with pleasure a triangle not noticed before, or a pair of similar triangles, or some other intimately known configuration. Examining an algebraic formula we may recognize a complete square, or some other familiar combination. Of course, we may also recognize, and it may be very useful to recognize, some more complex situation to which we cannot yet attach a name and for which we have not yet a formal definition, but which strikes us as familiar and important.

We have good reasons to be pleased when we have recognized a triangle in the proposed figure. In fact, we know several theorems and have solved various problems about triangles, and one or the other of these known theorems or former solutions may be applicable to our present problem. By recognizing a triangle, we establish contact with an extensive layer of our formerly acquired knowledge, some streak of which might be useful now. Thus, in general, recognizing may lead us to *remembering* something helpful, to mobilization of relevant knowledge.

11.10. Supplementing and regrouping

We have recognized a triangle in the figure and have succeeded in remembering a theorem about triangles that has some chance to be applicable to the present situation. Yet, to actually apply that theorem we must add some auxiliary line to our triangle, for instance, an altitude. Thus, in general, prospective useful elements just mobilized may be added to our conception of the problem to enrich it, to make it fuller, to fill in gaps, to supply its deficiencies, in a word, to *supplement* it.

Supplementing introduces new materials into our conception of the problem and is an important step in its organization. Yet sometimes we can make an important advance in organization without introducing any new material, just by changing the disposition of the elements already present, by conceiving them in new relations, by rearranging or *regrouping* them. By regrouping its elements, we change the "structure" of our problem's conception. Thus regrouping means *restructuring*.[4]

Let us state this consideration more concretely. The decisive step in the solution of a geometric problem may be the introduction of an

[4] *Cf.* Duncker, *loc. cit.* pp. 29–30.

appropriate auxiliary line. Yet sometimes we can take the decisive step without introducing any new line, just by conceiving the lines already present in a new fashion. For example, we may notice that certain lines form a pair of similar triangles. In noticing this familiar configuration, we recognize hitherto unobserved relations between the elements of the figure, we see the elements differently grouped, we see a new structure, we see the figure as a better arranged, more harmonious, more promising whole—we have restructured the problem material.

Regrouping may involve a change in emphasis. Elements and relations which were in the foreground before the regroupment may now surrender their privileged place and recede into the background; they may even recede so far that they practically drop out from the conception of the problem. For better organization we must now and then reject things which we thought relevant some time ago. Yet, on the whole, we add more than we reject.

11.11. Isolation and combination

When we are examining a complex whole, our attention may be attracted now by this detail and then by another. We concentrate on a certain detail, we focus on it, we emphasize it, we single it out, we distinguish it from its surroundings, in one word, we *isolate* it. Then the spotlight shifts to another detail, we isolate still another detail, and so on.

After examining various details and revaluing some of them, we may feel the need of visualizing again the situation as a whole. In fact, after the revaluation of some details, the appearance of the whole, the "vue d'ensemble," the "Gestalt" may have changed. The combined effect of our reassessment of certain details may result in a new mental picture of the whole situation, in a new, more harmonious *combination* of all the details.

Isolation and combination may advance the solution in complementing each other. Isolation leads to decomposing the whole into its parts, a subsequent combination reassembles the parts into a more or less different whole. Decomposed and recombined, again decomposed and again recombined, our view of the problem may evolve toward a more promising picture.

11.12. A diagram

A diagrammatic summary of the foregoing sections is offered by Fig. 11.1, which the reader should take for what it is worth. Nine terms are arranged in a square; one occupies the center of the square, four

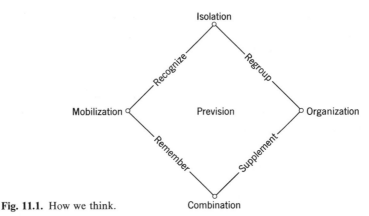

Fig. 11.1. How we think.

others the four vertices, and four more terms are written along the sides.

Mobilization and *organization* are represented by opposite ends of the horizontal diagonal of the square. In fact, these are complementary activities. Mobilization is extracting relevant items from our memory, organization is connecting such items purposefully.

Isolation and *combination* are represented by opposite ends of the vertical diagonal. In fact, these are complementary activities. Isolation is selecting a particular detail from the surrounding whole, combination is assembling dispersed details to a meaningful whole.

The sides adjoining the corner assigned to mobilization are labeled *recognize* and *remember*. In fact, mobilization of items relevant to the problem often starts from recognizing some element given with the problem and consists in remembering connected elements.

The sides adjoining the corner assigned to organization are labeled *supplement* and *regroup*. In fact, organization means supplementing the conception of the problem, making it fuller by adding new details and filling the gaps; and it also means regrouping the whole conception.

As we read the terms along the sides of the square, from left to right, we proceed from mobilized details to the organized whole; a detail just recognized, carefully isolated and focused, may induce a regroupment of the whole conception. Also, a detail remembered which fits into a combination is suitably added to the conception and supplements the whole.

Prevision is the center of our activity aimed at the solution, as the corresponding point is the center of our symbolic square. We keep on mobilizing and organizing, isolating and recombining, recognizing and remembering all sorts of elements, regrouping and supplementing our conception of the problem, just to foresee the solution, or some feature of the solution, or a bit of the path leading to it. If prevision comes to us abruptly, in a flash, we call it inspiration, or illuminating idea; our central desire is to have such an idea.

The mental operations surveyed in Fig. 11.1 take more specific forms when applied to special material. Thus, correspondingly to the four sides of the square, we list four mental operations important in solving mathematical problems:

Recognize:	Regroup:
use definitions	transform the problem
Remember:	Supplement:
known theorems and problems	introduce auxiliary elements

There is another point. The problem solver's moves are prompted or accompanied by feelings of relevancy and proximity, and feelings gauging the goodness of his guess. In discussing this we have mentioned incidentally that more sophisticated people have more differentiated feelings concerning such points. I do not wish to suppress here a rather speculative remark:[5] some such shades of feeling may be connected with the mental operations surveyed in Fig. 11.1.

We cheer up when our conception of the problem appears *well balanced* and *coherent, complete* with all details, and all details are *familiar*. If we have *distinct details* in a *harmonious whole,* the idea of the solution appears *near*. What we express with these terms is, it seems to me, that certain *activities considered above are well progressing,* or have already reached their goal.

Our conception of the problem appears well balanced when we do not feel the need of *regrouping* it, and appears as coherent when we have no trouble in *remembering* its details, but any detail easily recalls the others. When there is no need of *supplementing* it, the conception appears as complete, and it appears as familiar when all details have been *recognized.* Distinctness of details comes from foregoing *isolation* of, and concentration on, each detail, and the harmony of the whole conception results from successful *combination* of the details. We say that the idea is near when we feel that we are well progressing toward fuller *prevision.*

Wishing to arrange these favorable signs of our progress systematically, we place them so that their relative positions are the same as those of the corresponding terms in the square of Fig. 11.1. Thus, we arrange seven terms so as the four sides of that square and the three important points on its vertical diagonal are disposed. See the scheme:

[5] *Cf.* HSI, Signs of progress 4, p. 184.

Well isolated:
distinct details

Well recognized: Well grouped:
familiar   well balanced

Prevision promising:
idea near

Well remembered: Well supplemented:
coherent   complete

Well combined:
harmonious whole

11.13. The part suggests the whole

A whistling boy passed me in the street and I caught one or two measures of a melody which I like very much but had not heard for a long time. Suddenly that melody filled by mind, ousting completely whatever worries or idle thoughts I had before.

This little event is a good illustration of the "association of ideas," a phenomenon already described by Aristotle and by many authors after him. Bradley gives a good description: "Any part of a single state of mind tends, if reproduced, to reinstate the remainder." In fact, in my case, one measure brought back the whole impact of that melody and then, by and by, the remaining measures. Here is another description which lacks essential details but is easy to remember: "The part suggests the whole." Let us regard this short sentence as a convenient abbreviation of Bradley's more precise formulation.

Notice the important words "tends" and "suggests." The statements
"The part suggests the whole"
"The part tends to reinstate the whole"
"The part has a chance to reinstate the whole"
may be acceptable, but the sentence
"The part reinstates the whole"
is certainly inacceptable as an expression for the "law of association": there is no necessity of recall, just a chance, a tendency. We also know something about the strength of that tendency; a part more in the focus of attention suggests the whole stronger; several parts jointly suggest the whole stronger than any one of them singly. These additions are important if we wish to understand the role of association in the problem solver's mental experience.

Let us consider a strongly schematized example. A mathematical problem can be quickly solved by the application of a certain decisive theorem D, but it is very difficult to solve it without D. At the outset, the problem solver does not even suspect that the theorem D is relevant to his problem, although he is quite well acquainted with the theorem D itself. How can he discover the decisive role of D? There are various cases.

The case is relatively simple if the proposed problem and the theorem D have a common component part. The problem solver, after having tried this and that, will come upon that component part, isolate and focus it, and then that common part has a chance to recall or "reinstate" the whole theorem D.

The case is less simple if the original conception of the problem and the decisive theorem D have no common component. Still, if there is another theorem C, also known to the problem solver, that has some component in common with the problem and another component in common with D, the problem solver may attain D by first contacting C and then passing from C to D.

Of course, the chain of associations may be still longer; the proposed problem may be in associative contact with A, A with B, B with C, and finally C with D. The longer the chain the longer must the problem solver "shake the bag" or "shake the sieve" till the decisive D eventually falls out.

Shaking the bag or the sieve is a metaphorical way to describe the problem solver's mental experience (see the quotation prefixed to this chapter). The foregoing sections summarized by Fig. 11.1 attempted to describe this experience somewhat less metaphorically. There is a quite plausible interpretation of the activities described; through them, the problem solver seeks to establish desirable associative contacts.

In fact, in recognizing an element, the problem solver places it in a context with which it has strong associative contact. Any newly mobilized element, added to the problem's conception, offers chances to attract further elements with which it is in associative contact. When the problem solver isolates and focuses an element, the attention spent on it gives it more chance to bring in associated elements. A regroupment may bring together elements which could exercise more associative attraction jointly than anyone could singly.

It is, however, hardly possible to explain the problem solver's mental experience by association *alone;* there must be something else besides associative attraction to distinguish between relevant and irrelevant, desirable and undesirable, useful and useless associated elements and combinations.[6]

Examples and Comments on Chapter 11

11.1. *Your experience, your judgment.* The aim of this book is to improve your working habits. In fact, however, only you yourself can improve your own habits. You should find out the difference between what you are usually doing and what

---

[6] *Cf.* Duncker, *loc. cit.* p. 18.

you ought to do. This chapter was written to help you to see better what you are usually doing.

The following exercises, ex. 11.2–11.6, ask you to illustrate passages of the foregoing text. In the first place, try to find illustrations from your own work—such illustrations as come to your mind spontaneously have the best chance to be illuminating. Try to judge with an open mind whether the descriptions in the text or the illustrations in the solutions agree with your experience.

11.2. *Mobilization.* Recall your work on some problem of geometry where the figure, originally almost empty, became more and more filled by auxiliary elements as the solution progressed.

11.3. *Prevision.* Can you recall a case in which, at a pretty definite moment, you became suddenly convinced that the solution will succeed?

11.4. *More parts suggest the whole stronger.* Can you agree, judging by your own experience?

11.5. *Recognizing.* Can you recall a case in which recognizing an element (noticing its formerly unnoticed familiar role) appeared as the turning point of the solution?

11.6. *Regrouping.* Can you recall a case in which regrouping the figure appeared as the key to the solution?

11.7. *Working from inside, working from outside.* Establishing contacts between the proposed problem and his previous experience is certainly an essential part of the problem solver's performance. He can try to discover such contacts "from inside" or "from outside." He may remain within the problem, examining its elements till he finds one that is capable of attracting some usable element from outside, that is, from his previously acquired knowledge. Or he may go outside the problem, examining his previously acquired knowledge until he finds some element applicable to his problem. Working from inside, the problem solver scans his problem, its component parts, its aspects. Working from outside, he surveys his existing knowledge, and ransacks the provinces of knowledge that are most likely to be applicable to the present problem. The two parts of Fig. 11.2 attempt to give visual expression to "inside" and "outside" work.

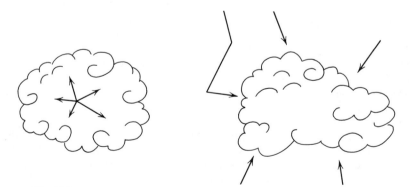

Fig. 11.2. Working from inside, working from outside—to pierce the clouds.

11.8. *Heuristic rat's maze.* Figure 11.3 may represent trails in wooded hilly country made by woodcutters without much care, and then abandoned; the point *E* marks the entrance. Figure 11.3 may also represent a maze in which rats are made to run in a psychological experiment.

But Fig. 11.3 may also symbolize the problem solver's activity when he is work-ing in a certain way. After a straightforward beginning, he follows a curving trail until he reaches an (actual or imagined) impasse. Then he turns back and starts retracing his steps, but noticing a side path he follows it to another impasse which makes him turn back. And so he goes on, attempting several trails, retrac-ing many of his steps, noticing new issues, exploring his problem and proceeding on the whole, we hope, in the right direction.

11.9. *Progress.* As the solution progresses, the problem solver's conception of the problem continually changes; especially, he collects more and more material connected with the problem. Let us tentatively assume that we can

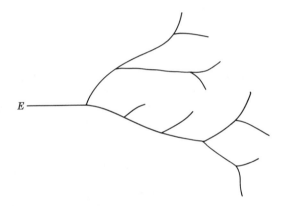

Fig. 11.3. Trials, trails, impasses, and side issues.

somehow measure the extent of the material collected at each instant, and that this extent is somehow proportional to the appropriateness of the problem solver's conception of his problem. Of course, this assumption is crude and unlikely, but it enables us to represent graphically the progress of the solution.

In a coordinate system, see Fig. 11.4, we take the time as abscissa and the "measure" of the problem's conception at the instant considered as the corre-sponding ordinate. The resulting curve represents the problem's conception as function of the time; it gives visual expression to the solution's growth in the problem solver's mind.

Let us assume that the solution develops without setbacks; correspondingly, Fig. 11.4 displays a rising curve, representing a never decreasing function of the time. The curve begins (from the left) with a slowly, almost uniformly rising dotted line, which intends to symbolize the "prehistory," the subconscious growth of the problem. The point *C*, from which the curve is traced in full, marks the beginning of *conscious* work. The slope of the curve at any point represents the pace of progress at the corresponding instant. The pace varies; it is slowest at the point S, which is a point of momentary *stagnation*; the tangent to the curve at the point S is horizontal. The pace is fastest at the point I where the slope is a maximum; I, a point of inflection, marks the emergence of the decisive idea, the instant of *inspiration*. (The point S is also a point of inflection, but of the opposite kind; the slope at S is a minimum.)

The development of the solution in the problem solver's mind is a complex process which has an inexhaustible variety of aspects. Figure 11.4 does not pretend to exhaust these aspects, but it may add a little to our other discussions; it may complement, for instance, the suggestions of Fig. 7.8.

11.10. *You too.* Much of what I know, or imagine to know, about problem solving came to me through reflecting on relatively few suggestive experiences. In reading a book, having a discussion with a friend, talking with a student, or

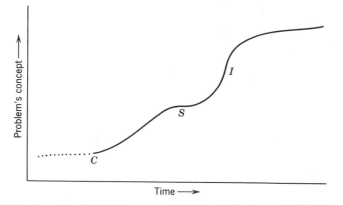

Fig. 11.4. Conscious beginning—stagnation—idea, inspiration, point of inflection.

observing the faces in an audience I suddenly recognized something and was tempted to say to myself: "You too, you are acting very much the same way." In fact I sometimes had this feeling "you too" in observing animals: dogs, birds, and once a mouse.

11.11. *Mice and men.* The landlady hurried into the backyard, put the mousetrap on the ground (it was an oldfashioned trap, a cage with a trapdoor) and called to her daughter to fetch the cat. The mouse in the trap seemed to understand the gist of these proceedings; he raced frantically in his cage, threw himself violently against the bars, now on this side and then on the other, and in the last moment he succeeded in squeezing himself through and disappeared in the neighbour's field. There must have been on that side one slightly wider opening between the bars of the mousetrap. The landlady looked disappointed, and so did the cat who arrived too late. My sympathy from the beginning was with the mouse, and so I found it difficult to say something polite to the landlady, or to the cat; but I silently congratulated the mouse. He solved a great problem, and gave a great example.

That is the way to solve problems. We must try and try again until eventually we recognize the slight difference between the various openings on which everything depends. We must vary our trials so that we may explore all sides of the problem. Indeed, we cannot know in advance on which side is the only practicable opening where we can squeeze through.

The fundamental method of mice and men is the same; to try, try again, and to *vary the trials* so that we do not miss the few favorable possibilities. It is true that men are usually better in solving problems than mice. A man need not throw himself bodily against the obstacle, he can do so mentally; a man can vary his trials more and learn more from the failure of his trials than a mouse.

CHAPTER 12

THE DISCIPLINE OF THE MIND

Method consists entirely in properly ordering and arranging the things to which we should pay attention.

DESCARTES: *Œuvres,* vol. X, p. 379; Rules for the Direction of the Mind, Rule V.

12.1. How we ought to think

Chapter 11 attempted to describe the typical mental behavior of the problem solver. Yet, is the typical also rational? We may behave so, but should we behave so?

These questions, in their vague generality, can hardly be answered, but they serve to indicate the trend of this chapter. Guided by the mental experience of the problem solver, which we surveyed in chapter 11, we shall attempt to list such mental operations (steps, procedures, etc.) as are typically useful for solving problems and, in listing them, we shall try to assign the proper position to each in the problem solving process.

We shall express the typically useful operations of problem solving in a concise, condensed form by "stereotyped" questions and recommendations. The reader should realize that these questions and recommendations can be interpreted in two different ways: either as quotations from a soliloquy of the problem solver, or as addressed by an understanding teacher to a student of his whom he wishes to help.[1]

---

[1] The reader is urged to compare the exposition of the present chapter with the parallel passages of HSI; for the foregoing see pp. 1–5, Purpose. The "stereotyped" questions and recommendations, which I consider as an essential part of my method, were introduced in my paper no. 2, see the Bibliography.

12.2. Focusing the aim

When you have a problem it often comes to your mind, and it may do so so often that it becomes an obsession. Yet you should not only think of your problem in some vague way, you should face it, you should see it clearly, you should ask yourself: *What do you want?*

There are many opportunities to ask this question in the course of the solution. When you become too deeply engaged in some side issue which may be irrelevant after all, when your thoughts start wandering, it may be advisable to ask yourself: *What do you want?* and bring your aim again into focus.

The aim of a problem to find is the unknown; to focus this aim, you should ask: *What is the unknown?* The aim of a problem to prove is the conclusion, and so the appropriate form of the question is: *What is the conclusion?*

Having clearly seen your aim, the thing you would like to have, you should take stock of the things in your possession which you could possibly use to attain your aim, and so you should ask yourself: *What have you?*

In fact, if you wish to join two points, to find your way from one point to the other, it may help you to look at the points alternately, first at one, then at the other, and so you often have the opportunity to ask the questions in succession: *What do you want? What have you got?*

Adapted to problems to find, the questions are: *What is the unknown? What are the data? What is the condition?* Adapted to problems to prove: *What is the conclusion? What is the hypothesis?*

What do these questions prompt us to do? To turn our attention to the points mentioned. According to Descartes (see the motto of this chapter), method consists in paying attention to all relevant points, one after the other, in appropriate sequence. Now, there is little doubt that the principal parts of a problem to find (unknown, data, and condition) and those of a problem to prove (conclusion and hypothesis) are relevant. They are even so important that it seems appropriate to consider them early; when you have well understood the problem as a whole, pay attention to its principal parts.

12.3. Assessing the prospects

A problem solver earnestly concerned with his problem keenly feels the proximity of the goal and the pace of his progress toward the goal; he keenly feels any change affecting the prospects of his plan. Now and then it is desirable to go a little beyond mere feelings, to see more clearly our position in relation to the problem, to diagnose the problem, to assess the prospects, and such is the tendency of the following questions.

Some problems are hopeless. If the problem in hand is hopeless, we should not get too deeply involved in it, and so we ask: *Is there any answer to this question?* Is there a clear answer, a sensible answer? *If there is an answer, can I find it?*

When we are dealing with a problem to find, we should ask: *Is there a solution?* We may ask more elaborately: *Is there just one solution, or several solutions, or no solution?* Is the condition just enough to determine the unknown, or does it demand too little, or too much?

When we are dealing with a problem to prove, the appropriate question is: *Is the theorem true or false?* We may ask more elaborately: *Is the theorem true? Or is a stronger hypothesis needed to imply the conclusion? Is the theorem sharp? Or is a weaker hypothesis enough to imply the conclusion?* And so on.

In fact, we cannot give a definitive answer to any of these questions before we have finished our work and solved the problem. Yet the questions do not really aim at a definitive answer, they demand only a provisional answer, a guess. In trying to guess right we may *clarify our position* in relation to the problem, and this is the desirable effect. We have stated the foregoing questions in a short, colloquial form. More cautiously, we should ask: Is it *likely* that there is an answer, a solution? The theorem stated may be true or false: what is *more likely?*

How soon should we ask the foregoing questions? There is (and there should be) no hard and fast rule. Very often they naturally follow the questions of sect. 12.2 concerned with the principal parts of the problem.

12.4. Wanted: an approach

The end suggests the means; the consideration of the aim (of the unknown, of the conclusion) may suggest an approach. One question leads to another: *What do you want? What is the unknown? How can you find this kind of unknown? From what data can you derive this kind of unknown?* And these questions may initiate a regressive approach; if we perceive "data" from which the unknown of the proposed problem could be derived, we may choose them as target for an auxiliary problem and so we may start working backward [see ex. 8.1(2)].

For a problem to prove, the corresponding questions are: *What do you want? What is the conclusion? How can you derive this kind of conclusion? From what hypothesis can you derive this kind of conclusion?*

Instead of emphasizing the unknown (the conclusion) we may emphasize the data (the hypothesis): *What are the data? What are such data good for? What can you derive from such data?* For problems to prove there is a corresponding string of questions: *What is the hypothesis? What is such a hypothesis good for? What can you derive from such a hypothesis?* These questions may initiate a progressive approach. (See ex. 8.1; we have discussed there, and we should remember, that, on the whole, regressive planning is preferable to progressive planning.)

Unfortunately, it can easily happen that we cannot design a serviceable plan, neither by regressive nor by progressive work. There are, however, other questions that could suggest an approach, and here are some that we may ask with advantage early in the game: *What kind of problem is it? Is it related to any known problem? Is it like any known problem?* Trying to classify our problem, trying to find relations and resemblances to known problems, we may perceive a familiar pattern applicable to our problem, and then we have something to begin with— we see the first stretch of a road that may lead us to the solution.

Trying to find a usable related problem we may survey those relations that are most often useful: *Do you know a related problem? Could you imagine a related problem? Do you know, or could you imagine, a problem of the same kind, or an analogous problem, or a more general, or a more special problem?* These questions, however, may lead us far from the proposed problem, so it is usually more profitable to ask them a little later when the problem is clarified and well fixed in our mind so that we do not risk losing sight of it when working at some distance from it.

12.5. Wanted: a more promising aspect

When you are handling material things (for instance, when you are about to saw a limb off a tree) you automatically put yourself in the most convenient position. You should act similarly when you are facing any kind of problem; you should try to put yourself in such a position that you can tackle the problem from the most accessible side. You turn the problem over and over in your mind; try to turn it so that it appears simpler. The aspect of the problem that you are facing at this moment may not be the most favorable: Is the problem as simply, as clearly, as suggestively expressed as possible? *Could you restate the problem?*

Of course you want to restate the problem (transform it into an equivalent problem) so that it becomes more familiar, more attractive, more accessible, more promising.

The aim of your work is to bridge the gap between what you want and what you have, to connect the unknown with the data, the conclusion with the hypothesis. Could you restate your problem so that unknown and data, hypothesis and conclusion appear *nearer to each other?*

Transform the conclusion, or transform the hypothesis, or transform both, but so that you bring them nearer to each other. Transform the unknown, or transform the data or the condition, or transform the whole problem, but so that you bring the unknown and the data nearer to each other.

As the solution progresses, new lines appear in the figure considered; new materials and new relations are added to the growing structure in the problem solver's mind. Each transformation of the problem is apt to bring in new elements. *Going back to definitions* is an important way of adding new material to the problem's conception.

For instance, our problem deals with the frustum of a pyramid (as in chapter 7). What is the frustum of a pyramid? How is it *defined?* The frustrum is the portion of a full pyramid that remains when a smaller pyramid is cut off from it by a plane parallel to its base. This answer brings two new solids to our attention, the full pyramid and the small pyramid, and we may find it advantageous to incorporate one or the other or both in the problem's conception.[2]

In going back to the definition of the elements given with the problem we introduce new elements, which in turn introduce still more new elements, and continuing thus we may further *unfold* the problem's concept. Such unfolding often brings us closer to the solution, but it need not; sometimes it may encumber the problem with unnecessary detail.

There are many ways of transforming problems which would deserve consideration; some are applicable only to certain particular kinds of problems, but others are rather general. (See ex. 12.1, 12.2.)

12.6. Wanted: relevant knowledge

The solution consists essentially in connecting the proposed problem with appropriate elements of our formerly acquired knowledge. When we are trying to restate the problem in a more promising form, we are in fact seeking a connection in starting from the problem, we are trying to pierce the clouds surrounding the problem by working at it "from inside." We may seek the connection from the other end, in trying to find some usable piece of knowledge in working at the problem "from outside."

It is obviously impossible to survey all our formerly acquired knowledge. Therefore, we should start exploring the parts of our knowledge that are most likely to be relevant to our problem.

If you are familiar with the domain to which your problem belongs, you know its "key facts," the facts you had most opportunity to use.

[2] *Cf.* HSI, Definition, pp. 85–92.

Make them ready as a good workman makes his favorite tools ready for use so that they are within easy reach.

If you have a problem to find, problems with the same unknown deserve your especial interest; such a problem may be the starting point for regressive planning (see chapter 8). If you have a problem to prove, theorems having the same conclusion as the theorem you are concerned with deserve your especial interest as possible starting points.

What are the key facts here? Is there a problem (especially a formerly solved problem) *with the same kind of unknown? Is there a theorem* (especially a formerly proved theorem) *with the same conclusion?* These questions have a good chance to extract some usable element from your previously acquired knowledge; it is advisable to start with them when you wish to collect pertinent facts. If they remain barren, however, you may be obliged to examine more complex or more recondite facts, or previously considered problems which have some other element in common with your problem, not just the unknown or the conclusion. There are, no doubt, elements in your knowledge which you could use for your present problem—but how are they related to it? How should you reach them? You may try generalization, specialization, analogy; you may ransack the whole province of knowledge to which your problem belongs.

Of course, the more extensive your knowledge is and the better it is organized, the more chance you have to find what you need. See ex. 12.3.

12.7. Wanted: reassessment of the situation

You are disappointed with the progress of your work. Various ideas you have had fizzled out, various routes you have tried led to an impasse. The figure in front of you, the whole conception of the problem in its present state, is perplexing and obscure, crowded but still incomplete; some essential element, some essential link is missing.

The trouble may be that you are involved in side issues and burdened with irrelevant material. Try to go back to the bare conception of the problem; look again at the unknown, the data and the condition, or the hypothesis and the conclusion. *Did you take into account the whole condition? Did you use all the data? Did you take into account the whole hypothesis,* did you use every part of it?

These questions are particularly relevant if you have previously convinced yourself that all the data and all the parts of the condition are needed to determine the unknown, or that the whole hypothesis is needed to derive the conclusion. Yet even if you have no such definite knowledge and you just suspect that all the data and all the clauses of

THE DISCIPLINE OF THE MIND

the condition, or of the hypothesis, may be essential, the foregoing questions are justified and may be helpful. They suggest that you should *try to use* a datum or clause which you have neglected so far, and so they may lead you to the missing link.

Or the trouble may be that you have not realized sufficiently the meaning of the essential terms of the problem. *Do you understand— have you visualized—all concepts essentially involved in the problem?* This question may prompt you to *go back to the definition* of some terms, and so it may suggest that you should unfold your problem; it has a chance to lead you to a more satisfactory restatement, to usable new elements.

12.8. The art of asking questions

In the foregoing sections we surveyed typical mental operations or "moves" of the problem solver. The description of each move culminated in a question (or advice—in *italics*) which may serve as an epitome, a condensed expression of the move.[3] It is important to understand how the problem solver (or the teacher) can use these questions.

Each of the collected questions, if asked at the right place and in the right time, may evoke the right answer, the right idea, a well-adapted move which can advance the solution. Thus the question may act as a *stimulant,* prompting the desirable reaction. These questions are *idea-needlers.*

Of course, under given circumstances you may not know what question to ask. Yet you may try several questions, one after the other, and arrive at a question that helps. Thus you can use the foregoing sections as a *repertory* of eligible questions, as a *checklist.*

Do not, however, use this checklist in a haphazard way, taking the questions at random, and do not use it mechanically, going through the questions in a fixed order. Instead, use this list of questions as an expert workman uses his *tool chest.* He takes a good look at the work he has to do and then he selects his tools. He may be obliged to try several tools until he finds the right one, but he does not pick his tools at random, or in a mechanically fixed order; he selects his tools *with judgement.* That is the way how you should select a question from those collected in this chapter when you are facing a problem.

Of course the workman probably has acquired his workmanship by long experience and by watching other workmen carefully. That is

[3] The questions concerned with "problems to find" are listed HSI pp. XVI–XVII (immediately inside the cover of the hard-cover edition). The reader is urged to study this list and the pertaining explanations and illustrations.

how you may master the use of the questions collected here. There is no hard and fast rule directing their use. Yet, if your personal experience of success and failure is behind these questions, and if you are keenly aware of your aim, there is a good chance that you will pick a good one.

Of course, the tools of a good workman are in good repair and well arranged in his tool chest. If you know the questions and the underlying moves described in this chapter, not from description but from experience, and if you have well understood their respective roles in the problem solving process, there is a good chance, I think, that you will handle your problems more professionally, less clumsily, and in a less haphazard way than most people.

Perhaps, any kind of mental discipline consists in the possession and the proper use of a set of appropriate questions. Yet how can we learn the art of asking questions? Has this art no rules?

Examples and Comments on Chapter 12

12.1. *Restate the problem.* The aim of our problem is to prove (or disprove) the statement: "If A is true then B is true." It may be advantageous to transform the problem and try to prove (or disprove) the equivalent contrapositive statement: "If B is false then A is false." See ex. 9.10.

Here is an analogous situation. Let x denote the unknown of a problem to find and $a, b, c, \ldots, l$ the data. (For instance, the unknown and the data may be measurements of various parts of a geometric figure.) It may be advantageous to interchange the unknown x and one of the data, say a. By so doing we pass from the original problem to a new one of which the unknown is a and the data $x, b, c, \ldots, l$. See ex. 2.33, 2.34, and 2.35.

We have considered here two types of transformations *independent of the subject matter*. The study of such types belongs properly to heuristics.

12.2. *Express it in mathematical language.* Descartes's great project which we discussed in sect. 2.1 may be (roughly) condensed into one piece of advice: Transform your problem, whatever it may be, into a mathematical problem by expressing it in the form of algebraic equations. Descartes's project failed, but we may revive it by generalizing it: *Express your problem in mathematical language.* The success of this advice depends, of course, on the scope of the mathematical language at our disposal. For instance, if we know and can use not only the symbols of algebra as Descartes, but also the symbols of the differential and integral calculus, we can treat many more problems.

"Mathematical language," if taken in a very wide sense, may include every kind of sufficiently clarified concept formation. In this very wide interpretation the advice "Express it in mathematical language" may be theoretically perfect, but it is practically pointless: it means no more than "try to be clear."

There is, however, a much narrower and even somewhat vague interpretation which is often useful. Graphs, diagrams, or geometric figures used as symbols form a sort of mathematical language. To *draw a figure,* to express the problem in the language of geometric figures, is often helpful. Some people have an urge to represent their ideas by some sort of geometric symbols. *Cf.* ex. 14.8.

12.3. *A well-stocked and well-organized store* of knowledge is an important asset of the problem solver. Good organization which renders the knowledge readily available may be even more important than the extent of knowledge. Anyhow, there are cases where too much knowledge may be an obstacle; it may prevent the problem solver from seeing a simple approach. Yet good organization can only help.

In a well-arranged store the items most frequently demanded are in the most accessible places, the items often used together are stored together, and labeling and details of arrangement are so planned that we can conveniently assemble any two, or more, related items.

A sensible arrangement of the books in your library or the tools in your tool chest may be helpful, but a sensible arrangement of the knowledge in your memory may be still more helpful and may deserve more care. Let us look at some points of organization important for the problem solver.

(1) In any subject matter there are some *key facts* (key problems, key theorems) which should be stored somehow in the forefront of your memory. When you are starting a problem, you should have some key facts around you, close at hand, just as an expert workman lays out his most frequently used tools around him when he starts working.

When you intend to prove a proposition of elementary plane geometry in the manner of Euclid, you may regard as key facts the four cases of congruence and the four cases of similarity of triangles. When you intend to reduce a problem of elementary geometry to equations in the Cartesian manner (chapter 2), you may regard as key facts the theorem of Pythagoras and the proportionality of the sides in similar triangles. You name the relevant key facts if you are experienced in dealing with the convergence of series, or with some other class of problems.

(2) Two questions which are again and again useful to the problem solver are: *By what data can you determine this kind of unknown? From what hypothesis can you derive this conclusion?* In view of the continual use of these questions, formerly solved problems with the same kind of unknown should be somehow "stored together," as should formerly proved theorems with the same conclusion.

(3) Do you know the city you live in? If you know it very well, you should be able to find the shortest route and the most convenient means of transportation between any two points of the city. Such is the desirable organization of knowledge; in the domain in which you work you should be able to find a practicable connection between any two points.

Surveying related problems may make for a better organization. Thus, the first part of this book has extensively surveyed problems related to each other by the common pattern of their solution. We may also survey strings of problems related by a common unknown, or by common data, or by analogy, and so on.

(4) Euclid wrote not only the "Elements" but also several other works. One

of them, the "Data," surveys the various data by which geometric objects can be determined. I like to believe that Euclid wrote the "Data" to help the problem solver by storing geometric knowledge in a *form readily available* to readers who often ask themselves: *by what data could I determine this kind of unknown?*

12.4. *By what data can you determine this kind of unknown?* List simple "problems to find," whose unknowns are described by one of the following sentences. (Capitals denote points.)

(1) ... find the point *P*.
(2) ... find the length *AB*.
(3) ... find the area of △*ABC*.
(4) ... find the volume of the tetrahedron *ABCD*.

12.5. *From what hypothesis can you derive this conclusion?* List simple theorems of plane geometry whose conclusions coincide with one of the following. (Capitals *A, B, C, ...* denote points.)

(1) ... then $AB = EF$.
(2) ... then $\angle ABC = \angle EFG$.
(3) ... then $AB:CD = EF:GH$.
(4) ... then $AB < AC$.

12.6. *Analogy: The triangle and the tetrahedron.* Here is a pair of problems, the first about the triangle, the second about the tetrahedron, which are analogous to each other:

Inscribe a circle in a given triangle.
Inscribe a sphere in a given tetrahedron.

List more pairs of problems, or pairs of theorems, similarly related to each other. Are the solutions, or proofs, also analogous, or how are they related to each other?

12.7. State a theorem about triangles analogous to the following theorem about tetrahedra:

A line joining the midpoints of two opposite edges of a tetrahedron passes through the center of gravity of any cross-section of the tetrahedron that is parallel to those two edges.

Could the theorem on the triangle help to prove the theorem on the tetrahedron? Also answer the corresponding questions dealing with the theorems in ex. 12.8 and ex. 12.9.

12.8. (Continued) Any plane passing through the midpoints of two opposite edges of a tetrahedron bisects its volume.

12.9. (Continued) In any tetrahedron the plane bisecting a dihedral angle divides the opposite edge into segments that are proportional to the areas of the adjacent faces including the dihedral angle.

12.10. *Attention and action.*

(1) Does method consists entirely in paying attention to all relevant points, one after the other, in appropriate sequence? (See the motto of this chapter.) I would not dare to assert that. Yet, certainly, a good part of the methodical work of the problem solver consists in focusing the relevant elements of his problem and their various combinations, one after the other.

The question, *"What is the unknown?"* and the advice, *"Look at the unknown!"* aim at the same effect: to turn the problem solver's attention to the unknown of his problem. Working methodically, the problem solver proceeds so as if he were directed by an inner voice:

Look at the problem as a whole.
Look at the unknown.
Look at the data.
Look at the condition.
Look at each datum separately.
Look at each clause of the condition separately.
Look especially at the datum that you have not used yet.
Look especially at the clause of the condition that you have not used yet.
Look at the combination of these two data.
And so on.

(2) Attention may initiate action.

Look at the unknown! What is the unknown? How can you find this kind of unknown? By what data can you determine this kind of unknown? Do you know— have you solved—a problem with this kind of unknown? The attention given the unknown induces the problem solver to search his memory for formerly solved problems with the same unknown. If the search succeeds, the problem solver may attempt to solve his problem in working backward (see chapter 8).

The case considered (recursive work initiated by attention paid to the unknown) is particularly frequent and useful, but attention given to any relevant element of the problem may lead to a profitable contact and hence to profitable action. For instance, attention to a term arising in the formulation of the problem may lead to *going back to the definition* of that term and so to a helpful restatement of the problem, to the introduction of usable new elements into the restated problem.

(3) The problem solver pays attention successively to various elements of the problem and to various combinations of these elements—he hopes to detect one that gives access to some profitable action—or to detect the one that gives access to the most profitable action. He hopes to find a bright idea that shows in a flash what there is to do.

12.11. *Productive thinking, creative thinking.* Thinking is productive if it produces the solution of the present problem, thinking is creative if it creates means to solve future problems. The greater the number and variety of the problems to which the means created are applicable, the greater is the creativity.

The problem solver may do creative work even if he does not succeed in solving his own problem; his effort may lead him to means applicable to other problems.

Then the problem solver may be creative *indirectly,* by leaving a good unsolved problem which eventually leads others to discovering fertile means.

I think that the Greeks who left us the problem of the trisection of an arbitrary angle did great creative work although they did not solve this problem, and inspite of the fact that, through the intervening centuries, this problem occasioned a tremendous amount of unproductive work. Yet this problem revealed a contrast: whereas an arbitrary angle can be bisected, only some special angles (such as $90°$) can be easily trisected by ruler and compasses; this led to the problem of dividing an angle into 5, 7, or 17 equal parts, contributed to the problem of solving equations by radicals, and eventually to the discoveries of Gauss, Abel, and Galois—which created means applicable to countless problems unsuspected by the Greek who first meditated on the trisection of an angle.

CHAPTER 13

RULES
OF DISCOVERY?

*Though it is difficult to prescribe any Thing in these Sorts of Cases,
and every Person's own Genius ought to be his Guide in these
Operations; yet I will endeavour to show the Way to Learners.*
NEWTON: *Universal Arithmetick* translated by Ralphson, 1769, p. 198.

*Since Arts are more easily learnt by Examples than Precepts, I have
thought fit to adjoin the Solution of the following Problems.*
NEWTON: *op. cit.*, pp. 177–178.

13.1. Rules and rules

As the work of the problem solver progresses, the face of the problem
continually changes. At each stage, the problem solver is confronted
with a new situation and a new decision: what should he do in this
situation, what should be the next step? *If* he possessed a perfect
method, an infallible strategy of problem solving, he could determine
the next step from the data of the present situation by clear reasoning,
on the basis of precise rules. Unfortunately there is no universal perfect method of problem solving, there are no precise rules applicable
to all situations, and in all probability there will never be such rules.

Yet there are rules and rules. Rules of conduct, maxims, and guidelines can be quite useful and reasonable without being as strict as the
rules of mathematics or logic. A mathematical rule is like a mathematical "line without breadth" separating black from white. Yet there
are quite reasonable rules which leave some latitude, some room for
secondary considerations; there is no sharp line of demarcation, and
sometimes there is no black and no white, just various shades of grey.

There seem to be attitudes, modes of thinking, habits of mind useful in many, and perhaps in most, problem solving situations. At any rate, the examples and discussions of the foregoing chapters seem to suggest such attitudes. Therefore, we should not ask: "Are there rules of discovery?" but we should put the question differently, perhaps so: "Are there maxims of some sort expressing attitudes useful in problem solving?"[1]

13.2. Rationality

We call an act or a belief *rational* if it is based on well-considered clear reasons and does not spring merely from less articulate and more obscure sources such as habit, unscrutinized impressions, feelings, or "inspiration." The assent we accord to a mathematical theorem after having critically examined its demonstration step by step, is the prototype of rational belief. From a certain point of view the main merit of the study of mathematical demonstrations is that it brings us nearer to that ideal rational behavior that befits man, the "rational being."

It is not obvious, however, how the problem solver should behave rationally. Let us consider his difficulty a little more concretely, let us visualize a frequently arising typical situation. It occurs to a problem solver as he is working at a certain problem A, that this problem is connected with another problem B. The study of problem B could possibly bring him nearer to his goal, the solution of his original problem A. On the other hand, however, the study of problem B may remain sterile and could result in loss of time and effort. And so the problem solver faces a decision: should he abandon for a time the study of his original problem A and switch to the study of the new problem B? His dilemma is to introduce or not to introduce B as an auxiliary problem. Can the problem solver arrive at a rational decision?

One important benefit that the problem solver may derive from problem B is that work on B may stir his memory and bring to the surface elements that he can use in solving his original problem A. What are the chances that work on B will have this desirable effect? It seems impossible to evaluate such chances merely on the basis of some clear, rational argument; to some extent, in some manner the problem solver must rely on his inarticulate feelings.

On the other hand, there may be articulate reasons for or against introducing B as an auxiliary problem; we have surveyed some such reasons in chapter 9. How should the problem solver take into account both his inarticulate feelings and his articulate reasons? He may—and

---

[1] The majority of the "rules" discussed in this chapter have been introduced, in a somewhat different formulation, in paper no. 13 of the author; see the Bibliography.

this seems to be a sensible procedure—carefully consider his articulate reasons for an appropriate time and refer to his feelings afterwards for the final decision. In fact, the careful consideration of his articulate reasons has had a chance to influence his feelings in the right direction; he could scarcely have acted more rationally.

At any rate, the problem solver must learn to balance inarticulate feelings against articulate reasons. And perhaps this is the most important thing he must learn. It seems to me that the main rule of conduct, the principal maxim of the problem solver should be:

> *Never act against your feelings, but try to see with an open mind clear reasons for or against your plans.*

13.3. Economy, but no predeterminable limitation

Tendency to economy needs no explanation. Everybody understands that you wish to save your assets, that you try to expend as little money, time, and effort as possible in performing a given task. Your mind may be your most important asset, and saving mental effort may be the most important kind of economy. *Do not do with more what can be done with less.* This is the general principle of economy; we observe it in problem solving if we try to attain the solution by bringing in *as little extraneous material as possible.*

Obviously, we have to scrutinize carefully the problem itself and such materials as immediately belong to it; we should first try to see a path to the solution without examining other things. If we fail to notice such a path, we examine materials which do not belong so immediately to the problem but are still close to it. If we fail again to notice some useful suggestion we may proceed to further details but—and this is the desirable general attitude—we are reluctant to spend time and effort on farfetched things as long as there is some hope that we can solve the proposed problem by starting from things closer to it. This sort of common sense *economy* should be expressed by the maxim:

Stay as close to the problem as possible.

Yet we cannot predict how close to the problem we shall be able to stay. Some superior being, possessing a perfect method of problem solving unattainable to us, could foretell with certainty how far away he will be obliged to go to collect the materials necessary to the solution —yet we cannot. In accordance with the principle of economy, we first explore the proposed problem itself; if this is not enough, we explore the immediate neighborhood of the problem. If even this is not enough, we explore a wider neighborhood; whenever our exploration fails to discover a path to the solution, we are obliged to go further. If you have made up your mind to solve the problem at any cost, you may pay a very high cost indeed; at any rate, you cannot limit the cost in advance. The resolute problem solver has to accept the principle of *absence of limitation,* counterpart of the principle of economy:

Yet be prepared to go as far away from the problem as circumstances may oblige you to go.

13.4. Persistence, but variety

"Genius is patience."

"Genius is one per cent inspiration and ninety-nine per cent perspiration."

One of these sayings is attributed to Buffon, the other to Edison, and both convey the same message: a good problem solver must be obstinate, he must stick to his problem, he must not give up.

What applies to the whole does not quite apply to the parts. Certainly, the problem solver examining some detail or some aspect of his problem should stick to it, should not give up too soon. Yet he should also try to assess the prospects and should not insist on squeezing an orange he has already squeezed completely dry.

Stick to the point examined till there is hope for some useful suggestion.

The work of the problem solver is largely a work of mobilization; he has to extract from his memory items applicable to his present problem. The item he needs to recall may be closer associated with, and more easily remembered through, a certain aspect or a certain detail of his problem than through other aspects or details. But the problem solver does not know in advance which detail or aspect will bring him nearer to his goal. Therefore, he should consider a variety of aspects or details, and he should certainly consider all the more basic or more promising ones.

To cover the whole territory without loss of time, the problem solver should not stay too long in, or return too soon to, the same spot. He should seek variety, he should see something new at each stage—a new point, or a new combination of points previously examined, or see in a new light points and combinations already considered. The aim is, of course, to see the whole problem in a new, more promising light.

In brief, variety is a necessary counterpart of persistence. As stated earlier, you should explore the points you consider with some persistence. *Yet try to examine some ground not yet covered at each step and try to perceive some useful suggestion in whatever you examine.*

The most obvious danger that this maxim warns you of, the greatest enemy of variety, is falling in a groove—repeating the same thing over and over again without change and without progress.

13.5. Rules of preference

If there are two approaches to the same problem which appear equally advantageous in other respects, but one of which seems easier than the other, it is natural to try the easier approach first. We see here a (rather trivial) *rule of preference* which we may state:

The less difficult precedes the more difficult.

In fact, this statement as it stands is incomplete: we should have added as a limitation or restriction *"ceteris paribus"* or "other things being equal." Now, let us note that this essential limitation, although not expressed, must be *understood* in this and in all subsequent similarly formulated rules of preference. Following are a couple equally obvious rules of this kind:

The more familiar precedes the less familiar.

An item having more points in common with the problem precedes an item having less such points.

These rules are obvious, but their application may be less obvious. The restriction *"ceteris paribus,"* especially, not expressed but understood, may require all the subtlety of the problem solver.

There are other less obvious, less general, more specific rules of preference. To examine them in an orderly fashion we should classify the elements involved. Here is a classification which may be incomplete, but at least the most conspicuous cases fit quite naturally into it:

(1) Materials inherent in the problem.
(2) Available knowledge.
(3) Auxiliary problems.

We shall take up the connected rules in the three following sections.

13.6. Materials inherent in the problem

When you start examining a problem, you do not know yet which details of the problem will turn out important. Hence there is the danger that you may emphasize too much an unimportant detail, and then you may get caught by it, you cannot get away from it. Therefore, begin by examining the problem as an undivided whole, do not concern yourself with details, let the whole of the problem work on your mind until you fully realize the point, the aim of the problem. *The whole precedes the parts.*

When you have the impression that you cannot profit much more by the consideration of the problem as a whole and you are about to go into details, observe that there is something like a hierarchy of details.

The highest ranking, nearest to the "center" of the problem, are the principal parts. (As we have discussed, the hypothesis and the conclusion are the principal parts of a problem to prove, the unknown, the data, and the condition those of a problem to find.) It is natural to begin the detailed consideration of the problem with the principal parts; you should see clearly, very clearly, the desired conclusion and the hypothesis from which it should follow—or the desired unknown, the available data, and the connecting condition. *The principal parts precede the other parts.*

One or the other of the principal parts may be subdivided: the hypothesis may consist of several assumptions, the conclusion of several assertions, and the condition of several clauses; the unknown may be a multipartite unknown having several components; and there may be several data which you may have lumped together in a first consideration. After the principal parts, their subdivisions deserve your next attention; you may examine each of the data, each of the unknowns, each clause of the condition, each assumption of the hypothesis, each assertion of the conclusion by itself. At any rate, other details of the problem may be considered as further removed from its center than the principal parts, which rank highest, and the subdivisions of the principal parts, which come next. There may be an order of precedence also among remoter parts of the problem. (Thus, a certain concept A may be involved in the statement of the problem, and another concept B may be involved in the definition of A; obviously B is further removed from the center of the problem than A.) Now, do not go further than necessary. Other things being equal (this limitation remains with us), you have more chance to make good use of a part closer to the center of the problem than of a more remote one. *Less remote parts precede more remote parts.*

13.7. Available knowledge

As we have repeatedly discussed, an essential (perhaps the most essential) performance of the problem solver is to mobilize the relevant elements of his knowledge and connect them with the elements of his problem. The problem solver can work at this task "from inside" or "from outside." He may remain inside his problem, unfolding it, scrutinizing its various parts, and hoping that such scrutiny will attract some usable piece of knowledge. Or he may go outside his problem, roaming over various regions of his knowledge and looking for usable pieces. In the foregoing section we have observed the problem solver working from inside, now we wish to see him working from outside.

Any piece of knowledge, any experience of our past could be usable for the solution of the problem at hand, but it is obviously impossible to pass in review all our knowledge or to recall our whole past point by point. Even if our problem is mathematical and we take into account only that relatively clear and well-ordered portion of our knowledge that consists of formerly solved problems and formerly proved theorems within a certain branch of mathematics, we cannot undertake to examine all that material item by item. We have to restrict ourselves, we have to pick out such items as have the best chance to be usable.

Let us consider, one after the other, problems to solve and problems to prove.

We have a problem to solve. We have already considered its principal parts, the unknown, the data, the condition. Now we are searching our memory for some formerly solved problem that could be helpful. It is natural to look for one that has something in common with our present problem, the unknown, or one of the unknowns, the data, or one of the data, some concept essentially involved, and so on. There is a possibility, more or less remote, that any such formerly solved problem may be helpful, but there are certainly too many for us to examine. Yet, among all the possible points of contact there is one that deserves more attention than the others—the *unknown*. (Especially, we may try to use a formerly solved problem that has the same kind of unknown as our present problem as the starting point of a regressive solution, of working backward; see chapter 8.) Of course, in special situations other points of contact may be preferable, but in general, *a priori*, other things being equal, we should first *look at the unknown*.[2] *Formerly solved problems having the same kind of unknown as the present problem precede other formerly solved problems.*

If we cannot find a sufficiently approachable formerly solved problem that has the same kind of unknown as the problem in hand we may look for one that has a *similar* kind of unknown—even such problems have a high priority, although not the highest priority.

If we have a problem to prove, the situation is similar. In searching our memory for a formerly proved theorem that could be helpful, we should *look at the conclusion*. *Formerly proved theorems with the same conclusion as the theorem that we are trying to prove precede other formerly proved theorems.*

The next best thing to a formerly proved theorem with the same conclusion may be one with a conclusion similar to the conclusion of the theorem we want to establish.

[2] HSI, pp. 123–129, Look at the unknown.

13.8. Auxiliary problems

One of the most crucial decisions confronting the problem solver is the choice of an appropriate auxiliary problem. He may look for such a problem working from inside, or working from outside, or (what is often the most sensible procedure) working alternately from inside and outside. Certain sorts of auxiliary problems, other things being equal, have more chance to be useful than others.

An auxiliary problem may advance the solution of the proposed problem in an inexhaustible variety of ways: it may yield material help, methodological help, stimulating influence, guidance, or practice. Yet whatever kind of help we are seeking, we have more chance *a priori* to obtain it from an auxiliary problem whose connection with the proposed problem is closer, than from another whose connection is looser. *Problems equivalent to the proposed one precede such problems as are more or less ambitious, and these precede the rest.* We can express the same thing in other terms: *Bilateral reduction precedes unilateral reduction which precedes looser connections* (see chapter 9).

13.9. Summary

Rationality. *Never act against your feelings, but try to see with an open mind clear reasons for or against your plans.*

Economy but no predeterminable limitation. *Stay as close to the problem as possible. Yet be prepared to go as far away from it as circumstances may oblige you to go.*

Persistence but variety. *Stick to the point examined until there is hope for some useful suggestion. Yet try to examine some ground not yet covered at each step and try to perceive some useful suggestion in whatever you examine.*

Rules of preference

The less difficult precedes the more difficult.

The more familiar precedes the less familiar.

An item having more points in common with the problem precedes an item having less such points.

The whole precedes the parts, the principal parts precede the other parts, less remote parts precede more remote parts.

Formerly solved problems having the same kind of unknown as the present problem precede other formerly solved problems.

Formerly proved theorems with the same conclusion as the theorem that we are trying to prove precede other formerly proved theorems.

Problems equivalent to the proposed one precede such problems as are more or less ambitious, and these precede the rest. Or: Bilateral reduction precedes unilateral reduction which precedes looser connections.

Yet add mentally to all rules of preference: OTHER THINGS BEING EQUAL.

Examples and Comments on Chapter 13

13.1. *The genius, the expert, and the beginner.* The genius acts according to the rules without knowing that there are rules. The expert acts according to the rules without thinking of the rules, but, if needed, he could quote the rule applicable to the case in question. The beginner, trying to act according to the rule, may learn its veritable meaning from success and failure.

These remarks are not new of course. St. Augustine, speaking of orators and the rules of rhetoric, put it so: "They follow the rules because they are eloquent, they are not eloquent because they follow the rules."

13.2. *Of plums and plans.* Shall I take this fruit? Is it ripe enough to be picked? Of course, if I leave it on the tree, it may still ripen and gain in flavor. On the other hand, if it remains on the tree, it may be attacked by birds or insects, knocked down by the wind, taken by the neighborhood children, or be spoiled or lost in some other way. Should I take it now? Has it quite the right color, shape, softness, smell, general appearance?

Color, shape, softness, and smell indicate the flavor, they do not guarantee it. When examining a fruit from my own garden, I can assess such indications quite reliably, at least I think so. When the fruit is not so familiar to me, my estimate is certainly much less reliable. Anyhow, estimating the taste of a fruit on the basis of its appearance can scarcely be considered as "fully objective." Such estimates depend to a large extent on personal experience which can scarcely be explained completely or argued exhaustively on an impersonal level.

Shall I take this step? Is my plan ripe enough to be carried through? Of course it is not certain that my plan will work. If I pondered it a little longer, I could see its chances a little better. On the other hand, I should do something sooner or later and, for the moment, I cannot think of a more reliable plan. Should I start carrying through that plan now? Does it look promising enough?

In picking plums or pursuing plans we may have some articulate reasons, but our decision will scarcely be based on reason alone. Our evaluation of the general aspect, of the promise of the fruit or of the problem situation, is bound to depend on feelings not fully analyzable.

13.3. *Style of work.* Anybody who intends to formulate "rules of discovery" should realize that different problem solvers work differently. Each good problem solver has his individual style.

Let us compare two problem solvers; one has the mentality of an "engineer," the other the mentality of a "physicist." They are trying to solve the same problem, but they are working differently as they have different interests. The engineer aims at a neat, short, efficient solution (the "least expensive," the "most marketable" solution). The physicist aims at the principle underlying the solution. The engineer is more inclined to "productive," the physicist to "creative," thinking (ex. 12.11). Accordingly they prefer different means to the same end.

Let us consider a half-concrete example. There seem to be two approaches to the problem that the engineer and the physicist are trying to solve. On the one hand, the proposed problem shows some analogy to a certain formerly solved problem A. On the other hand, the proposed problem seems to be amenable to treatment according to a general pattern B. There is a choice between these two approaches, between A and B. I am inclined to think that under such circumstances, other things being equal, the engineer prefers to investigate the concrete problem A and the physicist the general pattern B.

This example leads to the general statement: the style of work of the problem solver consists essentially in a *system of preferences,* of priorities. To the rules of preference summarized in sect. 13.9, the problem solver may add others (for instance: "General patterns precede particular facts"). Moreover, he may *emphasize* certain rules of preference more than others ("When there is a conflict, rule X has more weight than rule Y").

CHAPTER 14

ON LEARNING, TEACHING, AND LEARNING TEACHING[1]

*What you have been obliged to discover by yourself leaves a path in
your mind which you can use again when the need arises.*

G. C. LICHTENBERG: *Aphorismen.*

*Thus all human cognition begins with intuitions, proceeds from thence
to conceptions, and ends with ideas.*

I. KANT: *Critique of Pure Reason*, translated by
J. M. D. Meiklejohn, 1878, p. 429.

*I [planned to] write so that the learner may always see the inner
ground of the things he learns, even so that the source of the invention
may appear, and therefore in such a way that the learner may understand everything as if he had invented it by himself.*

G. W. VON LEIBNITZ: *Mathematische Schriften*, edited by Gerhardt,
vol. VII, p. 9.

14.1. Teaching is not a science

I shall tell you some of my opinions on the process of learning, on
the art of teaching, and on teacher training.

My opinions are the result of a long experience. Still, such personal
opinions may be irrelevant and I would not dare to waste your time by
telling them if teaching could be fully regulated by scientific facts and

---

[1] Sect. 14.1–14.7 were presented as an address at the 46th Annual Meeting of the Mathematical Association of America at Berkeley and printed previously; see the Bibliography,
paper no. 20 of the author.

theories. This, however, is not the case. Teaching is, in my opinion, not just a branch of applied psychology—at any rate, it is not yet that for the present.

Teaching is correlated with learning. The experimental and theoretical study of learning is an extensively and intensively cultivated branch of psychology. Yet there is a difference. We are principally concerned here with complex learning situations, such as learning algebra or learning teaching, and their long-term educational effects. The psychologists, however, devote most of their attention to, and do their best work about, simplified short-term situations. Thus the psychology of learning may give us interesting hints, but it cannot pretend to pass ultimate judgment upon problems of teaching.[2]

14.2. The aim of teaching

We cannot judge the teacher's performance if we do not know the teacher's aim. We cannot meaningfully discuss teaching, if we do not agree to some extent about the aim of teaching.

Let me be specific. I am concerned here with mathematics in the high school curriculum and I have an old fashioned idea about its aim: first and foremost, it should teach those young people to THINK.

This is my firm conviction; you may not go along with it all the way, but I assume that you agree with it to some extent. If you do not regard "teaching to think" as a primary aim, you may regard it as a secondary aim—and then we have enough common ground for the following discussion.

"Teaching to think" means that the mathematics teacher should not merely impart information, but should try also to develop the ability of the students to use the information imparted: he should stress know-how, useful attitudes, desirable habits of mind. This aim may need fuller explanation (my whole printed work on teaching may be regarded as a fuller explanation) but here it will be enough to emphasize only two points.

First, the thinking with which we are concerned here is not daydreaming but "thinking for a purpose" or "voluntary thinking" (William James) or "productive thinking" (Max Wertheimer). Such "thinking" may be identified here, at least in first approximation, with "problem solving." At any rate, in my opinion, one of the principal aims of the high school mathematics curriculum is to develop the students' ability to solve problems.

Second, mathematical thinking is not purely "formal"; it is not con-

---

[2] E. R. Hilgard, *Theories of Learning*, 2nd ed., Appleton-Century-Crofts, New York, 1956. *Cf.* pp. 485–490.

cerned only with axioms, definitions, and strict proofs, but many other things belong to it: generalizing from observed cases, inductive arguments, arguments from analogy, recognizing a mathematical concept in, or extracting it from, a concrete situation. The mathematics teacher has an excellent opportunity to acquaint his students with these highly important "informal" thought processes, and I mean that he should use this opportunity better, and much better, than he does today. Stated incompletely but concisely: let us teach proving by all means, but let us also teach guessing.

14.3. Teaching is an art

Teaching is not a science, but an art. This opinion has been expressed by so many people so many times that I feel a little embarrassed repeating it. If, however, we leave a somewhat hackneyed generality and get down to appropriate particulars, we may see a few tricks of our trade in an instructive sidelight.

Teaching obviously has much in common with the theatrical art. For instance, you have to present to your class a proof which you know thoroughly having presented it already so many times in former years in the same course. You really cannot be excited about this proof—but, please, do not show that to your class; if you appear bored, the whole class will be bored. Pretend to be excited about the proof when you start it, pretend to have bright ideas when you proceed, pretend to be surprised and elated when the proof ends. You should do a little acting for the sake of your students who may learn, occasionally, more from your attitudes than from the subject matter presented.

I must confess that I take pleasure in a little acting, especially now that I am old and very seldom find something new in mathematics; I may find a little satisfaction in re-enacting how I discovered this or that little point in the past.

Less obviously, teaching has something in common also with music. You know, of course, that the teacher should not say things just once or twice, but three or four or more times. Yet, repeating the same sentence several times without pause and change may be terribly boring and defeat its own purpose. Well, you can learn from the composers how to do it better. One of the principal art forms of music is "air with variations." Transposing this art form from music into teaching you begin by saying your sentence in its simplest form; then you repeat it with a little change; then you repeat it again with a little more color, and so on; you may wind up by returning to the original simple formulation. Another musical art form is the "rondo." Transposing the rondo from music into teaching, you repeat the same essential sentence several times with little or no change, but you insert between two repetitions some appropriately contrasting illustrative material. I hope that when you listen the next time to a theme with variations by Beethoven or to a rondo by Mozart you will give a little thought to improving your teaching.

Now and then, teaching may approach poetry, and now and then it may approach profanity. May I tell you a little story about the great Einstein? I listened once to Einstein as he talked to a group of physicists in a party. "Why have all the electrons the same charge?" said he. "Well, why are all the little balls in the goat dung of the same size?" Why did Einstein say such things? Just to make some snobs to raise their eyebrows? He was not disinclined to do so, I think. Yet, probably, it went deeper. I do not think that the overheard remark of Einstein was quite casual. At any rate, I learnt something from it: Abstractions are important; use all means to make them more tangible. Nothing is too good or too bad, too poetical or too trivial to clarify your abstractions. As Montaigne put it: The truth is such a great thing that we should not disdain any means that could lead to it. Therefore, if the spirit moves you to be a little poetical, or a little profane, in your class, do not have the wrong kind of inhibition.

14.4. Three principles of learning

Teaching is a trade that has innumerable little tricks. Each good teacher has his pet devices and each good teacher is different from any other good teacher.

Any efficient teaching device must be correlated somehow with the nature of the learning process. We do not know too much about the learning process, but even a rough outline of some of its more obvious features may shed some welcome light upon the tricks of our trade. Let me state such a rough outline in the form of three "principles" of learning. Their formulation and combination is of my choice, but the "principles" themselves are by no means new; they have been stated and restated in various forms, they are derived from the experience of the ages, endorsed by the judgment of great men, and also suggested by the psychological study of learning.

These "principles of learning" can be also taken for "principles of teaching," and this is the chief reason for considering them here—but more about this later.

(1) *Active learning.* It has been said by many people in many ways that learning should be active, not merely passive or receptive; merely by reading books or listening to lectures or looking at moving pictures without adding some action of your own mind you can hardly learn anything and certainly you can not learn much.

There is another often expressed (and closely related) opinion: *The best way to learn anything is to discover it by yourself.* Lichtenberg (an eighteenth century German physicist, better known as a writer of aphorisms) adds an interesting point: *What you have been obliged to discover by yourself leaves a path in your mind which you can use again when the need arises.* Less colorful but perhaps more widely applicable, is the following statement: *For efficient learning, the learner should discover by himself as large a fraction of the material to be learned as feasible under the given circumstances.*

This is the *principle of active learning* (Arbeitsprinzip). It is a very old principle: it underlies the idea of "Socratic method."

(2) *Best motivation.* Learning should be active, we have said. Yet the learner will not act if he has no motive to act. He must be induced to act by some stimulus, by the hope of some reward, for instance. The interest of the material to be learned should be the best stimulus to learning and the pleasure of intensive mental activity should be the best reward for such activity. Yet, where we cannot obtain the best we should try to get the second best, or the third best, and less intrinsic motives of learning should not be forgotten.

For efficient learning, the learner should be interested in the material to be learned and find pleasure in the activity of learning. Yet, beside these best motives for learning, there are other motives too, some of them desirable. (Punishment for not learning may be the least desirable motive.)

Let us call this statement the *principle of best motivation.*

(3) *Consecutive phases.* Let us start from an often quoted sentence of Kant: *Thus all human cognition begins with intuitions, proceeds from thence to conceptions, and ends with ideas.* The English translation uses the terms "cognition, intuition, idea." I am not able (who is able?) to tell in what exact sense Kant intended to use these terms. Yet I beg your permission to present my reading of Kant's dictum:

Learning begins with action and perception, proceeds from thence to words and concepts, and should end in desirable mental habits.

To begin with, please, take the terms of this sentence in some sense that you can illustrate concretely on the basis of your own experience. (To induce you to think about your personal experience is one of the desired effects.) "Learning" should remind you of a classroom with yourself in it as student or teacher. "Action and perception" should suggest manipulating and seeing concrete things such as pebbles, or apples, or Cuisenaire rods; or ruler and compasses; or instruments in a laboratory; and so on.

Such concrete interpretation of the terms may come more easily and more naturally when we think of some simple elementary material. Yet after a while we may perceive similar phases in the work spent on mastering more complex, more advanced material. Let us distinguish three phases: the phases of *exploration, formalization,* and *assimilation.*

A first *exploratory* phase is closer to action and perception and moves on a more intuitive, more heuristic level.

A second *formalizing* phase ascends to a more conceptual level, introducing terminology, definitions, proofs.

The phase of *assimilation* comes last: there should be an attempt to perceive the "inner ground" of things, the material learned should be mentally digested, absorbed into the system of knowledge, into the whole mental outlook of the learner; this phase paves the way to applications on one hand, to higher generalizations on the other.

Let us summarize: *For efficient learning, an exploratory phase should precede the phase of verbalization and concept formation and, eventually, the material learned should be merged in, and contribute to, the integral mental attitude of the learner.*

This is the *principle of consecutive phases.*

14.5. Three principles of teaching

The teacher should know about the ways of learning. He should avoid inefficient ways and take advantage of the efficient ways of learning. Thus he can make good use of the three principles we have just surveyed, the principle of active learning, the principle of best motivation, and the principle of consecutive phases; these principles of learning are also principles of teaching. There is, however, a condition: to avail himself of such a principle, the teacher should not merely know it from hearsay, but he should understand it intimately on the basis of his own well-considered personal experience.

(1) *Active learning.* What the teacher says in the classroom is not unimportant, but what the students think is a thousand times more important. The ideas should be born in the students' mind and the teacher should act only as midwife.

This is a classical Socratic precept and the form of teaching best adapted to it is the Socratic dialogue. The high school teacher has a definite advantage over the college instructor in that he can use the dialogue form much more extensively. Unfortunately, even in the high school, time is limited and there is a prescribed material to cover so that all business cannot be transacted in dialogue form. Yet the principle is: let the students *discover by themselves as much as feasible* under the given circumstances.

Much more is feasible than is usually done, I am sure. Let me recommend to you here just one little practical trick: let the students *actively contribute to the formulation* of the problem that they have to solve afterwards. If the students have had a share in proposing the problem they will work at it much more actively afterwards.

In fact, in the work of the scientist, formulating the problem may be the better part of a discovery, the solution often needs less insight and originality than the formulation. Thus, letting your students have a share in the formulation, you not only motivate them to work harder, but you teach them a desirable attitude of mind.

(2) *Best motivation.* The teacher should regard himself as a salesman: he wants to sell some mathematics to the youngsters. Now, if the salesman meets with sales resistance and his prospective customers refuse to buy, he should not lay the whole blame on them. Remember, the customer is always right in principle, and sometimes right in practice. The lad who refuses to learn mathematics may be right; he may be neither lazy nor stupid, just more interested in something else—there are so many interesting things in the world around us. It is your duty as a teacher, as a salesman of knowledge, to convince the student that mathematics *is* interesting, that the point just under discussion is interesting, that the problem he is supposed to do deserves his effort.

Therefore, the teacher should pay attention to the choice, the formulation, and a suitable presentation of the problem he proposes. The problem should appear as meaningful and relevant from the student's standpoint; it should be related, if possible, to the everyday experience of the students, and it should be introduced, if possible, by a little joke or a little paradox. Or the problem should start from some very familiar knowledge; it should have, if possible, some point of general interest or eventual practical use. If we wish to stimulate the student to a genuine effort, we must give him some reason to suspect that his task deserves his effort.

The best motivation is the student's interest in his task. Yet there are other motivations which should not be neglected. Let me recommend here just one little practical trick. Before the students do a problem, let them *guess the result,* or a part of the result. The boy who expresses an opinion commits himself; his prestige and self-esteem depend a little on the outcome, he is impatient to know whether his guess will turn out right or not, and so he will be actively interested in his task and in the work of the class—he will not fall asleep or misbehave.

In fact, in the work of the scientist, the guess almost always precedes the proof. Thus, in letting your students guess the result, you not only motivate them to work harder, but you teach them a desirable attitude of mind.

(3) *Consecutive phases.* The trouble with the usual problem material of the high school textbooks is that they contain almost exclusively merely routine examples. A routine example is a short range example; it illustrates, and offers practice in the application of, just one isolated rule. Such routine examples may be useful and even necessary, I do not deny it, but they miss two important phases of learning: the exploratory phase and the phase of assimilation. Both phases seek to connect the problem in hand with the world around us and with other knowledge, the first before, the last after, the formal solution. Yet the routine problem is obviously connected with the rule it illustrates and it is scarcely connected with anything else, so that there is little profit in seeking further connections. In contrast with such routine problems, the high school should present more challenging problems at least now and then, problems with a rich background that deserve further exploration, and problems which can give a foretaste of the scientist's work.

Here is a practical hint: if the problem you want to discuss with your class is suitable, let your students do some preliminary exploration: it may whet their appetite for the formal solution. And reserve some time for a retrospective discussion of the finished solution; it may help in the solution of later problems.

(4) After this much too incomplete discussion, I must stop explaining the three principles of active learning, best motivation, and consecutive phases. I think that these principles can penetrate the details of the teacher's daily work and make him a better teacher. I think too that these principles should also penetrate the planning of the whole curriculum, the planning of each course of the curriculum, and the planning of each chapter of each course.

Yet it is far from me to say that you must accept these principles. These principles proceed from a certain general outlook, from a certain philosophy, and you may have a different philosophy. Now, in teaching as in several other things, it does not matter much what your philosophy is or is not. It matters more whether you have a philosophy or not. And it matters very much whether you try to live up to your philosophy or not. The only principles of teaching which I thoroughly dislike are those to which people pay only lip service.

14.6. Examples

Examples are better than precepts; let me get down to examples—I much prefer examples to general talk. I am here concerned principally with teaching on the high school level, and I shall present you a few examples on that level. I often find satisfaction in treating examples at the high school level, and I can tell you why: I attempt to treat them so that they recall in one respect or the other my own mathematical experience; I am re-enacting my past work on a reduced scale.

(1) *A seventh grade problem.* The fundamental art form of teaching is the Socratic dialogue. In a junior high school class, perhaps in the seventh grade, the teacher may start the dialogue so:

"What is the time at noon in San Francisco?"

'But, teacher, everybody knows that' may say a lively youngster, or even 'But teacher, you are silly: twelve o'clock.'

"And what is the time at noon in Sacramento?"

'Twelve o'clock—of course, not twelve o'clock midnight.'

"And what is the time at noon in New York?"

'Twelve o'clock.'

"But I thought that San Francisco and New York do not have noon at the same time, and you say that both have noon at twelve o'clock!"

'Well, San Francisco has noon at twelve o'clock Western Standard Time and New York at twelve o'clock Eastern Standard Time.'

"And on what kind of standard time is Sacramento, Eastern or Western?"

'Western, of course.'

"Have the people in San Francisco and Sacramento noon at the same moment?"

"You do not know the answer? Well, try to guess it: does noon come sooner to San Francisco, or to Sacramento, or does it arrive exactly at the same instant at both places?"

How do you like my idea of Socrates talking to seventh grade kids? At any rate, you can imagine the rest. By appropriate questions the teacher, imitating Socrates, should extract several points from the students:

(a) We have to distinguish between "astronomical" noon and conventional or "legal" noon.

(b) Definitions for the two noons.

(c) Understanding "standard time": how and why is the globe's surface subdivided into time zones?

(d) Formulation of the problem: "At what o'clock Western Standard Time is the astronomical noon in San Francisco?"

(e) The only specific datum needed to solve the problem is the longitude of San Francisco (in an approximation sufficient for the seventh grade).

The problem is not too easy. I tried it on two classes; in both classes the participants were high school teachers. One class spent about 25 minutes on the solution, the other 35 minutes.

(2) I must say that this little seventh grade problem has various advantages. Its main advantage may be that it emphasizes an essential mental operation which is sadly neglected by the usual problem material of the textbooks: *recognizing the essential mathematical concept in a concrete situation.* To solve the problem, the students must recognize a *proportionality:* the time of the highest position of the sun in a locality on the globe's surface changes *proportionally* to the longitude of the locality.

In fact, in comparison with the many painfully artificial problems of the high school textbooks, our problem is a perfectly natural, a "real" problem. In the serious problems of applied mathematics, the appropriate *formulation* of the problem is always a major task, and often the most important task; our little problem which can be proposed to an average seventh grade class possesses just this feature. Again, the serious problems of applied mathematics may lead to practical action, for instance, to adopting a better manufacturing process; our little problem can explain to seventh graders why the system of 24 time zones, each with a uniform standard time, was adopted. On the whole, I think that this problem, if handled with a little skill by the teacher, could help a future scientist or engineer to discover his vocation, and it could also contribute to the intellectual maturity of those students who will not use mathematics professionally.

Observe also that this problem illustrates several little tricks mentioned in the foregoing: the students actively contribute to the formulation of the problem [*cf.* sect. 14.5(1)]. In fact, the exploratory phase which leads to the formulation of the problem is prominently important [*cf.* sect. 14.5(3)]. Then, the students are invited to guess an essential point of the solution [*cf.* sect. 14.5(2)].

(3) *A tenth grade problem.* Let us consider another example. Let us start from what is probably the most familiar problem of geometric construction: *construct a triangle, being given its three sides.* As analogy is such a fertile source of invention, it is natural to ask: what is the analogous problem in solid geometry? An average student, who has a little knowledge of solid geometry, may be led to formulate the problem: *construct a tetrahedron, being given its six edges.*

It may be mentioned here parenthetically that this problem of the tetrahedron comes as close as it can on the usual high school level to practical problems solved by "mechanical drawing." Engineers and designers use well executed drawings to give precise information about the details of three dimensional figures of machines or structures to be built: we intend to build a tetrahedron with specified edges. We might wish, for example, to carve it out of wood.

This leads to asking that the problem should be solved precisely, by straight-edge and compasses, and to discussing the question: which details of the tetrahedron should be constructed? Eventually, from a well conducted class discussion, the following definitive formulation of the problem may emerge:

Of the tetrahedron ABCD, we are given the lengths of its six edges

AB, BC, CA, AD, BD, CD.

Regard △ABC as the base of the tetrahedron and construct with ruler and compasses the angles that the base includes with the other three faces.

The knowledge of these angles is required for cutting out of wood the desired solid. Yet other elements of the tetrahedron may turn up in the discussion such as

(a) the altitude drawn from the vertex D opposite the base,
(b) the foot F of this altitude in the plane of the base;

(a) and (b) would contribute to the knowledge of the solid, they may possibly help to find the required angles, and so we may try to construct them too.

(4) We can, of course, construct the four triangular faces which are assembled in Fig. 14.1. (Short portions of some circles used in the construction are preserved to indicate that $AD_2 = AD_3$, $BD_3 = BD_1$, $CD_1 = CD_2$.) If Fig. 14.1 is copied on cardboard, we can add three flaps, cut out the pattern, fold it along three lines, and paste down the

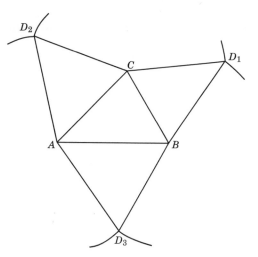

Fig. 14.1. Tetrahedron from six edges.

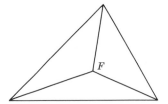

Fig. 14.2. An aspect of the finished product.

flaps; in this way we obtain a solid model on which we can measure roughly the altitude and the angles in question. Such work with cardboard is quite suggestive, but it is not what we are required to do: we should construct the altitude, its foot, and the angles in question with ruler and compasses.

(5) It may help to take the problem, or some part of it, "as solved." Let us visualize how Fig. 14.1 will look when the three lateral faces, after having been rotated each about a side of the base, will be lifted into their proper position. Fig. 14.2 shows the orthogonal projection of the tetrahedron onto the plane of its base, $\triangle ABC$. The point F is the projection of the vertex D: it is the foot of the altitude drawn from D.

(6) We may visualize the transition from Fig. 14.1 to Fig. 14.2 with or without a cardboard model. Let us focus our attention on one of the three lateral faces, on $\triangle BCD_1$, which was originally located in the same plane as $\triangle ABC$, in the plane of Fig. 14.1 which we imagine as horizontal. Let us watch the triangle BCD_1 rotating about its fixed side BC and let our eyes follow its only moving vertex D_1. This vertex D_1 describes an arc of a circle. The center of this circle is a point of BC; the plane of this circle is perpendicular to the horizontal axis of revolution BC; thus D_1 moves in a vertical plane. Therefore, the projection of the path of the moving vertex D_1 onto the horizontal plane of Fig. 14.1 is a straight line, perpendicular to BC, passing through the original position of D_1.

Yet there are two more rotating triangles, three altogether. There are three moving vertices, each following a circular path in a vertical plane—to which destination?

(7) I think that by now the reader has guessed the result (perhaps even before reading the end of the foregoing subsection): the three straight lines drawn from the original positions (see Fig. 14.1) of D_1, D_2, and D_3 perpendicularly to BC, CA, and AB, respectively, meet in one point, the point F, our supplementary aim (b), see Fig. 14.3. (It is enough to draw two perpendiculars to determine F, but we may use the third to check the precision of our drawing.) And what remains to do is easy. Let M be the point of intersection of D_1F and BC (see

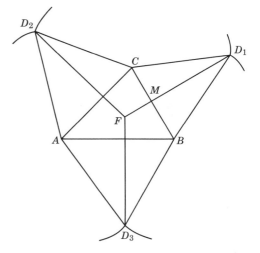

Fig. 14.3. The common destination of three travelers.

Fig. 14.3). Construct the right triangle *FMD* (see Fig. 14.4), with hypotenuse $MD = MD_1$ and leg *MF*. Obviously, *FD* is the altitude [our supplementary aim (a)] and $\angle FMD$ measures the dihedral angle included by the base $\triangle ABC$ and the lateral face $\triangle DBC$ which was required by our problem.

(8) One of the virtues of a good problem is that it generates other good problems.

The foregoing solution may, and should, leave a doubt in our mind. We found the result represented by Fig. 14.3 (that the three perpendiculars described above are concurrent) by considering the motion of rotating bodies. Yet the result is a proposition of geometry and so it should be established independently of the idea of motion, by geometry alone.

Now, it is relatively easy to free the foregoing consideration [in subsections (6) and (7)] from ideas of motion and establish the result by

Fig. 14.4. The rest is easy.

ideas of solid geometry (intersection of spheres, orthogonal projection).[3] Yet the result is a proposition of plane geometry and so it should be established independently of the ideas of solid geometry, by plane geometry alone. (How?)

(9) Observe that this tenth grade problem also illustrates various points about teaching discussed in the foregoing. For instance, the students could and should participate in the final formulation of the problem, there is an exploratory phase, and a rich background.

Yet here is the point I wish to emphasize: the problem is designed to deserve the attention of the students. Although the problem is not so close to everyday experience as our seventh grade problem, it starts from a most familiar piece of knowledge (the construction of a triangle from three sides), it stresses from the start an idea of general interest (analogy), and it points to eventual practical applications (mechanical drawing). With a little skill and good will, the teacher should be able to secure for this problem the attention of all students who are not hopelessly dull.

14.7. Learning teaching

There remains one more topic to discuss and it is an important topic: teacher training. In discussing this topic, I am in a comfortable position: I can almost agree with the "official" standpoint. (I am referring here to the "Recommendations of the Mathematical Association of America for the training of mathematics teachers," *American Mathematical Monthly,* **67** (1960) 982–991. Just for the sake of brevity, I take the liberty to quote this document as the "official recommendations.") I shall concentrate on just two points. To these two points I have devoted a good deal of work and thought in the past and practically all my teaching in the last ten years.

To state it roughly, one of the two points I have in mind is concerned with "subject matter" courses, the other with "methods" courses.

(1) *Subject matter.* It is a sad fact, but by now widely recognized, that our high school mathematics teachers' knowledge of their subject matter is, on the average, insufficient. There are, certainly, some well-prepared high school teachers, but there are others (I met with several) whose good will I must admire but whose mathematical preparation is not admirable. The official recommendations of subject matter courses may not be perfect, but there is no doubt that their acceptance would result in substantial improvement. I wish to direct your attention to a point which, in my considered opinion, should be added to the official recommendations.

---

[3] *Cf.* sect. 6.2(3).

Our knowledge about any subject consists of information and know-how. Know-how is ability to use information; of course, there is no know-how without some independent thinking, originality, and creativity. Know-how in mathematics is the ability to do problems, to find proofs, to criticize arguments, to use mathematical language with some fluency, to recognize mathematical concepts in concrete situations.

Everybody agrees that, in mathematics, know-how is more important, or even much more important, than mere possession of information. Everybody demands that the high school should impart to the students not only information in mathematics but know-how, independence, originality, creativity. Yet almost nobody asks these beautiful things for the mathematics teacher—is it not remarkable? The official recommendations are silent about the mathematical know-how of the teacher. The student of mathematics who works for a Ph.D. degree must do research, yet even before he reaches that stage he may find some opportunity for independent work in seminars, problem seminars, or in the preparation of a master's thesis. Yet no such opportunity is offered to the prospective mathematics teacher—there is no word about any sort of independent work or research work in the official recommendations. If, however, the teacher has had no experience in creative work of some sort, how will he be able to inspire, to lead, to help, or even to recognize the creative activity of his students? A teacher who acquired whatever he knows in mathematics purely receptively can hardly promote the active learning of his students. A teacher who never had a bright idea in his life will probably reprimand a student who has one instead of encouraging him.

Here, in my opinion, is the worst gap in the subject matter knowledge of the average high school teacher: he has no experience of active mathematical work and, therefore, he has no real mastery even of the high school material he is supposed to teach.

I have no panacea to offer, but I have tried one thing. I have introduced and repeatedly conducted a *problem solving seminar* for teachers. The problems offered in this seminar do not require much knowledge beyond the high school level, but they require some degree, and now and then a higher degree, of concentration and judgment—and, to that degree, their solution is "creative" work. I have tried to arrange my seminar so that the students should be able to use much of the material offered in their classes without much change; that they should acquire some mastery of high school mathematics; and so that they should have even some opportunity for practice teaching (in teaching each other in small groups). *Cf.* vol. 1, pp. 210–212.

(2) *Methods.* From my contact with hundreds of mathematics teachers I gained the impression that "methods" courses are often re-

ceived with something less than enthusiasm. Yet so also are received, by the teachers, the usual courses offered by the mathematics departments. A teacher with whom I had a heart to heart talk about these matters found a picturesque expression for a rather widespread feeling: "The mathematics department offers us tough steak which we cannot chew and the school of education vapid soup with no meat in it."

In fact, we should once summon up some courage and discuss publicly the question: Are methods courses really necessary? Are they in any way useful? There is more chance to reach the right answer in open discussion than by widespread grumbling.

There are certainly enough pertinent questions. Is teaching teachable? (Teaching is an art, as many of us think—is an art teachable?) Is there such a thing as the teaching method? (What the teacher teaches is never better than what the teacher is—teaching depends on the whole personality of the teacher—there are as many good methods as there are good teachers.) The time allotted to the training of teachers is divided between subject matter courses, methods courses, and practice teaching; should we spend less time on methods courses? (Many European countries spend much less time.)

I hope that people younger and more vigorous than myself will take up these questions some day and discuss them with an open mind and pertinent data.

I am speaking here only about my own experience and my own opinions. In fact I have already implicitly answered the main question raised; I believe that methods courses may be useful. In fact, what I have presented in the foregoing was a sample of a methods course, or rather an outline of some topics which, in my opinion, a methods course offered to mathematics teachers should cover.

In fact, all the classes I have given to mathematics teachers were intended to be methods courses to some extent. The name of the class mentioned some subject matter, and the time was actually divided between that subject matter and methods: perhaps nine tenths for subject matter and one tenth for methods. If possible, the class was conducted in dialogue form. Some methodical remarks were injected incidentally, by myself or by the audience. Yet the derivation of a fact or the solution of a problem was almost regularly followed by a short discussion of its pedagogical implications. "Could you use this in your classes?" I asked the audience. "At which stage of the curriculum could you use it? Which point needs particular care? How would you try to get it across?" And questions of this nature (appropriately specified) were regularly proposed also in examination papers. My main work, however, was to choose such problems (like the two problems I have here presented) as would illustrate strikingly some pattern of teaching.

(3) The official recommendations call "methods" courses "curriculum-study" courses and are not very eloquent about them. Yet you can find there one excellent recommendation, I think. It is somewhat concealed; you must put two and two together, combining the last sentence in "curriculum study courses" and the recommendations for Level IV. But it is clear enough: A college instructor who offers a methods course to mathematics teachers should know mathematics at least on the level of a Master's degree. I would like to add: he should also have had some experience, however modest, of mathematical research. If he had no such experience how could he convey what may be the most important thing for prospective teachers, the spirit of creative work?

You have now listened long enough to the reminiscences of an old man. Some concrete good could come out of this talk if you give some thought to the following proposal which results from the foregoing discussion. I propose that the following two points should be added to the official recommendations of the Association:

I. *The training of teachers of mathematics should offer experience in independent ("creative") work on the appropriate level in the form of a Problem Solving Seminar or in any other suitable form.*

II. *Methods courses should be offered only in close connection either with subject matter courses or with practice teaching and, if feasible, only by instructors experienced both in mathematical research and in teaching.*

14.8. The teacher's attitude[4]

As I have already said, my classes addressed to teachers were, to some extent, "methods" courses. In these classes I aimed at points of immediate practical use in the daily task of the teacher. Therefore, inavoidably, I had to express repeatedly my views on the teacher's daily task and on the teacher's mental attitude. My comments tended to assume a set form and eventually I was led to condense them into "Ten Commandments for Teachers"; see p. 116. I wish to add a few comments on these ten rules.

In formulating these rules, I had in mind the participants in my classes, teachers who teach mathematics on the high school level. Nevertheless, these rules are applicable to any teaching situation, to any subject taught on any level. Especially on the high school level, however, the mathematics teacher has more and better opportunities

---

[4] This section can be read independently of the foregoing (of which a few points will be repeated.) It is reprinted, with several modifications and the kind permission of the editor, from the *Journal of Education of the Faculty and College of Education, Vancouver and Victoria,* no. 3, 1959, pp. 61–69.

TEN COMMANDMENTS FOR TEACHERS

1. Be interested in your subject.
2. Know your subject.
3. Know about the ways of learning: The best way to learn anything is to discover it by yourself.
4. Try to read the faces of your students, try to see their expectations and difficulties, put yourself in their place.
5. Give them not only information, but "know-how," attitudes of mind, the habit of methodical work.
6. Let them learn guessing.
7. Let them learn proving.
8. Look out for such features of the problem at hand as may be useful in solving the problems to come—try to disclose the general pattern that lies behind the present concrete situation.
9. Do not give away your whole secret at once—let the students guess before you tell it—let them find out by themselves as much as is feasible.
10. Suggest it, do not force it down their throats.

to apply some of them than the teacher of other subjects; and this refers in particular to rules 6, 7, and 8.

On what authority are these commandments founded? Dear fellow teacher, do not accept any authority except your own well-digested experience and your own well-considered judgement. Try to see clearly what the advice means in your particular situation, try the advice in your classes, and judge after a fair trial.

Let us now consider the ten rules one by one, with especial attention to the task of the mathematics teacher.

(1) There is just one infallible teaching method: if the teacher is bored by his subject, his whole class will be infallibly bored by it.

This should be enough to render evident the first and foremost commandment for teachers: *Be interested in your subject.*

(2) If a subject has no interest for you, do not teach it, because you will not be able to teach it acceptably. Interest is a *sine qua non,* an indispensably necessary condition; but, in itself, it is not a sufficient condition. No amount of interest, or teaching methods, or whatever else will enable you to explain clearly a point to your students that you do not understand clearly yourself.

This should be enough to render obvious the second commandment for teachers: *Know your subject.*

Both interest in, and knowledge of, the subject matter are necessary for the teacher. I put interest first because with genuine interest you have a good chance to acquire the necessary knowledge, whereas some knowledge coupled with lack of interest can easily make you an exceptionally bad teacher.

(3) You may benefit a great deal from reading a good book or listening to a good lecture on the psychology of learning. Yet reading and listening are not absolutely necessary, and they are by no means sufficient; you should *know the ways of learning,* you should be intimately acquainted with the process of learning from *experience*—from the experience of your own studies and from the observation of your students.

Accepting hearsay as evidence for a principle is bad; paying lip service to a principle is worse. Now, there is one case where you cannot afford to be content with hearsay and lip service, there is one principle of learning that you should earnestly realize: the principle of active learning.[5] Try to see at least its central point: *The best way to learn anything is to discover it by yourself.*

(4) Even with some genuine knowledge and interest and some understanding of the learning process you may be a poor teacher. The case is unusual, I admit, but not quite rare; some of us have met with an otherwise quite competent teacher who was unable to establish "contact" with his class. In order that teaching by one should result in learning by the other, there must be some sort of contact or connection between teacher and student: the teacher should be able to see the student's position; he should be able to espouse the student's cause. Hence the next commandment: *Try to read the faces of your students, try to see their expectations and difficulties, put yourself in their place.*

The response of the students to your teaching depends on their background, their outlook, and their interests. Therefore, keep in mind and take into account what they know and what they do not know, what they would like to know and what they do not care to know, what they ought to know and what is less important for them to know.

(5) The four foregoing rules contain the essentials of good teaching. They form jointly a sort of necessary and sufficient condition: if you have interest in, and knowledge of, the subject matter and if, moreover, you can see the student's case and what helps or hampers his learning, you are already a good teacher or you will become one soon; you may need only some experience.

It remains to spell out some consequences of the foregoing rules,

---

[5] See sect. 14.4(1), 14.5(1). To be acquainted with the two other principles discussed in the foregoing is advisable.

especially such consequences as concern the high school mathematics teacher.

Knowledge consists partly of "information" and partly of "know-how." Know-how is skill; it is the ability to deal with information, to use it for a given purpose; know-how may be described as a bunch of appropriate mental attitudes; know-how is ultimately the ability to work methodically.

In mathematics, know-how is the ability to solve problems, to construct demonstrations, and to examine critically solutions and demonstrations. And, in mathematics, know-how is much more important than the mere possession of information. Therefore, the following commandment is of especial importance for the mathematics teacher: *Give your students not only information, but know-how, attitudes of mind, the habit of methodical work.*

Since know-how is more important in mathematics than information, it may be more important in the mathematics class how you teach than what you teach.

(6) First guess, then prove—so does discovery proceed in most cases. You should know this (from your own experience, if possible), and you should know, too, that the mathematics teacher has excellent opportunities to show the role of guessing in discovery and thus to impress on his students a fundamentally important attitude of mind. This latter point is not so widely known as it should be and, just for this reason, it deserves particular attention. I wish you would not neglect your students in this respect: *Let them learn guessing.*

Ignorant and careless students are liable to come forward with "wild" guesses. What we have to teach is, of course, not wild guessing, but "educated" "reasonable" guessing. Reasonable guessing is based on judicious use of inductive evidence and analogy, and ultimately encompasses all procedures of *plausible reasoning* which play a role in "scientific method."[6]

(7) "Mathematics is a good school of plausible reasoning." This statement summarizes the opinion underlying the foregoing rule; it sounds unfamiliar and is of very recent origin; in fact, I think that I should claim credit for it.

"Mathematics is a good school of demonstrative reasoning." This statement sounds very familiar—some form of it is probably almost as old as mathematics itself. In fact, much more is true: mathematics is coextensive with demonstrative reasoning, which pervades the sciences just as far as their concepts are raised to a sufficiently abstract and definite, mathematico-logical level. Under this high level there is no

---

[6] See chapter 15.

place for truly demonstrative reasoning (which is out of place, for instance, in everyday affairs). Still (it is needless to argue such a widely accepted point), the mathematics teacher should acquaint all his students beyond the most elementary grades with demonstrative reasoning: *Let them learn proving.*

(8) Know-how is the more valuable part of mathematical knowledge, much more valuable than the mere possession of information. Yet how should we teach know-how? The students can learn it only by imitation and practice.

When you present the solution of a problem, *emphasize* suitably the *instructive features* of the solution. A feature is instructive if it deserves imitation; that is, if it can be used not only in the solution of the present problem, but also in the solution of other problems—the more often usable, the more instructive. Emphasize the instructive features not just by praising them (which could have the contrary effect with some students) but by your *behavior* (a bit of acting is very good if you have a bit of theatrical talent). A well-emphasized feature may convert your solution into a *model solution,* into an impressive *pattern* by imitating which the student will solve many other problems. Hence the rule: *Look out for such features of the problem at hand as may be useful in solving the problems to come—try to disclose the general pattern that lies behind the present concrete situation.*[7]

(9) I wish to indicate here a little classroom trick which is easy to learn and which every teacher should know. When you start discussing a problem, let your students guess the solution. The student who has conceived a guess, or has even stated his guess, commits himself: he has to follow the development of the solution to see whether his guess comes true or not—and so he cannot remain inattentive.[8]

This is just a very special case of the following rule, which itself is contained in, and spells out, some parts of rules 3 and 6: *Do not give away your whole secret at once—let the students guess before you tell it— let them find out by themselves as much as is feasible.*

In fact, the credit for this rule is due to Voltaire who expressed it more wittily: "Le secret d'être ennuyeux c'est de tout dire." "The art of being a bore consists in telling everything."

(10) A student presents a long computation which goes through several lines. Looking at the last line, I see that the computation is wrong, but I refrain from saying so. I prefer to go through the computation with the student, line by line: "You started out all right, your first line is correct. Your next line is correct too; you did this and that.

---

[7] Do you want more details? Read the whole book.

[8] *Cf.* sect. 14.5(2).

The next line is good. Now, what do you think about this line?" The mistake is on that line and if the student discovers it by himself, he has a chance to learn something. If, however, I at once say "This is wrong" the student may be offended and then he will not listen to anything I may say afterwards. And if I say "This is wrong" once too often, the student will hate me and mathematics and all my efforts will be lost as far as he is concerned.

Dear fellow teacher, avoid saying "You are wrong." Say instead, if possible: "You are right, but. . . ." If you proceed so, you are not hypocritical, you are just humane. That you should proceed so, is implicitly contained in rule 3. Yet we can render the advice more explicit: *Suggest it, do not force it down their throats.*

Our last two rules, 9 and 10, tend in the same direction. What they jointly suggest is to leave the students as much freedom and initiative as possible under existing teaching conditions. Pressed for time, the mathematics teacher is often tempted to sin against the spirit of these rules, the *principle of active learning.* He may hurry to the solution of a problem without leaving enough time for the students to put the problem to themselves in earnest. He may name a concept or formulate a rule too soon, without sufficient preparation by appropriate material, before the students can feel the need for such a concept or rule. He may commit the celebrated mistake of *deus ex machina:* he may introduce some device (for instance, a tricky auxiliary line in a geometric proof) which leads to the result all right, but the students cannot see for their life how it was humanly possible to discover such a trick which appeared right out of the blue.

There are too many temptations to violate the principle. Let us, therefore, emphasize a few more of its facets.

Let your students ask the questions; or ask such questions as they may ask by themselves.

Let your students give the answers; or give such answers as they may give by themselves.

At any rate avoid answering questions that nobody has asked, not even yourself.

Examples and Comments on Chapter 14

First Part

14.1. Taking $122°25'41''$ west for the longitude of San Francisco answer the question (d) of sect. 14.6(1).

14.2. Following up a hint in sect. 14.6(8), prove the proposition suggested by Fig. 14.3 by solid geometry.

14.3. (Continued) Prove the proposition suggested by plane geometry.

14.4. Sect. 14.6(9) mentions some points discussed in the foregoing and illustrated by the problem presented in sect. 14.6(3) to 14.6(8). Do you notice more such points?

Second Part

14.5. *Why problem solving?* I am of the opinion that teaching "problem solving" could be an essential ingredient of various curricula rather different in other respects and should be, in fact, an essential ingredient of any useful high school mathematics curriculum. This opinion underlies the present book and my related writings and has already been explicitly stated in the foregoing (see sect. 5 of the preface and sect. 14.2). If the reader is not convinced by the foregoing chapters that this opinion has some merit, I cannot do much for him. Still, I wish to present a few comments on the role of problem solving in the high school curriculum.

(1) We are concerned here with the teaching of mathematics on the high school level and with the aims of such teaching. Responsible and realistic consideration of these aims should take into account the use that the students can be expected to make of what they are supposed to learn. Of course, there are different categories of students, and some will make more, and others less, use of the knowledge acquired in school, and some categories form a greater, and others a smaller, fraction of the student body. Reliable statistical data about these things would be very desirable, but they are scarcely available. The numerical proportions I shall use in what follows are rough estimates without serious statistical basis—I use them, in fact, just for the sake of concreteness.

(2) Let us consider such students as take some mathematics on the high school level (algebra, geometry, etc.) and, with regard to the future use of such study in their respective professions, let us distinguish three categories: mathematicians, users of mathematics, and nonusers of mathematics.

Let us draw the boundaries of the first category rather widely: Let us count as "mathematicians" or "producers of mathematics" also theoretical physicists, astronomers, and certain engineers in special research positions. Altogether they may form about 1% of the students. (The number of future Ph.D.'s in mathematics is closer to 0.1%.)

Engineers, scientists (also some social scientists), mathematics and science teachers, etc. are users (but, in general, not producers) of mathematics. Let us count as users of mathematics also such students as will not use mathematics in their profession, but who essentially need some mathematics in their studies (such is the case of numerous engineering graduates who become salesmen or managers). The number of all sorts of users of mathematics may be, say, 29% of the students.

Many of the remaining students could, but actually will not, use any mathematics beyond what they should have learned in primary school. It is a rough but not unrealistic estimate that 70% of the students will be nonusers of mathematics; almost all future businessmen, lawyers, clergymen, etc. belong to this category.

(3) We do not know in advance who will become what and so we do not know which students belong to which category. The mathematics class, therefore, should be so conducted that it conforms to two "principles":

First, each student should be able to derive some profit from his study irrespective of future occupation.

Second, such students as have some aptitude for mathematics should be attracted to it and not get disgusted with it by ill-advised teaching.

I take for granted that the reader accepts these two principles, at least to some extent. In fact, I think that to plan the curriculum without continuous earnest attention to these two principles would be unrealistic and irresponsible.

Let me hint briefly how the three categories of students here considered [see (2)] could gain something essential from the study of problem solving.

(4) The ability to solve mathematical problems needs, of course, some knowledge of the mathematical subject matter involved, but, moreover, it needs certain useful habits of mind, a certain general attitude which we are inclined to call "common sense" in everyday life. The teacher who wishes to serve equally all his students, future users and nonusers of mathematics, should teach problem solving so that it is about one-third mathematics and two-thirds common sense. It may not be too easy to implant common sense and useful mental habits, but if the mathematics teacher succeeds in implanting them he has rendered real service to his students, whatever their future occupation may be. This service is certainly the most significant thing he can do for those 70% of students who will have no use for technical mathematics in later life.

The 29% of students who will become users of mathematics need, as a preparation for subsequent studies, some technical proficiency (for instance, some facility in algebraic manipulation). Yet just the students with a good practical mind are reluctant to learn technicalities unless they are convinced that those technicalities serve a purpose, are good for something. The best the teacher can do to justify the teaching of technicalities is to demonstrate that they are efficacious in solving naturally arising interesting concrete problems.

Future mathematicians form only about 1% of the student body, but it is of paramount importance that they should be discovered: if they choose the wrong profession, their talent, needed by modern society in more than one way, may be wasted. The most important thing the high school teacher can do for this 1% is to awake their interest in mathematics. (To teach them a little more or less subject matter in high school hardly matters as, in any case, it can be only an infinitesimal fraction of the subject matter they have to know ultimately.) Now, problem solving is an important avenue to mathematics; it is not the only one, but it is connected with other important avenues (see ex. 14.6). Moreover, the teacher should treat a few problems which, although somewhat more difficult and time consuming, have real mathematical beauty and background (see chapter 15).

(5) I hope, as I said before, that arguments for teaching problem solving in the high school can be found in every part of the present book and of my related writings. A few specific points will be emphasized in the following.

14.6. *Problem solving and theory formation.* A devoted and well-prepared teacher can take a significant but not too complex problem and by helping the students discover its various aspects he can lead them through it, as through a gateway, into a whole theory. To prove that $\sqrt{2}$ is irrational, or that there are infinitely many prime numbers, are such significant problems. The former can be a gateway to a critical concept of the real number,[9] the latter to number theory.[10]

Such a procedure of the teacher parallels the historical evolution of science. The solution of a significant problem, the effort expended on, and the insight gained from, the solution may open the gate to a new science, or even to a new era of science. We should think of Galileo and the problem of falling bodies, or of Kepler and the problem of the orbit of Mars.

In his work quoted, M. Wagenschein offers an idea which deserves, in my opinion, the attention of all planners of curricula: instead of hurrying through all the details of a much too extended program, the teacher should concentrate on a few really significant problems and treat them leisurely and thoroughly. The students should explore all aspects of the problem accessible at their level, they should discover the solution by themselves, they should anticipate, led by the teacher, some consequences of the solution. In this way, a problem may become a representative example, a paradigm for a whole chapter of science. This is a first sketch of the idea of *paradigmatic teaching* about which every teacher seriously concerned with the curriculum should read more in the book quoted; see also ex. 14.11.

Let us note that the appropriate treatment of a single problem may be the gateway to, or the representative example of, a branch of science. It is in view of this and similar observations that I took the liberty to say in sect. 14.2 "thinking may be identified, at least in first approximation, with problem solving."

14.7. *Problem solving and general culture.* Many people think (and I am one of those many) that one of the essential tasks, and perhaps the most essential task, of the high school is to impart general culture. Let us not enter here, however, upon defining "general culture," because otherwise we could wrangle indefinitely about the "right" definition.

In teaching problem solving in the mathematics class we have an excellent opportunity to develop certain concepts and habits of mind which are, in my opinion, important ingredients of general culture; the box on p. 124 displays a list.[11] The list is not exhaustive; it contains only such more conspicuous, more tangible points as are, I hope, not beyond the reach of an average high school class. Most items of the list are explained at length in this book and in my related writings.[12] For additional remarks see ex. 14.8 and 14.10.

For a viewpoint which raises important questions about the relation between general culture and the teaching of mathematics see the work of Wittenberg, footnote 9.

[9] *Cf.* A. I. Wittenberg, Bildung und Mathematik (Klett Verlag 1963) pp. 168–253 *passim.*

[10] *Cf.* Martin Wagenschein, Exemplarisches Lehren im Mathematikunterricht, *Der Mathematikunterricht,* vol. 8, 1962, part 4, pp. 29–38.

[11] From *American Mathematical Monthly,* vol. 65, 1958, p. 103.

[12] For "Language of Formulas" see chapter 2, for "Reasonable Guessing" chapter 15. For Unknown, Data, and Condition, Generalization, Specialization, and Analogy, see the Index.

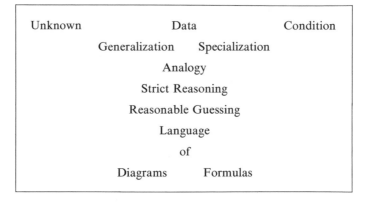

14.8. *Language of figures.* There are people who have an urge to represent their ideas by means of some sort of geometric symbols. In their mind some figures of speech which we all use have a tendency to develop into geometric figures. When they are thinking about their problems they are apt to find paper and pencil and start doodling: They may be struggling to express themselves in a language of geometric figures.

(1) There are many important nongeometric facts and ideas which are most appropriately expressed by geometric figures, graphs, or diagrams. By dots placed at appropriate elevations, high or low, the musical notation expresses the pitch of sounds, high or low. By means of geometric symbols (dots and connecting strokes) the chemical notation expresses the constitution of chemical compounds. Numbers and numerical relations can be expressed in various ways by geometrical objects and relations. Analytic geometry offers systematic means of translating numerical relations into geometrical relations and vice versa. Analytic geometry is, in a way, a dictionary of two languages, of the language of formulas and of the language of geometric figures: it enables us to translate readily from one language into the other. The ideas of analytic geometry underlie a great variety of graphs, diagrams, nomograms, etc. used in science, engineering, economics, and so on. Diagrams may also be useful in pure mathematics, and some can be well explained on the high school level; for an illustration not found in the usual textbooks, see ex. 14.9.

(2) Graphs and diagrams used in science are usually definite and precise in principle: the (idealized) geometric figure represents exactly the intended numerical relations. It is important to observe that graphical representations which are more or less vague or unprecise may still be useful. For instance, I believe that Fig. 11.2 has some suggestive value although it is scarcely more than a metaphor on paper, a figure of speech changed into a visible figure; Fig. 11.3 and 11.4 are of similar nature. Figures 15.1 to 15.5 have a clear mathematical meaning: they represent the set of all triangular shapes and certain subsets of this set. Yet their principal interest is that they hint something more, a process which we cannot yet conceive so clearly, the progress of an inductive argument.

Of two diagrams similar to the eye, one may be interpreted in an entirely definite mathematical meaning and the other in a rather vague metaphorical sense, and all gradations are possible between strict precision and poetical allusion. Flow charts, used for many purposes, can well illustrate the point.

(3) Geometry, our knowledge of space, has several aspects. Geometry can be conceived, as we know, as a science based on axioms. Yet geometry is also a skill of the eyes and the hands. Again, geometry can be considered as a part of physics (the most primitive part, say some physicists—the most interesting part, say some geometers). As a part of physics, geometry is also a field in which we can make intuitive or inductive discoveries and verify them subsequently by reasoning. To these aspects the foregoing has added one more: geometry is also the source of the symbols of a sort of language which can be colloquial or precise, and both ways helpful and enlightening.

There is a moral for the teacher: if you wish to instruct your students and not just hurry through the items of a curriculum dictated from above, do not neglect any of these aspects. Especially, do not insist too early or too much on the axiomatic aspect of geometry if you do not wish to disgust the future engineers and scientists among your students (or future artists and philosophers) who may be more attracted by the eye-knowledge of geometrical shapes, by space visualization, or by inductive discovery, or the powerful help to thinking afforded by diagrammatic representation.

14.9. *Rationals and irrationals.* What follows is only a rapid sketch of what should be very carefully done in the classroom—we are facing here what is perhaps the most delicate point of the high school mathematics curriculum. For the sake of brevity, I shall use a few terms and the notation of analytic geometry of which, however, very little knowledge is actually needed (just a little "graphing").

Let x and y denote, as usual, rectangular coordinates. The line with the equation $y = 1$ is the *number line* (just a "glorified yardstick"). See Fig. 14.5, which also displays the *lattice points,* that is, the points with integral coordinates. Figure 14.5 emphasises the lattice points along the number line ("milestones along a long straight road").

In Fig. 14.5 the number x is represented by the point $(x, 1)$ on the number line. We draw the straight line through the point $(x, 1)$ and the origin $(0, 0)$. Iff this

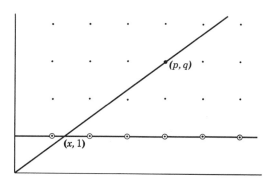

Fig. 14.5. Number line and lattice points.

line passes through the lattice point (p, q) (different from the origin!) the number x is rational:

$$\frac{x}{1} = \frac{p}{q}$$

from similar triangles.

The teacher should ask (but *not answer,* at least not answer for a considerable time) the following questions:

The origin $(0, 0)$ is a lattice point. Must every straight line passing through the origin pass also through another lattice point?

Of course, there are just two cases: a straight line through the origin does or does not pass through another lattice point; which case is more *likely?*

The teacher should let these questions and Fig. 14.5 "sink in" and only when the students have realized their import (perhaps after hours or weeks or months) should he start discussing the irrationality of $\sqrt{2}$, the approximation of irrationals by rationals (Fig. 14.5 may be used, after Felix Klein, as a gateway to continued fractions), etc.

14.10. *Strict reasoning.* Should we teach mathematical proofs in the high school? In my opinion, there is little doubt about the answer: Yes, we should, unless extremely adverse conditions compel us to lower the standard. Rigorous proofs are the hallmark of mathematics, they are an essential part of mathematics' contribution to general culture. The student who was never impressed by a mathematical proof missed a basic mental experience.

On what level of rigor should we teach mathematical proofs, and how? The answer to this question is not so simple; it is, in fact, beset with difficulties. Yet to ignore these difficulties and give some answer without much thought, just by following tradition, fashion, or prejudice, is not the kind of thing conscientious planners of the curriculum should do.

There are proofs and proofs, there are various ways of proving. The first thing we must realize is that certain ways of proving are more appropriately taught on a given level of age and maturity than other ways.

(1) A certain aspect of mathematical proofs was observed and described by Descartes with remarkable lucidity.

I quote the third of his *Rules for the Direction of the Mind:*[13] "About the objects of our study, we should not seek the opinion of others or our own conjectures, but only what we can see with clear and evident intuition or deduce with certitude, for there are no other ways to knowledge." In explaining this rule, Descartes considers the "two ways to knowledge," intuition and deduction, successively. Here is the beginning of his discussion of deduction:[14] "The evidence and certitude of intuition are required not only in propositions, but also in all kinds of reasoning. For instance, we intend to deduce that 2 and 2 equals 3 and 1. Then we must intuitively see not only that 2 and 2 make 4 and that 3 and 1 make 4, but also that from these two propositions that third one mentioned above necessarily follows."

[13] *Oeuvres,* vol. X, p. 366.
[14] *Oeuvres,* vol. X, p. 369.

A mathematical deduction appears to Descartes as a chain of conclusions, a sequence of successive steps. What is needed for the validity of deduction is *intuitive insight* at each step which shows that the conclusion attained by that step evidently flows and necessarily follows from formerly acquired knowledge (acquired directly by intuition or indirectly by previous steps of deduction).

(We know from chapter 7 that a graph with ramifications represents the structure of a proof more appropriately than a simple unramified chain, but this is not the point: If Descartes knew the diagrammatic representation we have studied in chapter 7, he would insist that each element of the diagram—such as represented by Fig. 7.18—must be supported by intuitive evidence.)

(2) Yet mathematics has several aspects. It can also be regarded as a game with symbols played according to arbitrarily fixed rules in which the principal consideration is to stick to the rules of the game. (This aspect is quite recent; fifty years ago most mathematicians and most philosophers would have thought that such an aspect of mathematics is revolting. Still, this aspect, which has been introduced under the influence of the great mathematician David Hilbert, is quite appropriate in certain studies of the foundations of mathematics.)

In this game with symbols, the symbols are meaningless (we are supposed to ignore their meaning if they have one). There are "proofs" in this game; a step of a proof consists in writing down a "well-formed" new formula (a combination of symbols complying with the rules). Such a step is considered as valid if the new formula is written down strictly in accordance with certain formulas introduced at the beginning ("axioms") with formulas written in previous steps, and with certain rules of inference fixed from the start. To be handled in this fashion, both the proofs and the propositions proved must be "atomized," decomposed into very small steps and minute component parts.

(3) Between the two extreme aspects considered in (1) and (2) there are others.[15] In fact, the concept of mathematical proof has evolved, changed from one era of science to the other. The study of this evolution and of the motives that led to it could have great interest for us teachers. By understanding how the human race has acquired a concept we could better see how the human child should acquire it. *Cf.* ex. 14.13.

A productive mathematician is, of course, free to prefer any aspect of mathematics; he should prefer the aspect most profitable for his work. Yet on the high school level our choice is not free, and if the choice is between (1) and (2) (between teaching proofs closer to one aspect or to the other), we can hardly hesitate.

I think that everybody prefers intuitive insight to formal logical arguments, including professional mathematicians. Jacques Hadamard, an eminent French mathematician of our times, expressed it so:[16] "The object of mathematical rigor is to sanction and legitimize the conquests of intuition, and there never was any other object for it." Yet, if we exclude professional mathematicians, almost nobody remains who would be in a position to properly appreciate formal argu-

---

[15] For a well-illustrated important investigation of the nature of proofs see Lakatos (quoted in the Bibliography).

[16] Émile Borel, *Lecons sur la théorie des fonctions,* 3rd ed. 1928; see p. 175.

ments. Intuition comes to us "naturally," formal arguments do not.[17] At any rate, intuition comes to us much earlier and with much less outside influence than formal arguments which we cannot really understand unless we have reached a relatively high level of logical experience and sophistication.

Therefore, I think that in teaching high school age youngsters we should emphasize intuitive insight more than, and long before, deductive reasoning. And when we present proofs, we should present them much closer to the idea of Descartes [see (1)] than to a certain idea of certain modern logicians [see (2)].

I have met with youngsters who had a definite interest, and probably some talent, for engineering or science, but refused to learn mathematics, and I have a hunch where this refusal comes from.

(4) Let me give an example. I consider the proposition: *Among three points on a straight line, there is just one that lies between the two others.*

Observe that this proposition says something essential about the nature of the straight line. If there are three points on a circle, no one plays a special role, no one is distinguished from the others by "betweenness."

Does this proposition about the three points on the straight line *need a proof?* In a university lecture on the foundations of geometry, a proof for this proposition starting from the axioms may be essential. Yet to present such a proof to a tenth grade high school class just starting the rational study of geometry is simply preposterous.

This is my opinion, which may be wrong. To have a defensible opinion you must picture to yourself the reaction of the class to such a proof. I imagine it so: The majority of the youngsters will be simply bored. Yet a less mediocre and less indifferent minority will feel more or less distinctly that the proof is superfluous and aimless. There may be one or two boys in the class who will be frankly disgusted and revolted. At any rate, I think that such would have been my own reaction if such a proof had been presented to me when I was of high school age. I do not pretend that I remember exactly the ideas of the teenager I was sixty years ago, and I certainly do not pretend that that teenager was always right. Still, I can vividly imagine my reaction to such a proof: It would have convinced me that my teacher is stupid—or that mathematics is stupid, or that both are stupid. And, at that point, I would have stopped listening to the explanations of the teacher—or, if obliged to listen, I would have listened with reluctance, suspicion, and contempt.

At any rate, I think that an adverse reaction to a proof of the nature outlined is natural and proper.

(5) There are many aspects of proofs. I think that the role of proofs in building up science is more complex than it is usually assumed, and there may be a question deserving a philosophical inquiry. We deal here, however, with another question: which aspect of proofs should be presented to beginners? This question seems to me easier, and I have a pretty firm opinion about it which I take the liberty to express.

In the first place, the beginner must be convinced that proofs deserve to be studied, that they have a purpose, that they are interesting.

---

[17] For an incidental remark expressing a very similar opinion see H. Weyl, *Philosophy of mathematics and natural science,* p. 19.

Proofs have a purpose in a law court. The defendant is suspected to be guilty, but this is just suspicion, there is a doubt. It must be proved either that he is guilty or that he is not. The purpose of a legal proof is to *remove a doubt,* but this is also the most obvious and natural purpose of a mathematical proof. We are in doubt about a clearly stated mathematical assertion, we do not know whether it is true or false. Then we have a problem: to remove the doubt, we should either prove that assertion or disprove it.

Now I can explain why I am so firmly convinced that the proof hinted at previously (about the three points on a line) is out of place in the high school. A high school age youngster who has understood that assertion about the three points can hardly doubt it. There is no doubt to remove, and so the proof appears useless, aimless, senseless. The case is aggravated if the proof starts from axioms, distinguishes several cases, and takes thirteen lines in the textbook. It may give the youngsters the impression that mathematics consists in proving the most obvious things in the least obvious way.

(6) Yet, as I have already said, the proof about the three points on the straight line is quite in order on the proper level. When we teach it in high school we commit the ugly and ridiculous pedagogical sin of the *confusion of levels.* See ex. 14.15.

On the research level, it can happen that a proposition appears intuitively obvious, we have strong plausible arguments for it, but no formal argument. It is in such a situation that a mathematician may do his best to discover a proof. For a preview of such situations on the high school level see ex. 14.11, and more examples in chapter 15.

(7) Before leaving this subject, I must sound a warning against another grave pedagogical error: overemphasizing trivial proofs. Crowding the pages of a textbook with pointless proofs which lack motivation and rewarding goals may make the worst impression on the best students, who have some gift of intuition which could be most useful in engineering or science or mathematics.

Also this ugly error may be due to a confusion of levels. Not for a high school age boy, but for a professional mathematician it may be necessary to check the formal justification for each step of a long argument. Such checking may be necessary, although it is not the most enjoyable part of the mathematician's work. Logic is the lady at the exit of the supermarket who checks the price of each item in a large basket whose contents she did not collect.

14.11. *Can a map be perfect?* A map is the representation of a part of the earth's surface on a flat piece of paper.

(1) To understand the situation before us, we generalize it and we describe more precisely the more general situation. (This transition from the particular to the general and from the more intuitive to the more abstract level is important; it is made here abruptly, but in the classroom it should be made gradually and carefully).

We consider the mapping of a surface S onto another surface S'. We consider a *one-to-one mapping,* that is, we suppose that to each point p of S there corresponds just one point p' of S', the *image* of p, and that, conversely, to each point of p' of S' there corresponds just one point p of S, the *original* of p'. We assume further that the mapping is "continuous": to the points of a "smooth" line on one surface there corresponds a set of points on the other surface forming a smooth line. Let L_1 and L_2 be lines on S intersecting in the point p at which they include the angle α. Let the lines L_1' and L_2' be the images of the lines L_1 and L_2, respectively. Then L_1' and L_2' intersect in the point p', the image of p, at which they include the angle α'; we consider α' as the image of α and α as the original of α'.

(In the particular case of geographical maps, S is a part of the earth's surface and S' the corresponding part of a plane. Let us think of important lines on the earth's surface—coast lines, rivers, boundary lines, roads, railway lines—which are represented by corresponding lines on the map.)

(2) Now we are in a position to give a clear definition. We call a mapping *perfect* if it satisfies two conditions:

(I)   All *lines* are reduced on the same scale.
(II) All *angles* are preserved.

Let us restate these two conditions with more detail.

(I) To the mapping there belongs a definite "scale" or fixed numerical ratio (for instance, $1:1\,000\,000$) in the following sense: If L', a line on the surface S', is the image of L, a line on the surface S, then the length of L' has that fixed numerical ratio ($1:1\,000\,000$ in our example) to the length of L, independently of the shape, size, and location of the lines.

(II) Each angle α' on S' is equal to the angle α on S of which α' is the image.

(3) Let us visualize the foregoing definition, let us see more concretely the details involved.

(3a) It is stated that the scale of a carefully done geographical map is $1:1\,000\,000$. This means that such is the scale *approximately*—but can the scale be the *same throughout the map exactly?* And, if such is the case, can the map also preserve the angles? That is the question.

(3b) If it is geometrically possible to map the surface S onto the surface S' on any fixed scale, it is, obviously, also possible to map a surface *geometrically similar* to S onto the surface S' on the scale $1:1$, that is, without reduction or enlargement. For example, let us assume that the planet on which we are living is an exact sphere. If any part of the earth's surface could be perfectly mapped onto a flat sheet of paper on the scale $1:1\,000\,000$, the corresponding part of a sphere of which the diameter is one millionth of the earth's diameter would be so mapped onto the same sheet that corresponding lines, original and image, would be of the same length throughout, and also corresponding angles would be equal.

(3c) We can roll a sheet of paper into a cylindrical or conical shape and, conversely, we can unroll the curved lateral surface of a cylinder, or of a cone, into a flat sheet. Such unrolling generates a perfect mapping of the curved cylindrical, or conical, surface onto a plane (imagine coastlines and rivers traced on the paper); lengths and angles are obviously preserved.

Yet could we similarly unroll into a plane a piece of a spherical surface, preserving all lengths and angles? We strongly suspect that it is not possible, and this suspicion may be based on experience, on observations we made in peeling apples or potatoes.

(4) Now we may perceive the kernel of our problem. Is it possible to map, with points corresponding one-to-one, a piece S of a spherical surface onto a piece S' of a plane so that *all lengths and all angles are preserved?*

We assume that (contrary to our expectation) such a mapping *is* possible and draw consequences from this assumption in the following (5) and (6).

(5) *Lengths are preserved.* Let p and q be two different points of S (of the sphere) and L any line on S connecting p and q; let p', q', and L' be the images in S' (in the plane) of p, q, and L, respectively. By our assumption, L and L' are of the same length. If L happens to be the *shortest line on the sphere* connecting p and q, shorter than any other connecting line, then, as lengths are preserved, L' must be shorter than any other line connecting p' and q', and so the shortest connecting line, *in the plane.* We know (the reader should know) that the shortest lines in the plane are straight lines, and the shortest lines on the sphere are arcs of great circles. The result of our consideration is: arcs of great circles on the spherical surface S are mapped onto segments of straight lines in the plane region S'. Especially, the sides of a spherical triangle which are arcs of great circles are mapped onto the sides of an ordinary triangle which are segments of straight lines.

(6) *Angles are preserved,* and so each angle of the spherical triangle just mentioned should be *equal* to the corresponding angle of the ordinary triangle. Yet this is impossible, since the sum of the three angles of an ordinary triangle is 180° whereas, as the reader should know, the sum of the three angles of a spherical triangle is *greater* than 180°.

A perfect mapping of the sphere onto the plane is impossible.

(7) The problem we have just solved may become a gateway both to practical applications (mapmaking) and to a great theory (a chapter of differential geometry centered in the "theorema egregium" of Gauss and reaching forward to general relativity). Here are a few points not too far beyond the high school level and closely connected with what we have just discussed; check them.

(7a) A plane (ordinary, Euclidean) triangle and a spherical triangle are so related that each side of one has the same length as the corresponding side of the other. Show that in this situation [considered in (5) and (6)] *each* angle of the spherical triangle is greater than the corresponding angle in the plane triangle. [That the sum of the angles of the first is greater than the sum of the angles of the second was the decisive remark of (6).]

(7b) The two conditions stated under (2) are not unrelated: (I) involves (II), that is if (I) is satisfied, (II) must be satisfied too.

(7c) Yet (II) does not involve (I). There are many mappings of the sphere onto the plane preserving all the angles, but in which the ratio of the length of a curve on the sphere to the length of the image curve in the plane is by no means constant [it cannot be constant, by virtue of the theorem proved in (5) and (6)].

(7d) There are mappings of the sphere onto the plane preserving all the areas (but they do not preserve the angles).

(7e) There are mappings of the sphere onto the plane preserving shortest lines, that is, mapping arcs of great circles onto segments of straight lines (but they do not preserve the angles).

(8) Yet I think that making explicite the role of continuity in the foregoing discussions and supplementing precise details in this respect would be too much above the level of the high school.

14.12. *What should we teach?* You, as a teacher, are paid by the community to teach the youngsters in your class. Therefore, your task is to teach what is profitable for the community and for the youngsters in your class.

You think that this advice does not amount to much. It may amount to more than you think. Just keep your task constantly in mind in short range as in long-range planning, in mapping out the next class period or the curriculum. Imagine that there is in your class a nice and clever boy, not yet spoiled by the school and not yet afraid of you who at any moment may ask you, honestly and naively: "But teacher, what is that good for?" If you try to picture to yourself what is in the mind of that nice boy and plan your teaching so that you can answer that critical question—or that he is constantly amused and challenged and has no opportunity to ask that critical question—you may become a better teacher.

I admit that the teacher's task is beset with temptations. For instance, we may be tempted to teach what is easy to teach, what is "teachable." Yet should we teach everything that is teachable? Is the teachable always profitable?

A clever trainer may teach a seal to balance a ball on its nose. But does such skill help the seal to catch more fish?

14.13. *The genetic principle.* Planning the curriculum involves more than choosing the facts and theories to be taught; we must also foresee in what sequence and by what methods those facts and theories should be taught. In this respect the "genetic principle" offers an important suggestion.

(1) The genetic principle of teaching can be stated in various ways, for instance: In teaching a branch of science (or a theory, or a concept) we should let the human child retrace the great steps of the mental evolution of the human race. Of course, we should not let him repeat in detail the thousand and one errors of the past, just the great steps.

This principle does not lay down a hard and fast rule; on the contrary, it leaves us much freedom of choice. What steps are great and what errors are negligible is a matter of interpretation. The genetic principle is a guide to, not a substitute for, judgement.

Yet in order to emphasize just this point, there may be some advantage in restating the principle more cautiously (and more vaguely): Having understood how the human race has acquired the knowledge of certain facts or concepts, we are in a better position to judge how the human child should acquire such knowledge. [We came very close to this formulation in ex. 14.10(3).]

(2) The genetic principle is supported by a biological analogy. The development of the individual animal retraces the evolutionary history of the race to which the animal belongs. That is, the embryo of the animal, as it passes through the successive stages of its development from the fertilized ovum to its adult form, resembles at each stage an ancestor of its race, and the sequence of its stages of development mirrors the sequence of its ancestors. If for "development of the individual animal" we say "ontogeny" and for "evolutionary history of the animal species" we say "phylogeny," we arrive at the concise form which the

German biologist Ernest Haeckel has given to his "fundamental biogenetic law": "Ontogeny recapitulates Phylogeny."

Such analogy is, of course, just a source of interesting suggestions and not a proof for the genetic principle of teaching, which itself should not be regarded as an "established principle" but just as a source of interesting suggestions.

(3) Thus, the genetic principle may suggest the principle of consecutive phases which we have discussed in sect. 14.4(3) and 14.5(3). In fact, in the historical development of various branches of science (of theories, of concepts) we may distinguish three phases. In an initial *exploratory* phase the first suggestive, but often incomplete or erroneous, ideas emerge from contact with the experimental material. In the next phase of *formalization* the material is ordered, appropriate terminology introduced, the laws recognized. In the last phase of *assimilation* the laws are seen in a broader context, extended, and applied.

Yet only the reading of the original works of great authors can really convince us of the genetic principle of teaching. Such reading may be like a brisk walk in the fresh air after the stale atmosphere of the textbooks. As James Clerk Maxwell wrote in the preface of his great *Treatise on Electricity and Magnetism:* "It is of great advantage to the student of any subject to read the original memoirs on that subject, for science is always most completely assimilated when it is in the nascent state."

(4) According to the genetic principle, the learner should retrace the path followed by the original discoverers. According to the principle of active learning, the learner should discover by himself as much as possible. A combination of the two principles suggests that the learner should *rediscover* what he has to learn—we have caught here a first glimpse of an important aspect of the educational process about which the reader should consult the two books of A. Wittenberg quoted in the bibliography.

14.14. *Lip service.* "General culture" is a catchword and, as a catchword, it is exposed to misuse. It is easy to pay lip service to "general culture." The most atrocious things can be done in the schools under the pretence that they are done for "general culture."

"General culture," "teaching to think," and "teaching problem solving" are catchwords which can be misinterpreted and misused inspite of the good substance behind them. Yet there is a difference in favor of "problem solving."

"Problem solving" can be explained not only in other general terms (equally liable to misinterpretation) but also by suggestive concrete examples (this book and my related writings attempt to present many such examples).

Moreover, lip service paid to problem solving can be more easily unmasked: "So, you are teaching problem solving—very interesting. What problems did you present to your class? What desirable mental attitudes do you expect to develop by such problems?"

14.15. *Confusion of levels.* "Contemporary mathematicians work much more with sets, operations, groups, fields, etc. than with oldfashioned geometry and algebra. Therefore, we must teach sets, operations, groups, and fields before those oldfashioned subjects."

This is an opinion. Here is a similar one:

"Contemporary American adults do many more miles driving a car than walking. Therefore, we must teach a baby to drive a car before he can walk."

14.16. *Isadora Duncan* was a celebrated dancer—as celebrated when I was young as Marilyn Monroe was in more recent years.

What is the connection between the lady and our subject? You see, it may seem such a good idea to let a team plan the curriculum and write the textbooks. The team consists of a university professor and a high school teacher. We may expect a splendid result in which the professor's mathematical perspective is coupled with the teacher's experience in handling high school classes. Yes, yes, but. . . .

When I was young everybody knew a certain story about Isadora Duncan. Allegedly, she offered something like marriage to Bernard Shaw: ". . . and think of the child who would have your brains and my looks." "Yes, yes," said Bernard Shaw "but what a calamity if the child had my looks and your brains."

Perhaps you too have seen some books recently which mirror the teacher's mathematical perspective coupled with the professor's experience in handling high school classes.

14.17. *Levels of knowledge.* In his "Treatise on the Improvement of the Mind" (Tractatus de Intellectus Emendatione), the philosopher Benedict Spinoza distinguishes four different levels of knowledge.[18] Spinoza exemplifies the four levels by four different ways of understanding the Rule of Three. In the following, (1) to (4), "rule" stands for any mathematical rule the reader learned some time ago and, if possible, for one that he has learned by stages, understanding it better at each stage.

(1) A student has learned a rule by heart, accepting it on authority, without proof, but he is able to use the rule, can apply it correctly. Then we say that he has *mechanical knowledge* of the rule.

(2) The student tried the rule on simple cases and convinced himself that it works correctly in all cases he has tested. He thus has *inductive knowledge* of the rule.

(3) The student has understood a demonstration of the rule. He has *rational knowledge* of the rule.

(4) The student conceives the rule clearly and distinctly and is so convinced of it that he cannot doubt that the rule is true. He has *intuitive knowledge* of the rule.

(5) I do not know whether the passage of Spinoza paraphrased in the foregoing has been noticed in the pedagogical literature. At any rate, the distinction between various levels of knowledge should be well understood by the teacher. Thus the curriculum requires of him that he should teach such and such chapter of mathematics in such and such grade. Yet what level of knowledge should the students attain? Is mechanical knowledge sufficient? Or should the teacher attempt to lead his students to intuitive knowledge? We have here two different aims, and it makes a great difference both for the teacher and for the students which one is adopted.

---

[18] See e.g. *Philosophy of Benedict Spinoza*, translated by R. H. H. Elwes, p. 7. The following is a very free paraphrase of Spinoza's text; especially, the names for the levels are added.

(6) When we consider the various levels of knowledge distinguished by Spinoza from the teacher's standpoint, we are led to several questions. How could we bring the students to this or that level? How could we test whether the students have attained this or that level? These questions are most difficult to answer in regard to the intuitive level.

(7) We could, of course, distinguish more than four levels of knowledge, and there is one level whose consideration may deserve the attention of teachers (and of learners, especially of ambitious learners who aspire to become scientists)— well anchored, well connected, well cemented, in one word *well-organized knowledge*.[19]

The teacher who aims at well-organized knowledge should be careful, in the first place, in introducing new facts.   A new fact should not appear out of nowhere, it should be motivated by, referred to, connected with the world around us and the existing knowledge, the daily experience, the natural curiosity of the student.

Moreover, when the new fact has been understood, it should be used to solve new problems, to solve old problems more simply, to shed light on things already known, to open new perspectives.

The ambitious learner should carefully study a new fact; he should turn it over and over, consider it under various aspects, scrutinize it from all sides, and try to fit it into his existing knowledge at the best place where it is most conveniently connected with related facts.   Then he will be able to see that new piece of knowledge with the least effort, the most intuitively.   Moreover, he should try to expand and enlarge any newly acquired knowledge by application, generalization, specialization, analogy, and in all other ways.

(8) As dedicated teachers we may find ways to anchor a new fact in the experience of the student, connect it with formerly learned facts, cement its knowledge by applications.   We can only hope that the student's well-anchored, well-connected, well-cemented, well organized knowledge will finally become intuitive.

14.18. *Repetition and contrast.*   If you like both music and teaching, you may observe various resemblances between them, and your observations, even if they are not scientifically perfect, may improve your teaching.   They may lead you to presenting the material you have to offer in a more artistic and more effective arrangement.

Why is that so?   Repetition and contrast play a role in all arts, also in the teacher's art, but their role is most conspicuous in music.   Hence the foreshadowing, the development, the repetition, alternation, and variation of themes in a musical composition may well suggest the analoguous treatment of themes in the classroom or in a literary composition.

14.19. *Inside help, outside help.*   In planning and writing this book I had in mind the task of the high school mathematics teacher and especially the following situation.   The teacher proposes a problem to his class; the students should learn from their own work, and the problem should be solved in class discussion.   This situation requires careful handling.   If the teacher helps too little, there will be

---

[19] *Cf.* ex. 12.3, and also the preface and the whole plan of G. Pólya and G. Szegö, *Aufgaben und Lehrsätze aus der Analysis.*

no progress. If the teacher helps too much, there will be no opportunity for the students to learn from their own work. How should the teacher avoid the horns of this dilemma? How much should the teacher help the students?

There is a better question; we should not ask "how much," we should ask "how." *How* should the teacher help the students? There are various ways to help.

(1) There are cases in which the teacher must ask several questions and repeat his questions several times until he succeeds in extracting a little work from the students. In the following dialogue a dotted line . . . . . . indicates the silence of the students. The discussion has been going on for some time as the teacher says:

"Tell me again: What is the unknown?"

'The length of the line *AB*.'

"How can you find this kind of unknown?"

. . . .

"How can you find the length of a line?"

. . . .

"From which data could you derive the length of a line?"

. . . .

"Did not we solve such problems before? I mean problems in which the unknown was the length of a line?"

'I think we did'.

"How did we proceed in such a case? From which data did we compute the unknown length?"

. . . .

"Look at the figure. You see the line *AB*, don't you? The length *AB* is unknown. Of which lines is the length known?"

'*AC* is given.'

"Good! Is there any other given line?"

'*BC* is also given.'

"Look at the lines *AB*, *AC*, and *BC*—look at their *configuration*. How would you describe this configuration?"

'*AB*, *AC*, and *BC* are the sides of △*ABC*.'

"What kind of triangle is △*ABC*?"

. . . .

Well, there are cases in which a teacher must be infinitely patient.

(2) A less patient teacher could proceed quite differently and tell the students right away: "Apply the theorem of Pythagoras to the right triangle △*ABC*."

(3) What is the difference between the two procedures (1) and (2)?

The most obvious difference is that (1) is long and (2) is short.

Yet we should also observe another difference: (1) offers more opportunity to the student than (2) to contribute to the solution something of his own.

Yet there is a more subtle difference.

The questions and suggestions offered by the teacher in procedure (1) could have *occurred to the student himself*. If you look at them closely, you may notice that several of these questions and suggestions are *tools* which the problem solver can use also for other problems, in fact, for wide classes of problems. These tools are at the disposal of everybody—of course, the more experienced, the "methodologically better prepared" problem solver handles them more deliberately and more skillfully.

Yet the action advised by the teacher in procedure (2) is not a tool a priori at the disposal of the problem solver—it is the solution itself, or it is almost that. It is a specific action suggested without reference to any general idea.

Let us call *inside help* such help as the problem solver earnestly concerned with his problem and familiar with methodological ideas has a good chance to find by himself. Let us call *outside help* such help as has little relation to methodological ideas—the problem solver has little chance to elicit such help by methodical work. I think that the most important difference between the procedures (1) and (2) is that the teacher offers inside help in the former and only outside help in the latter.

(4) If we accept the principle of active learning we must prefer inside help to outside help. In fact, the teacher should give outside help only in the last resort, when he has exhausted all obvious inside suggestions without result, or when he is pressed for time.

Outside help has very little chance to be instructive—appearing out of the blue, as *deus ex machina,* it can easily be disappointing.[20] Inside help may be the most instructive thing the teacher can offer; the student may catch on, he can realize that the question helps and that he could have put that question to himself by himself. And so the student may learn to use that question; the voice of the teacher may become for him an inner voice which warns him when a similar situation arises.[21]

To give inside help, the teacher may use all the "stereotyped" questions and recommendations collected in chapter 12—for this reason, chapter 12 may become for him the central chapter of the book. Of course, he must first become intimately acquainted with the situations to which those questions and recommendations are applicable. This book has been planned and written to help the teacher in this task.

14.20. In my class, when I get the impression that I have spoken too long without interruption and I should now ask a question of the audience, I am apt to remember a German jingle of which here is an approximate translation:

All are sleeping just one is preaching:
Such performance is called here "teaching."

14.21. *How difficult is it?* Both the scientist and the teacher may be led to this question, the one when he is struggling with a problem, the other when he is about to propose it to his class. In answering the question we must rely much more on "feeling" than on any clear argument. Still, now and then we arrive at assessing the degree of difficulty of a problem quite well. The scientist's feeling may be justified by the outcome of his research and the teacher's feeling by the outcome of an examination.

In most circumstances, clear arguments can contribute but little to the evalua-

---

[20] *Cf.* the end of sect. 3.2.

[21] *Cf.* HSI *passim,* especially sect. 17, Good questions and bad questions, pp. 22–23.

tion of the degree of difficulty of a problem, yet even that little deserves careful consideration.

(1) *Size of the region of search.* An offense has been committed (for instance, one of the children has broken a window) and we know that the offender is one of a set of n persons. Other things being equal, the difficulty to find the offender obviously increases with n. Generally speaking, we may expect that the difficulty of a problem increases with the size of the adequate region of search. (*Cf.* sect. 11.6.)

(2) *Number of items to be combined.* The students have to solve a problem which needs the application of n different rules introduced in the last chapter with which the students are much less familiar than with the foregoing chapters of the course. Under such circumstances, other things being equal, the difficulty of the problem obviously increases with n. Generally speaking, we may expect that the difficulty of a problem increases with the number of items which we have not combined before but which we have to combine to solve the problem.

(3) The foregoing considerations may help us to judge the difficulty of a problem a priori, before trying it. Judging the difficulty a posteriori, after having tried to solve the problem, involves, more or less explicitly, ideas of *statistics;* here is a schematic example. In an examination, of two problems proposed to 100 students, one was solved by 82 students, the other by 39 students. Obviously, the latter was more difficult *for this group* of 100 students. Will it also be more difficult for the next group of students? Yes, says the statistician, this has to be expected with such and such a degree of confidence provided that there are *no nonrandom differences* between the two groups. And there is the rub. In educational matters there are too many scarcely controllable circumstances which have a great influence so that the distinction between "random" and "nonrandom" becomes highly elusive. Thus the presentation of a certain point in the course, the insistence of the teacher on that point, the mood of the teacher, and many other unforseeable circumstances may incontrollably influence the outcome of an examination—they may influence it much more than that component of the situation about which we have desired to obtain information by statistical experiment. We have touched here but lightly on one of the many reasons which should make us cautious or even suspicious when we are dealing with educational statistics.

Of course, when the mathematician deals with a problem that was proposed two hundred or two thousand years ago yet still remains unsolved, he has a rather good "statistical" reason to suspect that the problem is hard. (The theory of numbers abounds in such problems.)

14.22. *Difficulty and educational value.* It is difficult to assess the difficulty, and still more difficult to assess the educational value, of a problem. Yet the teacher must try to assess both when he is about to propose the problem to his class.

The teacher can be assisted in his task by a classification of high school level problems. The credit for a classification of this kind is due to Franz Denk.[22] The following classification is somewhat different; it distinguishes four types of problems.

[22] Franz Denk, Werner Hartkopf, and George Pólya, pp. 39–42; see the Bibliography.

(1) *One rule under your nose.* The problem can be solved by straightforward mechanical application of a rule or by straightforward mechanical imitation of an example. Moreover, the rule to apply or the example to follow is thrust under the nose of the student; typically, the teacher proposes such problems at the end of the hour in which he has presented the rule or the procedure. A problem of this type offers practice but nothing else; it may teach the student to use that particular rule or procedure, but has little chance to teach him anything else. And there is the danger that, even of that single rule, the student will acquire just "mechanical," and not "insightful," knowledge.

(2) *Application with some choice.* The problem can still be solved by the application of a rule learned in class or the imitation of an example shown in class, yet it is not so immediately obvious which rule or example should be used; the student needs some mastery of the material covered in the last weeks and some judgement to find the usable item in a certain limited region of search.

(3) *Choice of a combination.* To solve the problem, the student must combine two or more rules or examples shown in class. The problem need not be too difficult if a somewhat similar (but not the same!) combination has been discussed in class. Of course, if the combination is quite novel, or if many pieces of knowledge must be combined, or pieces of knowledge from chapters wide apart, the problem may demand a higher degree of independence and may become quite difficult.

(4) *Approaching research level.* It is scarcely possible to draw a sharp line of demarcation between the kind of problem we have just considered [under (3)] and research problems. The examples and discussions of chapter 15 attempt to outline some characteristics desirable in "research problems on the classroom level."

On the whole, as the degree of difficulty increases along the lines presented in ex. 14.21(1) and (2), also the educational value of the problem has a chance to increase, especially if our educational aim is "teaching to think," and we judge values from the standpoint of this aim.

14.23. *Some types of problems.* For the use of teachers who wish to interrupt now and then the monotonous sequence of routine problems dished out by the textbooks, I collected elsewhere some types of nonroutine problems (MPR, v. 2, p. 160). I wish to add here one more problem of the "You may guess wrong" type (ex. 14.25) and one new type, the "red herring" problem. This latter type diverts the attention from the main point, from the most natural or most efficient procedure, by some conspicuous but irrelevant feature. "Red herring" problems should be sparingly used; they should be proposed only to students clever enough to see the joke and to learn better the relevant point by having brushed aside irrelevancies. See ex. 14.24.

14.24. Find the remainder of the division of the polynomial

$$x^3 + x^5 + x^7 + x^{11} + x^{13} + x^{17} + x^{19}$$

by the polynomial $x^2 - 1$.

14.25. Two spheres are tangent to each other. They are separated from each other by their *principal* common tangent plane that passes through their point of contact. They have infinitely many other common tangent planes which envelop their common *tangent cone*. This cone touches each sphere along a circle, and the part of the cone between these two circles is the lateral surface of a *frustum*.

Being given t, the slant height of the frustum, compute:

(1) the lateral area of the frustum

(2) the area of that piece of the "principal" tangent plane that lies within the tangent cone.

(Are there enough data to determine the unknowns?)

14.26. *A term paper.* For terminology and notation, see "Mathematics and Plausible Reasoning," vol. 1, pp. 137–138, ex. 33; you may also be able to use appropriate points from ex. 34–54 on pp. 138–141 and their solutions on pp. 249–256.

Consider

(a) a right prism,

(b) a right pyramid, and

(c) a right double pyramid.

The base of each of these three solids is a regular polygon with n sides, and each solid is so circumscribed about a sphere that the point of contact on each face of the solid is the center of gravity of that face.

(1) Find the ratio of the area of the base B to the area of the total surface S of the solid for (a), (b), and (c).

(2) Compute S^3/V^2 for (a), (b), and (c); V stands for volume.

(3) Collect in a table the numerical values of the ratios computed under (2) for $n = 3, 4, 5$, and 6.

(4) Describe the limiting case $n \to \infty$ for (a), (b), and (c). Find the limit of the ratios computed in (2) and add their numerical values to the table (3).

(5) Consider the (unsolved) problem: "Find the polyhedron with a given number F of faces and a given volume that has the minimum surface area."

Formulate a "plausible" conjecture for $F = 4, 6, 8, 12$, and 20; explain why it is plausible, and find indications for or against it in your foregoing work.

(6) Choose carefully a particular problem from the foregoing work that could be appropriately treated in a solid geometry class in the high school.

Formulate it clearly.

Note such "inside" questions and suggestions (from the list "How to Solve It," HSI, pp. XVI–XVII, for instance) [*cf.* chapter 12] as have a good chance to be efficiently used for the problem chosen.

Represent the progress of the solution of the chosen problem diagrammatically (as we did it in class for the volume of the frustum.) [*Cf.* chapter 7.]

(7) How would you "sell" (justify the choice of, explain the general interest of) the subject of this term paper to a class of bright students in a high school, or in a teacher-training institution? (Name the level, and be efficient but concise—please!)

(8) A *diagonal* of a convex polyhedron is a straight-line segment joining two vertices that, except for its two endpoints, lies completely inside the polyhedron (not on the surface). Let D denote the number of diagonals of the polyhedron.

Compute D

(a) for each of the five regular solids,

(b) for a polyhedron all F faces of which are triangles,

(c) for a general polyhedron, being given that it has F_n polygonal faces with n sides, $n = 3, 4, 5, \ldots,$

$$F_3 + F_4 + F_5 + \cdots = F$$

(9) (Optional) If the foregoing has suggested to you some mathematical idea which seems to you relevant, even if incomplete for the moment, note it here—but *very* clearly and concisely.

(The foregoing is a typical example of a "take-home final" such as I am accustomed to propose to my classes offered to high school teachers. For (5) see below ex. 15.35, for (8) ex. 15.13. To answer (6) and (7), some participants presented a dialogue between the teacher and the students similar to some dialogues in this book and especially in HSI. Some of these dialogues were very well done. The references in square brackets were added in print.)

14.27. *On talks at mathematical meetings. Zermelo's rules.* The situation of the speaker in a mathematical meeting resembles a little, and differs a great deal from, the situation of the teacher in the classroom. The speaker, as the teacher, intends to impart some information, but his audience consists of his peers and possibly his superiors, not of students. The situation of the speaker is not easy, also his performance is not often successful. This is not so much his fault, rather the fault of the vast expanse of mathematics. Any one mathematician can master only a small parcel of present day mathematics, and usually knows very little about the other small parcel that the next mathematician has mastered.

(1) Ernest Zermelo, whose name will be forever linked to the important "axiom of choice" of general set theory, liked to spend some of his time in coffee-houses. His conversation there was interspersed with sarcastic remarks about his colleagues. Commenting on an address which had great success at a recent mathematical meeting, he criticized the speaker's style and eventually condensed his disapproval into two rules according to which, he mockingly asserted, that address must have been constructed:

I. *You cannot overestimate the stupidity of your audience.*

II. *Insist on the obvious and glide nimbly over the essential.*[23]

Zermelo's personal remarks were often witty; very unjust on the whole, but striking and revealing about some particular point. So was the criticism implied by the two rules; I had to laugh and I could not forget the rules. Years later I realized that these rules, suitably interpreted, give often applicable sound advice.

(2) The speaker in a mathematical meeting usually treats his audience as if everybody knew everything about the subject he is discussing—especially each detail of his latest paper. Of course, just the contrary is the case, and the speaker should realize it. It would be better for him to overestimate than underestimate the lack of knowledge of the audience about the subject. In fact, the speaker

---

[23] In the original German: *I. Du kannst Deine Hörer nicht dumm genug einschätzen. II. Bestehe auf dem Selbstverständlichen und husche über das Wesentliche hinweg.*

could greatly profit from an interpretation of Zermelo's first rule: "You cannot overestimate the lack of knowledge of your audience about the things you are talking about."

(3) What is essential in the mathematician's work? Each detail of the proof is essential, but it is hardly possible to present adequately all details of a long and difficult proof to a mathematical meeting. Even if the speaker succeeded in hurrying through all the details, nobody would be able to follow. Therefore, "glide nimbly over the essential," over the details of the proof.

Yet even a long proof may hinge on a central remark which in itself is intuitive and simple. A good speaker should be able to detach from the proof some decisive remark and make it intuitive and obvious until everybody in the audience can understand it, take it home, and keep it for possible later use. In so doing the speaker succeeds in imparting useful information and, in fact, follows Zermelo's second rule: "Insist on the obvious and glide nimbly over the essential."

14.28. *Epilogue.* When I was young I devoured the novels of Anatole France. More than by the stories themselves I was attracted by the tone in which they are told, the voice of a sage who looks at all things human with a delicate irony mingled with pity.

Anatole France has a word about a subject we have been discussing: "Do not try to satisfy your vanity by teaching them great many things. Awake their curiosity. It is enough to open the minds, do not overload them. Put there just a spark. If there is some good inflammable stuff it will catch fire." (Le jardin d'Epicure, p. 200.)

There is a great temptation to paraphrase this passage: "Do not try to satisfy your vanity by teaching high school kids a lot of . . . just because you wish to make people believe that you understand it yourself. . . ." Yet let us resist temptation.

CHAPTER 15

GUESSING AND SCIENTIFIC METHOD [1]

Non-mathematical induction plays an essential role in mathematical research.

I. SCHUR: *Inaugural Dissertation,* Berlin 1901, Thesis I.

It will always be difficult, in any branch of knowledge, to describe with some approximation to the truth the method followed by the inventors. . . . Nevertheless, concerning the mental procedures of mathematicians there is a simple remark extensively confirmed by the history of science: observation has an important place and plays a great role in their procedures.

CHARLES HERMITE: *Œuvres,* vol. IV, p. 586.

Observation is the abundant source of invention in the world of subjective realities just so as it is in the world of phenomena perceptible by the senses.

CHARLES HERMITE: *Hermite et Stieltjes, Correspondance,* vol. I, p. 332.

15.1. Research problems on the classroom level

The teaching of mathematics should acquaint the students with all aspects of mathematical activity as far as possible. Especially it should give opportunity to the students for independent creative work—as far as possible.

The activity of the expert mathematician, however, differs a great deal and in several respects from the usual classroom activity. We shall be in a better position to see which differences deserve particular em-

---

[1] This chapter is dedicated to my friend and colleague, Charles Loewner.

phasis after a few examples. The following examples show that a good teacher can offer something approaching the experience of independent inquiry even to an average class by choosing appropriate problems and presenting them in the appropriate way.

15.2. Example

"Being given L, the length of the perimeter of an isosceles right triangle, compute its area A." This is the usual kind of problem printed in the usual textbook. It is not even a bad problem, just not too interesting when it is presented by itself, isolated from analogous problems. Compare it with the following presentation, and see the difference.

"In those legendary pioneer days" said the teacher "when land was plentiful and almost everything else very scarce, a man in the Middle West had hundreds of acres of flat grassland, but only one hundred yards of barbed wire. He intended to use the whole length of this wire to fence in a piece of his land. He thought of various shapes and wondered how many square yards would be the fenced in area."

"Well, which shape would you prefer? But remember, you have to compute the area, and so you rather choose some easy shape."

– A square.
– A rectangle with sides of 20 and 30 yards.
– An equilateral triangle.
– An isosceles right triangle.
– A circle.

"Very good. May I add a few shapes:

A rectangle with sides of 10 and 40 yards.
An isosceles triangle with sides of 42, 29, and 29 yards.
An isosceles trapezoid with sides of 42, 13, 32, and 13 yards.
A regular hexagon.
A semicircle."

"All these figures are *isoperimetric,* that is, of equal perimeter; the perimeter of each is supposed to be 100 yards. Compute the areas in square yards, and arrange the ten figures according to their areas, beginning with the largest and ending with the smallest. By the way, before computing you may try to guess which area will be the largest and which one will be the smallest."

This problem may be proposed as homework to an average high school class at an appropriate stage of the curriculum. The solution consists of the following list:

Circle	795	Rectangle 30, 20	600
Regular hexagon	722	Semicircle	594
Square	625	Equilateral triangle	481

Trapezoid 42, 13, 32, 13	444	Triangle 42, 29, 29	420
Isosceles right triangle	430	Rectangle 40, 10	400

"Are there any questions?"

15.3. Discussion

The point of our problem is to bring to the attention of the students the list of figures and areas which constitutes its solution; the observation of this list should suggest various remarks to the students. The more spontaneously these remarks emerge, the better it is. Yet, if they are slow in coming, the teacher may contribute to the discussion by well-placed, gently prodding questions, such as the following.

"Have you any comment on the list?"

"The circle leads the list. Have you any comment on this?"

"There are several triangles in the list, also several quadrilaterals. Which figure leads the list of quadrilaterals? What about the triangles?"

"Yes, it may be so as you say, but have you proved it?"

"If you have not proved it what reason have you to believe it?"

"A triangle may be considered as a degenerate quadrilateral, with one vanishing side (or with one angle of 180°). Does this remark contribute to your argument?"

Eventually the students should arrive, sooner or later, by their own means as far as possible, to remarks in the direction of the following.

The list suggests that of all plane figures of equal perimeter the circle has the maximum area.

The list suggests that of all quadrilaterals of equal perimeter the square has the maximum area.

The list suggests that of all triangles of equal perimeter the equilateral triangle has the largest area.

The list suggests that of all polygons with a given number n of sides and with a perimeter of given length the regular polygon has the largest area.

Another suggestion of the list: If two regular polygons have the same perimeter, the one with more sides has the larger area. (The more a polygon resembles the circle the larger seems to be its area.)

None of these statements is proved by the list, which yields, however, some kind of reason to believe in them, more or less.

Our experience may also suggest more general insights, such as the following: Verification in more cases gives us a stronger reason to believe.

There may be other relevant remarks, and some of the foregoing remarks may come more readily after more examples.

15.4. Another example

"The Greeks knew" said the teacher "a remarkable proposition on the area of the triangle which we call today 'Heron's formula' and express by the equation

$$A^2 = s(s - a)(s - b)(s - c)$$

where A stands for the area of the triangle, a, b, and c for the lengths of the three sides, and

$$s = \frac{a + b + c}{2}$$

for the semiperimeter.

"The proof for Heron's formula is not quite simple, and I do not wish to enter into it today. In absence of a proof, however, we cannot be certain that the equation as it stands is correct—my memory may have failed me as I wrote it down. Could you check the formula? How could you check it?"

– I would try it on the equilateral triangle.

In this case $a = b = c$, $s = 3a/2$, and the formula yields the correct result. "What else could we do?"

– I would try it on a right triangle.

– I would try it on an isosceles triangle.

In the first case $a^2 = b^2 + c^2$, in the second case $b = c$, and in both cases the formula yields, after some algebraic manipulation, the correct result. (The reader should work out the details.)

"Do you like it?"

– Yes, it checks.

"Could you think of a further particular case that we should check?"

"What about the degenerate triangle? I mean the extreme, or limiting, case in which the triangle collapses into a straight line segment."

In this case $s = a$ (or b, or c) and the formula obviously yields the correct result.

– Please, teacher, in how many cases must we check a formula, if we want to be sure that it is right?

The reader may picture to himself the discussion started by the last question.

15.5. Graphic representation of the inductive argument's progress

What is, and what is not, accomplished by the successive verifications of the proposed formula in sect. 15.4? Each verification deals with a certain triangular shape, and so a survey of all such shapes may contribute to elucidating our question.

Let x, y, and z denote the sides of a variable triangle listed in order of increasing length, so that

$$0 < x \leqq y \leqq z$$

Then we must have

$$x + y > z$$

Now, as only the shape of the triangle matters, and not its size, we may assume that

$$z = 1$$

Thus we have three inequalities

(1) $$x \leqq y, \quad y \leqq 1, \quad x + y > 1$$

Let us now represent the triangle with sides x, y, and 1, or triangle $(x, y, 1)$ for short, by the point (x, y) in a plane the rectangular coordinates in which are x and y. Each one of the three inequalities (1) restricts the point (x, y) to a half-plane (including the boundary line in the first two cases, excluding it in the third case). Jointly the three inequalities (1) characterize a set of points—the common part or *intersection* of the three half-planes. This intersection is a triangle, see Fig. 15.1, with vertices $(1, 1)$, $(0, 1)$, and $(\frac{1}{2}, \frac{1}{2})$ [including the vertex $(1, 1)$ and the two adjacent sides, but excluding the two other vertices and the third side]. This triangular area represents the totality of triangular shapes; the point (x, y) represents the triangle $(x, y, 1)$ and different points represent different shapes.

How are the particular cases considered in sect. 15.4 located in Fig. 15.1?

First, we have verified the proposed formula for the equilateral triangle. Such a triangle is $(1, 1, 1)$—we mark with this symbol the corresponding point $(1, 1)$ in Fig. 15.2.

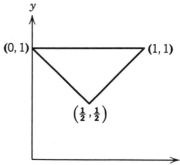

Fig. 15.1. The totality of triangular shapes.

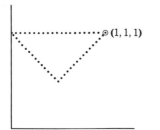

Fig. 15.2. Verified for the equilateral tri-angle.

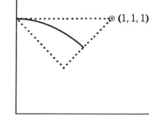

Fig. 15.3. . . . and for right triangles.

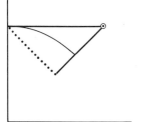

Fig. 15.4. . . . and for isosceles triangles.

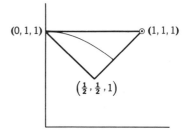

Fig. 15.5. . . . and for degenerate tri-angles.

Then we verified the formula for right triangles. If $(x, y, 1)$ is a right triangle, its greatest side 1 must be its hypotenuse and, therefore,

$$x^2 + y^2 = 1$$

Hence the right triangles are represented in Fig. 15.3 by a circular arc (of the unit circle).

Then we considered isosceles triangles. We should here distinguish between two kinds of isosceles triangles: either the two longer sides are equal and so

$$y = 1$$

or the two shorter sides are equal and so

$$x = y$$

Hence, the points representing isosceles triangles fill two boundary lines in Fig. 15.4 (fully drawn there—they were just dotted in the foregoing Fig. 15.2 and 15.3).

Finally, for a degenerate (collapsing) triangle $(x, y, 1)$

$$x + y = 1$$

Such "triangles" are represented by the third boundary line, fully drawn in Fig. 15.5 (but just dotted in the foregoing Figs. 15.2, 15.3, and 15.4).

Surveying the sequence of diagrams from Fig. 15.2 to Fig. 15.5, we may visualize the progress of the inductive argument. At the beginning, in Fig. 15.2, one point was enough to represent the extent of verification. Then more and more full lines appear in the diagram, indicating more and more classes of cases in which the verification has succeeded.

The points representing the triangular shapes for which the formula in question has been verified, are distributed along lines. Yet the formula remains unverified for the "bulk" of the triangular shapes represented by the area not covered by those lines. Still, since the formula has been verified along the whole boundary and also along a crossing line, we may reasonably expect that it will turn out correct in all cases. The part suggests the whole and suggests it strongly.

15.6. A historic example

We are going to investigate a problem of solid geometry by following in the footsteps of two great mathematicians. I shall tell their names later, but not too early, otherwise a good point of my story could be spoiled.

(1) *Analogy suggests a question.* A polyhedron is enclosed by plane faces just so as a polygon is enclosed by straight sides; polyhedra in space are analogous to polygons in a plane. Yet polygons are simpler, more accessible than polyhedra, a question about polygons has a good chance to be much easier than the corresponding question about polyhedra. When we know a fact about polygons, we should try to discover an analogous fact about polyhedra; in doing so we have a good chance to hit upon a stimulating question.

For instance, we know that the sum of the angles in a triangle is the same for all triangles: it is, independently of the size and shape of the triangle, equal to 180°, or two right angles, or π (in radians; we shall prefer this last way of measuring angles). More generally, the sum of angles in a polygon with n sides is $(n - 2)\pi$. Now, let us try to discover an analogous fact about polyhedra.

(2) *We try to exhaust the possibilities.* Our goal, however, is not quite clear. We wish to find out something about the sum of the angles in a polyhedron—but what angles?

Each edge of the polyhedron is associated with a *dihedral* angle, included by the two faces adjacent to that edge. Each vertex of the polyhedron is associated with a *solid* angle included by all the faces (three or more) adjacent to that vertex. Which kind of angle should we consider? Has the sum of all angles of the same kind some simple property? What about the sum of the six dihedral angles in a tetrahedron? What about the sum of the four solid angles in a tetrahedron?

It turns out that none of the two last named sums is independent of the shape of the tetrahedron (see ex. 15.14). How disappointing! We expected the tetrahedron to behave like a triangle.

Still, we may be able to save our original idea. We have not yet exhausted all possibilities. There are, in a polyhedron, angles of still another kind (this kind is the most familiar, in fact): each face enclosed by n sides (by n edges of the polyhedron) contains n interior angles. Let us call such angles *face angles,* and let us try to find the *sum of all face angles* of the polyhedron. Let $\Sigma\alpha$ denote the desired sum; see Fig. 15.6.

(3) *We observe.* If we see no other approach to our problem, we can always attack it experimentally: we take a few polyhedra and we compute $\Sigma\alpha$ (the sum of the face angles) for each. We may begin with the cube, see Fig. 15.7a. Each face of the cube is a square; the sum of the four angles in a square is 2π. As there are six faces, $\Sigma\alpha$ is

$$6 \times 2\pi = 12\pi$$

for the cube. We can handle just as easily the tetrahedron and the octahedron, see Fig. 15.7b and c.

The three polyhedra we have examined so far are regular. Let us examine, for a change, some nonregular polyhedron, for instance a 5-prism (a prism with pentagonal base, see Fig. 15.7d). This prism has two kinds of faces: there are five parallelograms and two pentagons. Therefore, $\Sigma\alpha$ is

$$5 \times 2\pi + 2 \times 3\pi = 16\pi$$

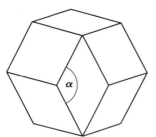

Fig. 15.6. Face angle.

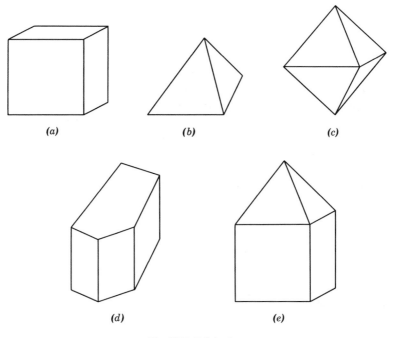

(a)   *(b)*   *(c)*

(d)   *(e)*

Fig. 15.7. Polyhedra.

for the 5-prism. Then let us take a polyhedron which we see less often in a classroom, see Fig. 15.7e: a pyramid is placed on a cube as "roof"; the resulting "tower" has 9 faces, 5 squares, and 4 triangles; hence its $\Sigma\alpha$ equals

$$5 \times 2\pi + 4 \times \pi = 14\pi$$

We collect our observations in Table I; to render more recognizable the polyhedra considered, let us note for each the number of its faces, F.

TABLE I

Polyhedron	F	$\Sigma\alpha$
Cube	6	12π
Tetrahedron	4	4π
Octahedron	8	8π
5-Prism	7	16π
Tower	9	14π

Do you notice something worth noticing—some law or pattern or regularity?

(4) *We observe, guided by an idea.* It is not surprising that, at this stage of the game, we see nothing striking in the observational material hitherto collected; observation without some leading idea seldom produces noteworthy results.

In reflecting on our own procedure, we may find our way out of this impasse. In (3), we have repeatedly computed the grand total of the face angles Σα by taking first the sum of those angles that belong to the same face—we know this sum precisely, in fact the knowledge of this sum was the starting point of our inquiry. Let us now take, for a change, the sum of those angles that have the same corner of the polygon as vertex. We do not know this sum precisely, but we know that it is less than 2π, a full plane angle. (We now restrict ourselves explicitly to *convex* polyhedra; the fact quoted is intuitive, but you may look up Euclid XI 21 for a proof.) Let V denote the number of vertices of the polyhedron considered; we perceive that the grand total of the face angles

$$\Sigma\alpha < 2\pi V$$

Let us verify this relation on our collected material! We extend our Table I into Table II.

TABLE II

Polyhedron	F	$\Sigma\alpha$	V	$2\pi V$
Cube	6	12π	8	16π
Tetrahedron	4	4π	4	8π
Octahedron	8	8π	6	12π
5-Prism	7	16π	10	20π
Tower	9	14π	9	18π

In fact, throughout Table II, $2\pi V$ is larger than $\Sigma\alpha$, and we can hardly fail to observe that the difference is constant

$$2\pi V - \Sigma\alpha = 4\pi$$

Is this a coincidence? A mere coincidence seems unlikely, and so we can hardly resist the temptation to guess that the relation observed holds not only in the few cases we have examined, but generally for all convex polyhedra. Thus we arrive at the *conjecture*

$$(?) \qquad \Sigma\alpha = 2\pi V - 4\pi$$

The query in parenthesis in front of the stated relation should remind us that it is not proved, merely conjectured.

(5) *We test our conjecture.* Our observation, guided by a lucky remark, produced a remarkable conjecture—but is it true?

Let us check a few more cases. There are two more regular polyhedra to consider, the dodecahedron and the icosahedron, with $F = 12$ and $F = 20$, respectively. Furthermore, we could consider a general prism, the n-prism, the base of which is a polygon with n sides, then the n-pyramid with a base of the same nature, and the n-double-pyramid; the last consists of two n-pyramids standing on opposite sides of a common base (which is *not* a face of the resulting double pyramid). The reader can easily extend Table II to the solids mentioned.

EXTENSION OF TABLE II

Polyhedron	F	$\Sigma\alpha$	V	$2\pi V$
Dodecahedron	12	36π	20	40π
Icosahedron	20	20π	12	24π
n-Prism	$n + 2$	$(4n - 4)\pi$	$2n$	$4n\pi$
n-Pyramid	$n + 1$	$(2n - 2)\pi$	$n + 1$	$(2n + 2)\pi$
n-Double-pyramid	$2n$	$2n\pi$	$n + 2$	$(2n + 4)\pi$

The conjectural relation (?) is verified in all cases examined, which is gratifying but does not amount to a proof.

(6) *Reflections on the procedure followed.* In computing $\Sigma\alpha$ we have used the same procedure several times: we started by taking the sum of such angles as belong to the same face. Why not apply this procedure *generally*?

To follow up this reflection we introduce suitable notation. Let

$$s_1, s_2, s_3, \ldots, s_F$$

denote the number of sides of the first, the second, the third, . . . , the last face, respectively. With this notation

$$\Sigma\alpha = \pi(s_1 - 2) + \pi(s_2 - 2) + \cdots + \pi(s_F - 2)$$
$$= \pi(s_1 + s_2 + \cdots + s_F - 2F)$$

Now, $s_1 + s_2 + s_3 + \cdots + s_F$ is the total number of all the sides of all F faces. In this total each edge of the polyhedron is counted exactly twice (as it is adjacent to exactly two faces) and so

$$s_1 + s_2 + \cdots + s_F = 2E$$

where E stands for the number of the edges of polyhedron. Hence we obtain

$$(!) \qquad \Sigma\alpha = 2\pi(E - F)$$

We have found here a second expression for $\Sigma\alpha$, but there is an essential difference: the first expression, (?) in subsection (4), was just conjectured, but (!) is proved. If we eliminate $\Sigma\alpha$ from (?) and (!) we obtain the relation

$$(??) \qquad F + V = E + 2$$

which, however, we have not proved and, therefore, we let (??) precede it. In fact, (??) is just as doubtful as (?); through the well-proven (!), each of the two relations (?) and (??) follows from the other, and so they stand and fall together, they are equivalent.

(7) *Verifications.* Both the familiar relation (??) and the less familiar relation (?) were discovered by Euler, who did not know that Descartes found relations (?) and (!) before him. We know about Descartes' work on this subject from a few brief sentences found among his unpublished manuscripts and printed about a century after Euler's death.[2]

Euler devoted two memoirs and a short remark of a third memoir to the subject.[3] The remark is concerned with the sum of the solid angles in a tetrahedron [this sum depends on the shape, as we have said in subsection (2)]. On the whole, the preceding exposition follows Euler's first memoir where he tells how he was led to his discovery, but offers no formal proof, only a variety of verifications. We wish to follow Euler also in this respect. By collecting the material from our foregoing tables and adjoining to it E, the number of edges, we obtain Table III.

TABLE III

Polyhedron	F	V	E
Tetrahedron	4	4	6
Cube	6	8	12
Octahedron	8	6	12
Dodecahedron	12	20	30
Icosahedron	20	12	30
Tower	9	9	16
n-Prism	$n + 2$	$2n$	$3n$
n-Pyramid	$n + 1$	$n + 1$	$2n$
n-Double-pyramid	$2n$	$n + 2$	$3n$

The conjectured relation (??) is verified throughout Table III, which is gratifying but, of course, does not amount to a proof.

[2] Descartes, *Œuvres,* vol. X, pp. 265–269.
[3] Euler, *Opera Omnia,* ser. 1, vol. 26, pp. XIV–XVI, 71–108, and 217–218.

(8) *Reflections on the result obtained.* In his second memoir, Euler attempted a proof for (??). His attempt failed, however; there is an essential gap in his proof. Yet, in fact, the foregoing considerations have brought us quite close to a proof; we just need to realize how far we have advanced.

Let us try to realize the full meaning of our result (!). Especially, let us see what happens when the polyhedron *varies.* Let us imagine that the polyhedron changes *continuously;* its faces are gradually inclining so that their lines and points of intersection, the edges and vertices of the polyhedron, are continuously changing, yet so that the "general plan," or "morphological structure," of the polyhedron, the connection between its faces, edges and vertices, remains unchanged. Then also the *numbers F, E,* and *V* (of faces, edges, and vertices respectively) remain *unchanged.* Such a change may affect each face angle α individually, but, by virtue of the well-proven (!), it cannot affect the face angles collectively, that is, it must leave $\Sigma\alpha$, the sum of all face angles, unchanged. And here we may see an opportunity to profit by such a change: the change could reduce the proposed polyhedron to some more accessible form for which we might more easily compute the (unchanged!) $\Sigma\alpha$.

Indeed, let us choose one of the faces of the polyhedron as "base." We place this base horizontally and we stretch it (whereas the other faces shrink) so that eventually the whole polyhedron can be orthogonally *projected onto its base;* Fig. 15.8 shows the result (*a*) for a cube and (*b*) for a "general" polyhedron. The result is a collapsed polyhedron, flattened into two superposed polygonal sheets (with the same rim): the lower sheet (the "stretched base") is undivided, the upper sheet is divided into $F - 1$ subpolygons, F being the number of faces of the original polyhedron. We let r denote the number of sides of the polygonal rim enclosing both sheets.

We compute $\Sigma\alpha$ for the flattened polyhedron (we know that it has the same value for the original, unflattened polyhedron). The total sum consists of three parts.

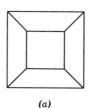 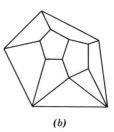

Fig. 15.8. Flattened polyhedra.   (*a*)   (*b*)

The sum of the angles in the lower sheet (the "stretched base") is $(r - 2)\pi$.

The sum of the angles on the rim of the upper sheet is the same.

There remain the angles in the interior of the upper sheet; they surround the $V - r$ interior vertices and so their sum is $(V - r)2\pi$.

From the three component parts, we obtain

$$\Sigma\alpha = 2(r - 2)\pi + (V - r)2\pi$$
$$= 2\pi V - 4\pi$$

This proves our conjecture (?) and, therefore, also our conjecture (??).

15.7. Scientific method: guess and test

The foregoing examples suggest general insights. Of course, such insights will emerge more spontaneously and with better documentation from more work on these examples and on similar ones (see the Examples and Comments at the end of this chapter). Yet, even what we have discussed may lead to some remarks tending toward the following.

Observation may lead to discovery.

Observation should disclose some regularity, pattern, or law.

Observation has more chance to yield worthwhile results if guided by some good remark, by some insight [as in sect. 15.6(4)].

Observation yields only tentative generalizations, conjectures, not proofs.

Test your conjecture: examine particular cases and consequences.

Any particular case or consequence that is verified (turns out to be true) adds to the credit of the conjecture.

Distinguish carefully between suggestion and proof, between conjecture and fact.

Do not neglect analogies: they may lead to discovery [as the analogy between polygons and polyhedra, sect. 15.6(1)].

Examine extreme cases [such as degenerate triangles and collapsed polyhedra, sect. 15.4 and sect. 15.6(8)].

These remarks would deserve to be stated more precisely, with more details, with more system and, especially, with much broader documentation (*cf.* MPR). Yet even as they stand, as they may emerge from examples such as the preceding and from well-directed classroom discussions, they can give students on the high school level a basic insight into the nature of science. Philosophers, past and present, have offered and do offer widely divergent views about the nature of science, of "scientific method," of "induction," etc. Yet what do scientists actually do? They devise hypothetical explanations and submit their hypotheses to the test of experience. If you want a description of scientific method in three syllables, I propose:

GUESS AND TEST

15.8. Some desirable features of "research problems"

The problems just presented differ from the usual routine problems in several respects. I wish to emphasize just three points.

(1) The schoolboy receives his problem ready made from the textbook or from the teacher and, in many cases, the teacher cares little and the textbook less whether the boy takes any interest or not in the problem. On the contrary, the mathematician's most crucial step may be to *choose* his problem; to spot, to invent a problem that is attractive and worthwhile but not beyond his force. In sect. 15.2 and 15.4 the teacher proceeds so that the students should *have some share in proposing the problem* [*Cf.* sect. 14.5(1)].

(2) Most textbook problems have little connection; they serve to illustrate just one rule and offer some practice in its application. After having rendered this service they may sink into oblivion. Yet the problems of sect. 15.2 and 15.6 have a *rich background;* they suggest challenging problems which suggest further challenging problems, until the ramifications of the original problem cover a wide domain. (Such ramifications will be discussed to some extent in the Examples and Comments at the end of the chapter.)

(3) In many classrooms "guessing" is taboo, whereas in mathematical research "First guess then prove" is almost the rule. In the foregoing problems, observation, conjectures, inductive arguments, in short, *plausible reasoning* play a prominent role.

(4) Although point (1) (participation of the students in proposing the problem) is not unimportant, the other two points are more momentous. Problems with a *background* connected with the world around us, or with other domains of thought, and problems involving *plausible reasoning,* challenging the judgment of the students,have more chance to lead them to intellectual maturity than the problems that fill the textbooks and serve only to practice this or that isolated rule.

15.9. Conclusion

As I see them, examples and remarks such as those presented in this chapter are communicable on the high school level and they can do, let us say, three things for the students.

First, they may give the student a taste of mathematics in the making, of independent creative work.

Second, and this is even more important, as it may serve a larger fraction of the student population: they contribute not only to the understanding of mathematics, but also to that of the other sciences. In fact, they give a reasonably good first idea of "inductive research" and "scientific method."

Third, they reveal an aspect of mathematics which is as important as it is rarely mentioned; mathematics appears here as a close relative to the natural sciences, as a sort of "observational science" in which observation and analogy may lead to discoveries. This aspect should especially appeal to prospective users of mathematics, to future scientists and engineers.

And so I hope that mathematical discovery, scientific method, and the inductive aspect of mathematics will not be so completely neglected in the high schools of the future as they are in the high schools of today.

Examples and Comments on Chapter 15

First Part

15.1. Among the several conjectures suggested by the list of sect. 15.2 and stated in sect. 15.3, are there some that you can prove? Pick out some readily accessible assertion and prove it.

15.2. [Sect. 15.4] Devise further means to check Heron's formula.

15.3. [Sect. 15.5] In Fig. 15.5, an arc of the unit circle divides the triangle (the points of which represent triangular shapes) into two parts, one above, the other below, the arc. What is the difference between the shapes represented by those two parts?

15.4. [Sect. 15.6(8)] Try to devise a *more definite* transition from the "general" convex polyhedron to the "flattened" one.

15.5. We consider a convex polyhedron with F faces, V vertices, and E edges. Let

F_n stand for the number of those faces that have precisely n sides, and

V_n for the number of those vertices in which precisely n edges meet.

Tell the value of ΣF_n and that of ΣV_n; the sign Σ is used here as an abbreviation for $\sum\limits_{n=3}^{\infty}$. (Of course, only a finite number among the F_n are different from 0, and the same holds for V_n; hence, in fact, Σ denotes a finite sum. The sign Σ will be used in the same sense in some of the following connected problems.)

15.6. (Continued) Express the number of the face angles in several different ways.

15.7. (Continued) By appropriately chosen diagonals ("face diagonals") which do not cross each other, dissect each face of a polyhedron into triangles. Express in several different ways the number of triangles into which the total surface of the polyhedron is so divided.

15.8. Show that

$$E \geqq \frac{3F}{2}, \qquad E \geqq \frac{3V}{2}$$

Can the case of equality be attained in the first inequality, and under what circumstances? Analogous question for the second inequality.

15.9. Show that in any convex polyhedron the average of the face angles is never less than $\pi/3$ but always less than $2\pi/3$.

15.10. Show that in any convex polyhedron there must be a face with less than 6 sides.

15.11. Being given V, the number of vertices of a convex polyhedron, find the maximum of F and the maximum of E. Under what condition are these maxima attained?

15.12. Being given F, the number of faces of a convex polyhedron, find the maximum of V and the maximum of E. Under what condition are these maxima attained?

15.13. When a straight-line segment connects two vertices of a convex polyhedron, there are three possible cases: the connecting segment may be an edge, or a face diagonal, or a diagonal. The last case arises iff, except the two endpoints, no point of the segment lies on the surface of the polyhedron. Let D denote the number of diagonals of the polyhedron; E, F, V, F_n, and V_n will be used in the same meaning as earlier.

(1) Find D for the five regular polyhedra.

(2) Find D for the n-prism, the n-pyramid, and the n-double-pyramid.

(3) Express D in terms of F when all faces of the polyhedron are polygons with the same number of sides $n = 3, 4, 5, \ldots$.

(4) Express D generally.

Illustrate each general case by examples. Be careful; questions may be misleading.

15.14. [Sect. 15.6(2)] We consider a tetrahedron and let $\Sigma\delta$ stand for the sum of its six dihedral angles and $\Sigma\omega$ stand for the sum of its four solid angles.

Compute these two sums in the following three limiting cases:

(1) The tetrahedron collapses into a triangle, three edges become sides of the triangle, three other edges line segments drawn from an interior point of the triangle to its vertices.

(2) The tetrahedron collapses into a convex quadrilateral, its six edges become the four sides and the two diagonals of the quadrilateral.

(3) One vertex of the tetrahedron goes to infinity, the three edges converging to it become parallel and perpendicular to the opposite face.

(A sphere with radius 1 is described about the vertex of a polyhedral angle as center. That part of the surface of the sphere that falls within the polyhedral angle is a spherical polygon. The area of this spherical polygon measures the "solid angle.")

15.15. (Continued) Examine the answer to ex. 15.14. Compare the two sums examined. Do they vary in the same way? Are their variations related?

15.16. For a polyhedron with F faces, V vertices, and E edges, we let $\Sigma\delta$ denote the sum of its E dihedral angles and $\Sigma\omega$ the sum of its V solid angles. Compute the two sums for a cube.

15.17. (Continued) Compute the two sums for two different easily accessible (degenerate) cases of the n-pyramid.

15.18. (Continued) Compute the two sums in accessible (limiting) cases of the n-prism and the n-double-pyramid.

15.19. (Continued) In all cases considered, compare the two sums with F, V, and E; observe the variation of the quantities compared; which variations appear most closely related?

15.20. (Continued) If you have found a rule that is supported by all your observations, try to prove it.

Second Part

15.21. Try to guess the answers to the following questions.

Of all triangles inscribed in a given circle which one has the largest area?

Of all quadrilaterals inscribed in a given circle which one has the largest area?

Of all polygons with a given number n of sides inscribed in a given circle which one has the largest area?

15.22. Try to guess the answers to the following questions.

Of all the triangles circumscribed about a given circle which one has the smallest area?

Of all the quadrilaterals circumscribed about a given circle which one has the smallest area?

Of all the polygons with a given number n of sides circumscribed about a given circle which one has the smallest area?

15.23. *The Principle of Non-Sufficient Reason.* The "usual" answers given to ex. 15.21 and 15.22 are correct.[4] We shall not discuss the proof here. We wish to explore why do people so often guess correctly in similar situations.

Of course, we cannot expect to find a very definite answer. Yet I think that the following expresses feelings voiced by many people.

Why are the regular polygons so popular? The circle is the most "perfect," the most symmetric plane figure; it has infinitely many axes of symmetry, it is symmetric with respect to each of its diameters. Of all polygons with a given number of sides the regular polygon is the "nearest in perfection" to the circle: it is the most symmetric, it has more axes of symmetry than the others. And so we expect that the inscribed regular polygon will "fill" the circle better (and the circumscribed regular polygon will "hug" the circle tighter) than any other polygon with the same number of sides.

Analogy plays a role too. The regular polygon yields the extremum in the isoperimetric problem (sect. 15.3, ex. 15.1) which is similar to the foregoing problems.

---

[4] Concerning ex. 15.21 see MPR vol. 1, pp. 127–128.

There are still other plausible arguments. We are going to discuss one that is rather subtle but deserves special attention. We are facing here problems with several unknowns where the condition is the same for each unknown; the condition does not favor any vertex of the polygon more than any other vertex, no side more than any other side. Hence we may expect that, in the polygon that satisfies the condition and so solves the problem, all sides will be equal and all angles will be equal. Thus we expect that the regular polygon will be the solution.

Underlying this expectation is a principle of plausible inference which we may try to express as follows:

"No one should be favored of eligible possibilities among which there is no sufficient reason to choose."

We may term this the "Principle of Non-Sufficient Reason." This principle is of some importance in problem solving; quite often it enables us to forecast the solution, or choose the procedure leading to the solution. In a mathematical context, a more specific formulation of the principle may be advantageous:

"Unknowns that play the same role in the condition may be expected to play the same role in the solution."

Or shorter: "No difference in the conditions, no difference in the results." Or: "Expect the same values for unknowns subject to the same conditions."

In geometrical problems, the principle favors symmetry (as we have seen.) Hence we may sometimes find the following formulations of the principle of nonsufficient reason more suggestive (although, in fact, they are less clear):

"We expect that any symmetry found in the data and condition of the problem will be mirrored by the solution."

"Symmetry should result from symmetry."

To some extent, the "symmetry found in the data and condition" should be mirrored not only by the "solving object" but also by the "solving procedure."[5]

Of course, we should not forget that the principle is merely heuristic and should not take plausibility for certainty.[6]

The principle of nonsufficient reason plays a certain role also in not purely mathematical questions.[7]

A striking example which flies in the face of the principle can be concisely described if we use some algebraic terminology. The problem is: Find n quantities of which the n elementary symmetric functions are given. The principle of nonsufficient reason makes us expect that those n quantities will be equal. Yet the n roots of an algebraic equation of which the coefficients are given "at random" should be expected to be all different from each other.

15.24. *The ass of Buridan.* An ass was very hungry when he found himself facing two equal and equally appetizing haystacks, one to his left, the other to his right, and himself in such a position in the middle that there was perfect symmetry.

---

[5] See HSI, pp. 199–200 (Symmetry), and ex. 5.13 for the terminology.

[6] *Cf.* MPR, vol. 1, pp. 186–188, ex. 40 and 41. See also a paper of the author "On the role of the circle in certain variational problems" *Annales Univ. Scient. Budapest. Sectio Math.* v. 3–4, 1960–61, pp. 233–239.

[7] See J. M. Keynes, *A treatise on probability*, pp. 41–64.

Attracted equally by both haystacks the ass could not decide between them and died of hunger.

Poor ass—he fell a victim to the principle of nonsufficient reason.

15.25. Of all polyhedra inscribed in a given sphere and having a given number V of vertices, which one has the largest volume?

Try to guess the answer for $V = 4$, 6, and 8.

15.26. Of all polyhedra circumscribed about a given sphere and having a given number F of faces, which one has the smallest volume?

Try to guess the answer for $F = 4$, 6, and 8.

15.27. Given the radius r of a sphere, compute the volume of the inscribed cube.

15.28. Consider the sphere with radius r as the globe. Inscribe a regular hexagon in the equator. The six vertices of this hexagon, the north pole, and the south pole are the eight vertices of a double pyramid. Compute its volume.

Any remarks?

15.29. Given the radius r of a sphere, compute the volume of the circumscribed regular octahedron.

15.30. A right prism with regular hexagonal base is circumscribed about a sphere with radius r, which we regard as the globe. The surface of the prism touches the sphere in six points equally spaced along the equator, in the north pole, and in the south pole. Compute the volume of the prism.

Any remarks?

15.31. Compare the solids considered in the foregoing (ex. 15.27 with ex. 15.28, ex. 15.29 with ex. 15.30) and try to find a plausible explanation.

15.32. Here is a plausible assumption: of two polyhedra inscribed in the same sphere and having the same number V of vertices, the one with more faces and edges has more chance to "fill" the sphere. Granted this, what kind of polyhedron do you expect to be the solution of ex. 15.25?

15.33. Here is a plausible assumption: of two polyhedra circumscribed about the same sphere and having the same number F of faces, the one with more vertices and edges has more chance to "hug" the sphere tighter. Granted this, what kind of polyhedron do you expect to be the solution of ex. 15.26?

15.34. Are there any more remarks suggested by ex. 15.31?

15.35. Of all polyhedra having a given surface area and a given number F of faces, which one has the largest volume?

Try to guess the answer for $F = 4$, 6, and 8.

15.36. Solve the system, giving all solutions:

$$2x^2 - 4xy + 3y^2 = 36$$
$$3x^2 - 4xy + 2y^2 = 36$$

What about the principle of nonsufficient reason?

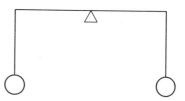

Fig. 15.9. Equal weights at equal distances.

15.37. Solve the system, giving all solutions:

$$6x^2 + 3y^2 + 3z^2 + 8(yz + zx + xy) = 36$$
$$3x^2 + 6y^2 + 3z^2 + 8(yz + zx + xy) = 36$$
$$3x^2 + 3y^2 + 6z^2 + 8(yz + zx + xy) = 36$$

15.38. Solve the system, giving all solutions:

$$x^2 + 5y^2 + 6z^2 + 8(yz + zx + xy) = 36$$
$$6x^2 + y^2 + 5z^2 + 8(yz + zx + xy) = 36$$
$$5x^2 + 6y^2 + z^2 + 8(yz + zx + xy) = 36$$

15.39. *The principle of nonsufficient reason in physics. Or, nature ought to be predictable.* The beginning of Archimedes' work "On the Equilibrium of Planes or the Centres of Gravity of Planes"[8] deals with the equilibrium of the lever. (The lever is a rigid horizontal bar supported at one point which is called the fulcrum; the weight of the bar is regarded as negligible.) Archimedes considers the case where the midpoint of the lever is the fulcrum and the two endpoints carry equal weights, see Fig. 15.9; he admits as evident that in this situation of perfect symmetry there is equilibrium; in fact, his first postulate states: "Equal weights at equal distances are in equilibrium." Indeed, the lever is in the situation of Buridan's ass: it has no sufficient reason to incline to one side rather than to the other.

We can penetrate here a little deeper. Let us see what happens if somebody contradicts Archimedes' postulate and proposes a different rule: in the situation of Fig. 15.9 the right-hand weight will sink. Well, if this prediction turns out correct for me as I am looking at the lever, it must turn out wrong for my friend who, facing me, is looking at the lever from the opposite side: the rule which contradicts Archimedes' postulate *cannot be generally valid.* We may now perceive a deeper source of the confidence with which we are inclined to accept Archimedes' postulate: we want nature to be predictable.

15.40. *Choosing n points on a spherical surface.* We restate ex. 15.25 and ex. 15.26 as the first two in a series of analogous problems.

On the surface of a given sphere choose n points so that

(1) the *inscribed polyhedron* of which the vertices are the n points should have *maximum volume;*

(2) the *circumscribed polyhedron* the n faces of which touch the sphere in the n points should have *minimum volume;*

(3) the shortest of the $n(n-1)/2$ distances between the n points should

---

[8] See his Works, edited by T. L. Heath, p. 189.

be a maximum (a "maximum minimorum": this is the "problem of the *n* misanthropes*");

(4) each point carrying a unit electric charge, the *n* mutually repelling charges should be in the *most stable electrostatic equilibrium.*

(5) On the surface of the sphere, there is an arbitrary continuous mass distribution the density of which is observed at *n* points. Choose the *n* points so that on the basis of those *n* observations the total mass can be most accurately estimated. (This is the "problem of the *n reporters*" of a worldwide news agency, or the problem of *best interpolation.* For a line segment, Gauss has solved the analogous problem, in a certain sense, by his celebrated mechanical quadrature.)

In all five foregoing problems, if *n* = 4, 6, 8, 12, or 20, the vertices of a certain inscribed regular polyhedron deserve consideration, although they may not yield the solution as certain foregoing examples show. *Cf.* L. Fejes-Tóth, *Lagerungen in der Ebene, auf der Kugel und im Raum.*

If the *n* points are chosen at random (the *n* brightest fixed stars on the sky appear to be so chosen when *n* is not very large), the average distance of a point from its next neighbor, its second next neighbor, its third next neighbor, and so on, can be computed. See a paper of the author, *Vierteljahrsschrift der Naturforschenden Gesellschaft in Zürich,* v. 80, 1935, pp. 126–130.

Third Part

15.41. *More problems.* Consider some research problems similar to, but different from, the problems discussed in this chapter. Pay especial attention to such questions as the following: Does the problem fit, and where does it fit, into the curriculum? Is the problem instructive? Has it a worthwhile background? Does it illustrate some idea of general interest? Does it afford opportunity for inductive, plausible reasoning? Or opportunity for a challenging proof on the level of the class? How should it be presented to the class?

15.42. In sect. 15.4 we tested a general formula by discussing particular cases. Where have you opportunity for a discussion of this kind? Carry through such a discussion in a few more cases. What is the merit of such a discussion?

15.43. The main purpose of sect. 15.5 is to illustrate graphically an aspect of inductive reasoning. Could the students benefit from this section in other respects?

15.44. *Periodic decimals.* The three decimal expansions

$$\tfrac{1}{6} = 0.166666666\ldots$$
$$\tfrac{1}{7} = 0.142857142\ldots$$
$$\tfrac{1}{8} = 0.125$$

belong to three different types. The one representing $\tfrac{1}{8}$ is a *terminating* decimal expansion, the two others are *infinite.* In fact, they are recurring, repeating, or *periodic decimals;* with the standard notation, they are written in the form

$$\tfrac{1}{6} = 0.1\overline{6}$$
$$\tfrac{1}{7} = 0.\overline{142857}$$

The bar is placed over the repetend or *period,* that is, the sequence of digits that must be infinitely often repeated. The period of $\frac{1}{6}$ is of length 1, that of $\frac{1}{7}$ of length 6; generally, the number of digits in a period is called its *length.* The decimal expansion of $\frac{1}{7}$ is *purely periodic* that for $\frac{1}{6}$ is *mixed:* The first does not and the second does contain an initial sequence of digits that does not belong to the period. Here is one more example for each of the three types

$$\tfrac{39}{44} = 0.8\overline{863} \qquad\qquad \tfrac{19}{27} = 0.\overline{703} \qquad\qquad \tfrac{19}{20} = 0.95$$

Find out by observation as much as you can about the three types of decimal expansions, about the length of their periods, about the distribution of the digits in the period, or whatever else you find worth noticing. Try to prove or disprove the guesses to which observation leads you.

Choose the fractions you wish to observe, or look at the decimal fractions into which the numbers mentioned under (1) through (7) are expanded.

(1) $\frac{1}{7}, \frac{2}{7}, \frac{3}{7}, \frac{4}{7}, \frac{5}{7}, \frac{6}{7}, \frac{7}{7}$

(2) $\frac{1}{7}, \frac{10}{7}, \frac{100}{7}, \ldots$

(3) All proper fractions with a denominator that is relative prime to the numerator and is less than 14.

(4) All proper fractions with denominator 27 whose numerator is relative prime to 27.

(5) $\frac{1}{3}, \frac{1}{7}, \frac{1}{11}, \frac{1}{37}, \frac{1}{41}, \frac{1}{73}, \frac{1}{101}, \frac{1}{239}$

(6) $\frac{1}{9}, \frac{1}{99}, \frac{1}{999}, \frac{1}{9999}$

(7) $\frac{1}{11}, \frac{1}{101}, \frac{1}{1001}, \frac{1}{10001}$

At any rate observe and try to understand thoroughly that

$$7.00000\ldots = 6.99999\ldots$$

$$0.50000\ldots = 0.49999\ldots$$

15.45. (Continued) Observe that

$$\tfrac{1}{9} = 0.111111\ldots \qquad\qquad \tfrac{1}{11} = 0.090909\ldots$$

$$\tfrac{1}{27} = 0.037037\ldots \qquad\qquad \tfrac{1}{37} = 0.027027\ldots$$

$$\tfrac{1}{99} = 0.01010101\ldots \qquad\qquad \tfrac{1}{101} = 0.00990099\ldots$$

$$\tfrac{1}{271} = 0.0036900369\ldots \qquad\qquad \tfrac{1}{369} = 0.0027100271\ldots$$

and EXPLAIN.

15.46. (Continued) Starting from decimal expansions and passing from the base 10 to the base 2 we arrive at binary expansions. Here is an example:

$$\tfrac{1}{3} = 0.01010101\ldots$$

if we interpret the right-hand side in the binary system, that is, *if* we take the foregoing equation in the meaning

$$\frac{1}{3} = \frac{1}{2^2} + \frac{1}{2^4} + \frac{1}{2^6} + \frac{1}{2^8} + \cdots$$

Examine binary expansions in the same way as decimal expansions have been examined in ex. 15.44 and 15.45.

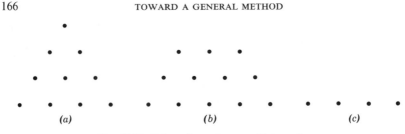

Fig. 15.10. Triangular and trapezoidal numbers.

15.47. (Continued) Assess the educational merit of the project of research sketched in ex. 15.44, 15.45, and 15.46.

15.48. *Trapezoidal numbers:* Fig. 15.10a represents the triangular number

$$1 + 2 + 3 + 4 = 10$$

cf. ex. 3.38 and Fig. 3.8. We could call analogously the number

$$3 + 4 + 5 = 12$$

represented by Fig. 15.10b a "trapezoidal" number. If we include extreme cases in our definition (which is often desirable) we must regard also the numbers represented by (a) and (c) in Fig. 15.10 as "trapezoidal." Yet then each positive integer would be "trapezoidal" (since it can be represented by one row of points as in Fig. 15.10c) and the definition would be pointless. Thus we are led to the following definition.

Let $t(n)$ denote the number of *trapezoidal representations* of the positive integer n, that is, the number of representations of n as sum of consecutive positive integers. Here are a few examples:

$$6 = 1 + 2 + 3$$
$$15 = 7 + 8 = 4 + 5 + 6 = 1 + 2 + 3 + 4 + 5$$

When

$$n = 1, 2, 3, 6, 15, 81, 105$$
$$t(n) = 1, 1, 2, 2, \ \ 4, \ \ 5, \ \ \ 8$$

Find a "simple expression" for $t(n)$ by appropriate observation followed, if possible, by a proof.

15.49. (Continued) Fig. 15.11 offers graphic help to survey your observations. Let us call the expression of n in the form

$$n = a + (a + 1) + (a + 2) + \cdots + (a + r - 1)$$

(a sum of r terms) a trapezoidal representation of n *with r rows*. Iff n admits a trapezoidal representation with r rows, the point with abscissa n and ordinate r is marked with a dot in Fig. 15.11.

If $t(n) = 1$, the only trapezoidal representation of n must be the "trivial" representation with $r = 1$. List the numbers n in Fig. 15.11 for which $t(n) = 1$.

What is $t(p)$ if p is a prime number?

15.50. (Continued) Let $s(n)$ denote the number of representations of the positive integer n as sum of consecutive *odd* positive integers. Find an expression for $s(n)$.

Examples

$$15 = 3 + 5 + 7$$
$$45 = 13 + 15 + 17 = 5 + 7 + 9 + 11 + 13$$
$$48 = 23 + 25 = 9 + 11 + 13 + 15 = 3 + 5 + 7 + 9 + 11 + 13$$

When

$$n = 2, 3, 4, 15, 45, 48, 105$$
$$s(n) = 0, 1, 1, \ \ 2, \ \ 3, \ \ 3, \ \ \ 4$$

15.51. Assess the project of research sketched in ex. 15.48 and 15.49.

15.52. Consider three plane figures:

(1) a square with a vertical diagonal,

(2) the circle (with radius a) circumscribed about it,

(3) the square circumscribed about the circle with a vertical side.

The vertical diagonal of (1) divides each figure in two symmetric *halves.* Rotating about their common vertical axis of symmetry, the three plane figures describe three solid figures:

(1) a double cone,

(2) a sphere,

(3) a cylinder.

Compute, in all three cases,

    V, the volume of the solid figure,

    S, the surface area of the solid figure,

    A, the area of the plane figure,

    L, the perimeter of the plane figure,

    X_A, the distance of the center of gravity of the half-area of the plane figure from the axis of revolution,

    X_L, the distance of the center of gravity of the half perimeter of the plane figure from the axis of revolution.

Group the eighteen quantities in a 3×6 table, OBSERVE, and try to explain what you have observed.

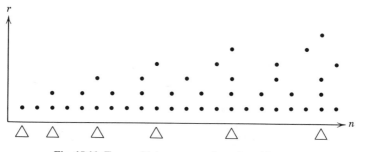

Fig. 15.11. Trapezoidal representation of n with r rows.

15.53. Observe that

$$\sqrt{2} - 1 \qquad\qquad\qquad = \sqrt{2} - \sqrt{1}$$
$$(\sqrt{2} - 1)^2 = 3 - 2\sqrt{2} \quad = \sqrt{9} - \sqrt{8}$$
$$(\sqrt{2} - 1)^3 = 5\sqrt{2} - 7 \quad = \sqrt{50} - \sqrt{49}$$
$$(\sqrt{2} - 1)^4 = 17 - 12\sqrt{2} = \sqrt{289} - \sqrt{288}$$

try to generalize, and prove your guess.

15.54. Quite often it matters little what your guess is, but it always matters a lot how you test your guess.

15.55. *Fact and conjecture.* The following story of which I cannot guarantee the authenticity is about Sir John and a janitor. Sir John, who is a member of the Royal Society, can be expected to draw a sharp distinction between a fact and a conjecture; the janitor, who is employed in the Royal Society building, unexpectedly drew such a distinction on a certain occasion.

One day Sir John arrived somewhat late to a meeting of the Royal Society and was visibly in a hurry. He had to deposit his hat in the cloakroom and receive a check with a number for it. The janitor who was in charge of the cloakroom that day said obligingly: "Please, do not wait for the check, Sir, I shall give you your hat without a check." Sir John went thankfully to the meeting without the check, but was somewhat worried about the fate of his hat. When, however, he entered the cloakroom after the meeting, the janitor, without hesitation, gave him his hat. Sir John was visibly pleased, and I do not know what prompted him to ask the janitor: "How did you know that this is my hat?" I do not know either what happened to the janitor—perhaps he found Sir John's tone unduly patronizing— at any rate he answered rather sharply: "Sir, I do not know whether it is your hat; it is the hat you gave me."

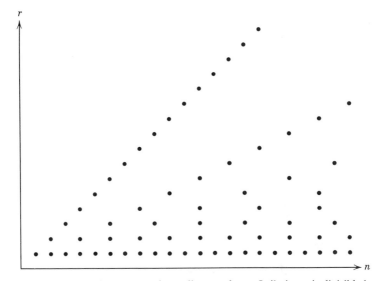

Fig. 15.12. For the observant reader, a diagram due to Leibnitz: n is divisible by r.

SOLUTIONS

Chapter 7

7.1. The connecting lines exhibited by Fig. 7.10 indicate the relation

$$F = Q + 4T + 4P$$

Obviously

$$Q = a^2h, \qquad T = \frac{h}{2} \cdot \frac{b-a}{2} \cdot a, \qquad P = \frac{h}{3}\left(\frac{b-a}{2}\right)^2$$

Introduce the connecting lines indicating these relations into Fig. 7.10 and satisfy yourself that the foregoing relations yield the same final expression for F as the approach considered in the text.

7.2. See Fig. S7.2, (3) and (4); the latter represents the mental situation just before the dramatic last line in Volume I, p. 36.

No solution: **7.3, 7.4, 7.5.**

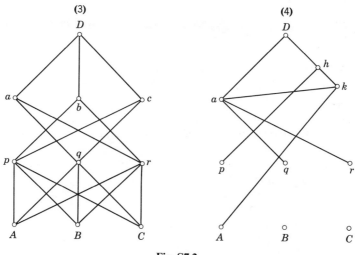

Fig. S7.2.

Chapter 8

8.3. (Stanford 1956.)   From Fig. 8.3

$$AB' = \frac{b}{\sqrt{3}}, \qquad AC' = \frac{c}{\sqrt{3}}$$

Apply the law of cosines to $\triangle B'AC'$:

$$3x^2 = b^2 + c^2 - 2bc \, \cos\left(\alpha + \frac{\pi}{3}\right)$$

Apply the law of cosines to $\triangle ABC$ to express $bc \cos \alpha$, set $bc \sin \alpha = 2T$ where T is the area of $\triangle ABC$, and obtain

$$6x^2 = a^2 + b^2 + c^2 + 4\sqrt{3}\,T$$

T is symmetric in a, b, and c.

8.4. The cross-section is

(1) a rectangle with area $2yz = \dfrac{2hx\sqrt{a^2 - x^2}}{a}$,

(2) a right triangle with area $\dfrac{xz}{2} = \dfrac{h(a^2 - y^2)}{2a}$,

(3) a segment of a circle.

Plan (2) is preferable (rational function of y) and the desired volume equals

$$\tfrac{1}{2}\int_{-a}^{a} xz \, dy = \frac{2a^2h}{3}$$

No solution: **8.1, 8.2, 8.5, 8.6, 8.7, 8.8.**

Chapter 9

9.5. See sect. 4.4.   The theorem A, stated in sect. 4.4(1) and illustrated by Fig. 4.4 is proved by the help of the weaker theorem B, stated in sect. 4.4(2) and illustrated by Fig. 4.3.

9.6. The problems A and B are equivalent.   There is a distinct advantage in passing from A to B since we arrive so at smaller numbers.   Repeating this passage, we obtain successively the following pairs of numbers:

$$(437, 323) \qquad (323, 114) \qquad (209, 114) \qquad (114, 95) \qquad (95, 19)$$
$$(76, 19) \qquad (57, 19) \qquad (38, 19) \qquad (19, 19)$$

Thus all the common divisors of 437 and 323 are 1 and 19, the divisors of 19 which is their greatest common divisor.   The procedure used in this example is generally applicable, is of basic importance, and is called *Euclid's algorithm* (see Euclid VII 2).

9.7. (1) Under certain important and frequently arising circumstances the condition of B is wider than that of A.   (Here is a simple case: $f(x)$ is defined, continuous, and possesses a derivative in the closed interval $a \leqq x \leqq b$; moreover it is known that the maximum of $f(x)$ is not attained for $x = a$ nor for $x = b$.)

(2) Finding the roots of the equation $f'(x) = 0$ appears in most cases as the more familiar problem; moreover we know means to remove those roots which do not yield the maximum of $f(x)$.

9.8. (1) Theorem B is stronger; it immediately implies A.

(2) B is easier to prove than A, because B adds to A a precise detail on which we can start to work: if we have only the less complete statement A, we have still to discover that precise detail, or some equivalent fact; the stronger theorem B is more accessible than A, because it is more explicit than A. This is a typical case (*cf.* HSI, pp. 114–121, especially p. 121, Induction and mathematical induction, especially point 7).

(3) Let A denote any vertex of the triangle and M the midpoint of the opposite side; prove first that $\triangle MGO \sim \triangle AGE$, from which $MO \parallel AE$ follows.

9.9. (1) A is a problem to prove, B is a problem to find which is more ambitious than A: we can foresee from the outset that a complete solution of B will either prove or disprove the assertion stated by A.

(2) A is a problem on limits, B on algebraic inequalities, and so B appears more elementary.

(3) We omit the case $\epsilon \geq 1$ which is easier. In the case $\epsilon < 1$ we have a chain of equivalent conditions:

$$\sqrt{x + 1} < \epsilon + \sqrt{x}$$
$$1 < \epsilon^2 + 2\epsilon\sqrt{x}$$
$$\left(\frac{1 - \epsilon^2}{2\epsilon}\right)^2 < x$$

That is, from a certain value of x onward, the quantity $\sqrt{x + 1} - \sqrt{x}$ is less than ϵ, an arbitrary (arbitrarily small!) positive quantity. This proves A.

9.10. (1) The two propositions with which A and B are concerned are equivalent (by contraposition) and so are the two problems A and B themselves.

(2) The statement, "n is composite," affirms the existence of two integers a and b such that $n = ab$, $a > 1$, $b > 1$. The statement "n is prime" denies that n is composite (we can disregard here the case $n = 1$) and this "negative" statement offers "less foothold." Hence B appears more tractable.

(3) If $n = ab$

$$2^n - 1 = (2^a)^b - 1$$

is divisible by $2^a - 1$.

9.11. Set, for $m = 1, 2, 3, \ldots,$

$$a_{3m-2} = 2m^{-1/3}, \qquad a_{3m-1} = a_{3m} = -m^{-1/3}$$

For a generalization of the present problem see *American Mathematical Monthly,* vol. 53, 1946, pp. 283–284, problem 4142. It seems impossible to set up helpful additional requirements, like (III) and (IV), without some idea, without some premonition or anticipation of the solution.

9.12. If $p_1, p_2, \ldots, p_l$ are different primes, and $n = p_1^{a_1}p_2^{a_2} \cdots p_l^{a_l}$ then

$$\tau(n) = (a_1 + 1)(a_2 + 1) \cdots (a_l + 1)$$

9.13. Ex. 1.47 and sect. 1.3(1); sect. 2.5(2), 2.5(3); sect. 3.2 and sect. 3.1.

9.15. *Why?* Here are two more tasks analogous to ex. 9.6–9.10.

(I) Compare the two problems:

A. Find $\sqrt[4]{100}$.

B. Find $\sqrt{100}$.

†(II) Let $f(x)$ and $g(x)$ stand for some given functions. Compare the two problems:

A. Prove that $\displaystyle\int_a^b f(x)\,dx \geqq \int_a^b g(x)\,dx$.

B. Prove that $f(x) \geqq g(x)$ for $a \leqq x \leqq b$.

Wherefrom? Auxiliary problems that originate, overtly or covertly, from one or the other of the four sources named seem to be the most usual. An instructive example: generalization, specialization, and analogy contribute jointly to the solution of ex. 3.84; see Vol. 1, p. 97, p. 191, and HSI ex. 20, pp. 238, 241–242, 252–253.

No solution: **9.1, 9.2, 9.3, 9.4, 9.14.**

Chapter 10

10.2.

(1)

```
_ _ _ E _ _ _ _ _ _
_ _ _ E _ _ _ A _ _
_ W _ E _ _ _ A _ _
_ W _ E _ H _ A _ _
_ W _ E _ H _ A _ T
```

(2)

$$\int \sqrt{x^7 - x^4}\,dx =$$
$$\int x^2 \sqrt{x^3 - 1}\,dx =$$
$$\int \sqrt{x^3 - 1} \cdot x^2 dx =$$
$$\tfrac{1}{3}\int \sqrt{x^3 - 1} \cdot 3x^2\,dx =$$
$$\tfrac{1}{3}\int (x^3 - 1)^{1/2}\,d(x^3 - 1)$$

No solution: **10.1.**

Chapter 11

11.2. Fig. 1.1 to 1.6; Fig. 7.1 to 7.6, summarized by Fig. 7.8.

11.3. See the two examples quoted in the solution of ex. 11.2: in the first, it may be Fig. 1.6; in the second, it may be Fig. 7.3.

11.4. Ex. 10.2.

11.5. Recognizing in Fig. 4.2 the abscissas $x_1, x_3, \ldots, x_n$ as *roots* of the desired polynomial $f(x)$.

11.6. The transition from Fig. 1.11 to Fig. 1.12. Or, in MPR, vol. II, p. 144, the transition from Fig. 16.2 to Fig. 16.3.

No solution: **11.1, 11.7, 11.8, 11.9, 11.10, 11.11.**

Chapter 12

12.4.

(1) The point P lies on two given straight lines; by construction, ...

The point P lies on a given straight line, and on a given circle; by construction, ...

The point P lies on two given circles; by construction, ...

(2) Given the lengths of two sides of $\triangle ABC$, and the included angle, the third side being AB, ...

Given the length of the side BC and two angles of $\triangle ABC$, ...

Given, in the right triangle $\triangle ABC$, the lengths of the two legs AC and BC, ...

(3) Given the base and the height, ...

Given three sides of $\triangle ABC$, ...

[Also all sets of data listed under (2)].

(4) Given the area of the base $\triangle ABC$ and the length of the altitude from the vertex D, ...

12.5.

(1) If $\triangle ABC \cong \triangle EFG$, ...

If $ABFE$ is a parallelogram, ...

(2) If $\triangle ABC \sim \triangle EFG$, ...

If $\angle ABC$ and $\angle EFG$ are corresponding angles, obtained by cutting two parallels by a transversal, ...

If $\angle ABC$ and $\angle EFG$ are inscribed in the same circle and intercept the same arc, ...

And so on; see ex. 8.8.

(3) If the figure $ABCD$ ... is similar to the figure $EFGH$ ..., the points being mentioned in the order as they correspond to each other, ...

If $AE \parallel BF \parallel CG \parallel DH$, and A, B, C, D are collinear and also E, F, G, H are collinear ... (A less heavy statement of the theorem is the following: "Corresponding parts of straight lines cut by parallels are proportional.")

(4) If, in $\triangle ABC$, $\angle C < \angle B$, ...

12.6. See ex. 1.47, sect. 2.5(2), ex. 2.10, ex. 2.13, ex. 2.51, sect. 14.6(3), also HSI pp. 37–46, especially p. 38 (Analogy, especially sect. 3) and MPR, vol. 1, pp. 45–46. There are many others.

12.7. A median of a triangle that ends in the midpoint of the side AB, passes through the midpoint of any segment intercepted by the two other sides on a line parallel to AB.

(In the tetrahedron, the cross sections considered are parallelograms. The plane passing through an edge of the tetrahedron and the midpoint of the opposite edge yields a useful intersection.)

12.8. A median of a triangle bisects its area.

(For the tetrahedron, the theorem stated follows from the principle of Cavalieri, since the plane in question bisects the area of each cross section considered in ex. 12.7.)

12.9. The bisector of an interior angle of a triangle divides the opposite side into segments that are proportional to the adjacent sides.

No solution: **12.1, 12.2, 12.3, 12.10, 12.11.**

Chapter 13

No solution: **13.1, 13.2, 13.3.**

Chapter 14

14.1. Approximately $9\frac{2}{3}$ minutes after noon. The longitude of the "central" meridian for Western Standard Time is 120° west.

14.2. Since

$$BD = BD_1, \qquad CD = CD_1$$

the point D belongs to the intersection of two spheres, one with center B and radius BD_1, the other with center C and radius CD_1. The plane of this circle of intersection is perpendicular to the line BC joining the centers of the spheres, and therefore it is perpendicular to the horizontal plane of $\triangle ABC$. Hence the orthogonal projection F of D onto the horizontal plane lies on the line through D_1 perpendicular to BC. Of course, D_2 and D_3 are analogously situated.

14.3. The point F is the *radical center* (defined in textbooks of analytic geometry) of the three circles indicated by short arcs in Fig. 14.1, and D_1F, D_2F, and D_3F are the three radical axes meeting in F.

14.4. Action and perception in sect. 14.6(4).

14.11. The following answers are sketchy.

($7a$) Using the law of cosines both of ordinary and of spherical trigonometry, we are led to prove the inequality

$$\frac{b^2 + c^2 - a^2}{2bc} > \frac{-\cos b \cos c + \cos a}{\sin b \sin c}$$

in supposing that $0 < a < \pi, 0 < b < \pi, 0 < c < \pi$, and that the three segments of length a, b, and c, respectively, form a triangle. The inequality can be derived by appropriate manipulation from the fact that the function $(\sin x)/x$ steadily decreases as x increases from 0 to π, and this fact itself can be made plausible by geometric considerations.

($7b$) By continuity, a "very small" spherical triangle is "almost plane," and so it is "almost congruent" to its image in the plane, having the same sides; hence the corresponding angles are "almost" equal.

($7c$) The stereographic projection of the sphere (from the north pole onto the plane of the equator) preserves the angles.

($7d$) The theorem of Archimedes on the surface area of a spherical zone yields a simple area preserving mapping.

($7e$) Projection of a hemisphere onto a plane from the center of the sphere.

14.24. The desired remainder is a polynomial of degree not exceeding 1, of the form $ax + b$. Take the problem as solved, suppose that the quotient $q(x)$ of the division has been found. Then you have the identity:

$$x^3 + x^5 + \cdots + x^{17} + x^{19} = (x^2 - 1)q(x) + ax + b$$

which yields, for $x = 1$ and $x = -1$, two equations to determine the two unknowns a and b:

$$7 = a + b, \qquad -7 = -a + b$$

Hence $b = 0$, $a = 7$, and the desired remainder is $7x$.

The prime numbers 3, 5, . . . 17, 19 are conspicuous but irrelevant; if they are replaced by any seven *odd* positive integers the result remains the same. We see this clearly *after* having solved the problem, yet when we are looking at the problem the first time, those prime numbers could lead us on a wild goose chase.

14.25.

(1) πt^2,

(2) $\pi t^2/4$.

No solution: **14.5–14.10, 14.12–14.23, 14.26, 14.27, 14.28.**

Chapter 15

15.1. "Being given the perimeter, find the figure with maximum area." This is the "isoperimetric problem"; it can be proposed for various classes of figures; here are a few references:

triangles: MPR, vol. 1, p. 133, ex. 16;

rectangles: HSI, pp. 100–102, Examine your guess 2;

quadrilaterals: MPR, vol. 1, p. 139, ex. 41;

polygons with a given number of sides: MPR, vol. 1, p. 178;

all plane figures: MPR, vol. 1, pp. 168–183.

For an introduction to some ideas of Jacob Steiner and to connected physical problems see *Modern Mathematics for the Engineer*, second series, edited by E. F. Beckenbach, pp. 420–441.

15.2. Symmetry in a, b, and c; test by dimensions.

15.3. Acute triangles above, obtuse triangles below, the arc.

15.4. Central projection of the polyhedron (of its inside) onto one of its faces w (the "window"); choose as center of projection a point outside the polyhedron, but sufficiently near to an inner point of w. *Cf.* MPR, vol. 1, p. 53, ex. 7.

15.5. $\Sigma F_n = F, \qquad \Sigma V_n = V$

15.6. $\Sigma n F_n = \Sigma n V_n = 2E$

15.7. Use ex. 15.5, ex. 15.6, the definition of sect. 15.6(2), and the final result of sect. 15.6(8)

$$\sum (n - 2) F_n = 2E - 2F = \sum \frac{\alpha}{\pi} = 2V - 4$$

15.8. From ex. 15.6 and 15.5

$$2E = \Sigma n F_n \geqq 3\Sigma F_n = 3F$$
$$2E = \Sigma n V_n \geqq 3\Sigma V_n = 3V.$$

There is equality in the first line iff each face is a triangle, and in the second line iff just three edges end in each vertex.

15.9. *First proof.* Take here for granted (but prove afterwards for yourself) the lemma.

Lemma. If a set of quantities can be subdivided into nonoverlapping subsets so that the average of the quantities in each subset is less than a, then the average of the quantities in the whole set is also less than a.

This lemma remains valid if the relation "less than" expressed by the sign $<$ is replaced by any of the relations expressed by $\leqq$, $>$, $\geqq$. We apply the lemma twice, to (1) and (2).

(1) The average of the angles in a face with n sides is

$$\frac{(n-2)\pi}{n} = \left(1 - \frac{2}{n}\right)\pi \geqq \frac{\pi}{3}$$

(2) The sum of the face angles with common vertex is $<2\pi$, their number $\geqq 3$, and so their average is $<2\pi/3$.

Second proof. By sect. 15.6(6) and ex. 15.6 the average of all face angles is

$$\frac{\Sigma\alpha}{2E} = \frac{2\pi(E-F)}{2E} = \pi\left(1 - \frac{F}{E}\right) \geqq \frac{\pi}{3}$$

by ex. 15.8. Equality is attained iff all faces are triangles.

On the other hand, by Euler's theorem proved in sect. 15.6(8),

$$\frac{\Sigma\alpha}{2E} = \frac{2\pi V - 4\pi}{2E} = \frac{\pi V}{E} - \frac{2\pi}{E} \leqq \frac{2\pi}{3} - \frac{2\pi}{E}$$

by ex. 15.8.

15.10. *First proof.* The average of the angles in a face with n sides is

$$\frac{(n-2)\pi}{n} = \left(1 - \frac{2}{n}\right)\pi \geqq \frac{2\pi}{3}$$

if $n \geqq 6$. If all faces had six sides or more, the average of all face angles would be $\geqq 2\pi/3$, which is impossible by ex. 15.9.

Second proof. From Euler's theorem, ex. 15.8 and ex. 15.6,

$$12 = 6F - 2E + 6V - 4E$$
$$\leqq 6F - 2E$$
$$= \Sigma(6-n)F_n$$
$$12 \leqq 3F_3 + 2F_4 + F_5$$

and so at least one of the three numbers F_3, F_4, and F_5 must be positive.

15.11. (1) If there is a face that has n sides where $n > 3$, we can divide it into $n - 2$ triangles by diagonals, and so we can replace it by $n - 2$ (which is >1) triangular faces without changing V. Therefore, F cannot be a maximum unless all F faces are triangles.

(2) From 15.8, $2E \geq 3F$ with equality iff $F_4 = F_5 = \cdots = 0$, $F = F_3$. Now,

$$F + V = E + 2 \geq \tfrac{3}{2}F + 2$$
$$V \geq \tfrac{1}{2}F + 2$$
$$F \leq 2(V - 2)$$

and

$$E = F + (V - 2) \leq 3(V - 2)$$

with equality iff each face is a triangle.

15.12. By analogy, both solutions of ex. 15.11 apply (with proper interpretation) to the present case and yield

$$V \leq 2(F - 2), \qquad E \leq 3(F - 2)$$

with equality iff just three edges start from each vertex of the polyhedron.

The inequalities derived have interesting applications. For instance, we combine the second inequality just obtained with ex. 15.8 as follows:

$$\frac{E + 6}{3} \leq F \leq \frac{2E}{3}$$

For $E = 6$ this yields

$$4 \leq F \leq 4$$

that is, the case of the tetrahedron. For $E = 7$ it yields, however,

$$\tfrac{13}{3} \leq F \leq \tfrac{14}{3}$$

and so F cannot be an integer! We are so driven to the conclusion that there exists no convex polyhedron with 7 edges—a fact already noticed by Euler.

15.13.

(1)   0   4   3   100   36

for   tetra-, hexa-, octa-, dodeca-, icosahedron

respectively.

(2) $n(n - 3)$ 0 $1 + \dfrac{n(n - 3)}{2}$

for the n-prism, n-pyramid, n-double pyramid

respectively.

(3) $\dfrac{(F - 2)(F - 4)}{8}$, $\dfrac{(F^2 - 5F + 2)}{2}$, $\dfrac{(9F^2 - 42F + 8)}{8}$

for $n =$ 3,   4,   5

respectively; $n > 5$ is impossible, see ex. 15.10; the . . . after 5 is misleading.

(4) In terms of F_n, cf. ex. 15.6, 15.7,

$$D = \frac{V(V - 1)}{2} - E - \sum \frac{n(n - 3)}{2} F_n$$
$$= 1 - \tfrac{1}{4}\Sigma(2n - 3)(n - 2)F_n + \tfrac{1}{8}[\Sigma(n - 2)F_n]^2$$

15.14.

	(1)	(2)	(3)
$\Sigma\delta$	3π	2π	$\dfrac{5\pi}{2}$
$\Sigma\omega$	2π	0	π

15.15. In the cases observed in ex. 15.14, both sums change in the same direction—the change of $\Sigma\omega$ is double of the change of $\Sigma\delta$—and

$$2\Sigma\delta - \Sigma\omega = 4\pi$$

Is this relation true for all tetrahedra?

15.16. *Cf.* ex. 15.19.

15.17. Generalize the cases (1) and (3) of ex. 15.14; in both the result is the same. *Cf.* ex. 15.19.

15.18. *Cf.* ex. 15.19.

15.19.	$2\Sigma\delta - \Sigma\omega$	F	V	E
Tetrahedron	4π	4	4	6
Cube	8π	6	8	12
n-Pyramid	$(2n-2)\pi$	$n+1$	$n+1$	$2n$
n-Prism	$2n\pi$	$n+2$	$2n$	$3n$
n-Double-pyramid	$(4n-4)\pi$	$2n$	$n+2$	$3n$

Neither V nor E, only F increases consistently when $2\Sigma\delta - \Sigma\omega$ increases.

15.20. $2\Sigma\delta - \Sigma\omega = 2\pi F - 4\pi$

For a proof, express the spherical area (the solid angle) associated with a vertex in terms of the dihedral angles associated with the edges that diverge from that vertex. Remember that the area of a spherical triangle with angles α, β, and γ is the "spherical excess" $\alpha + \beta + \gamma - \pi$, and derive hence an expression for the area of a spherical polygon. You obtain so (use ex. 15.5, 15.6)

$$\Sigma\omega = 2\Sigma\delta - \Sigma\pi(n-2)V_n = 2\Sigma\delta - 2\pi(E - V)$$

15.21. The usual answers are: the equilateral triangle, the square, the regular polygon with n sides.

15.22. Usually, the same shapes are guessed as in ex. 15.21.

15.25. The usual answers are: the regular tetrahedron, the regular octahedron, the cube.

15.26. The usual answers are: the regular tetrahedron, the cube, the regular octahedron.

15.27. A diagonal of the cube is a diameter of the sphere. Hence, if a is the length of an edge of the cube

$$(2r)^2 = 3a^2$$

(*Cf.* HSI, pp. 7–14, the main example of Part I.) Therefore, the required volume is

$$a^3 = \frac{8\sqrt{3}r^3}{9}$$

15.28. Let A denote the area of an equilateral triangle with side r. The required volume is

$$\frac{2 \cdot 6Ar}{3} = \sqrt{3}r^3$$

The answer to the case $V = 8$ of ex. 15.25, which appeared as a quite plausible guess, turned out to be wrong. (The answers to the cases $V = 4$ and 6 are correct, however; for the first, cf. MPR, vol. 1, p. 133, ex. 17.)

15.29. Let A be the area and h the altitude of a face of the octahedron. Then the required volume is (divide the octahedron into 8 tetrahedra)

$$V = \frac{8Ar}{3} = \frac{8r}{3} \frac{h^2}{\sqrt{3}}$$

The sphere touches, by symmetry, each face at its center. Hence h is, in a right triangle, the length of the hypotenuse which is divided by the altitude (of length r) into two pieces, of lengths $h/3$ and $2h/3$, respectively. Thus

$$\frac{h}{3} \cdot \frac{2h}{3} = r^2$$

By eliminating h, we find

$$V = 4\sqrt{3}r^3$$

15.30. Volume of the prism is

$$6 \frac{r^2}{\sqrt{3}} 2r = 4\sqrt{3}r^3$$

In answering the case $F = 8$ of ex. 15.26 we hardly expected this. (The answers to cases $F = 4$ and 6 are correct.)

15.31.

	F	V	E
Cube	6	8	12
Hexagonal double pyramid	12	8	18
Regular octahedron	8	6	12
Hexagonal right prism	8	12	18

The nonregular polyhedron that "beats" (or "ties with") the regular polyhedron agrees with it, of course, in the *prescribed* number of elements (V in the first case, F in the second) but is more complicated otherwise (has greater F and E in the first case, greater V and E in the second). Should this circumstance account for the observed failure of the principle of nonsufficient reason?

15.32. One which has only triangular faces, see ex. 15.11.

15.33. One which has only three-edged vertices, see ex. 15.12.

15.34. Between the cube and the octahedron there is a reciprocal relation, and we can observe the same relation between the corresponding rival nonregular polyhedra; cf. MPR, vol. 1, p. 53, ex. 3 and 4. This suggests a conjecture: between the *solutions* of the problems of ex. 15.25 and ex. 15.26 for the same number of data the same (topological) *reciprocal* relation will hold; cf. ex. 15.32 and 15.33.

15.35. The usual answers are the same as to ex. 15.26, and the whole situation is the same; the usual answer is correct for $F = 4$ and $F = 6$, incorrect for $F = 8$. *Cf.* MPR, vol. 1, p. 188, ex. 42.

15.36. (Stanford 1962.) We are required to find the points of intersection of two congruent ellipses symmetrical to each other with respect to the line $x = y$. Subtraction of the equations yields $x^2 = y^2$. Of the four points of intersection

$$(6, 6) \qquad (-6, -6) \qquad (2, -2) \qquad (-2, 2)$$

two do, and two do not, comply with the principle of nonsufficient reason. *Cf.* ex. 6.22.

15.37. By appropriate subtractions $x^2 = y^2 = z^2$. Of the eight solutions

$$(1, 1, 1) \qquad (-1, -1, -1)$$
$$(3, -3, -3) \quad (-3, 3, 3) \quad (-3, 3, -3) \quad (3, -3, 3) \quad (-3, -3, 3) \quad (3, 3, -3)$$

two do, and six do not, comply with the principle of nonsufficient reason.

15.38. (Stanford 1963.) Same solutions as in ex. 15.37, but it is less easy to establish that $x^2 = y^2 = x^2$.

15.41. See ex. 15.42–15.53.

15.42. (1) We can discuss the result of any problem "in letters" in this manner by checking particular cases; *cf.* sect. 2.4(3); ex. 2.61; HSI, Can you check the result? 2, p. 60; MPR, vol. 2, pp. 5–7, sect. 2, pp. 13–14, ex. 3–7; also paper no. 11 of the author, quoted in the Bibliography; etc.

(2) In checking its particular cases, we familiarize ourselves with the formula, we understand its "structure" better. Moreover, such a discussion may illustrate several important general ideas. We may learn that the merit of a formula consists in its generality, in its applicability. Then, we may also learn plausible, inductive reasoning—assessing the likelihood of a general assertion by examining its particular cases. In brief, the teacher who neglects discussions of the kind presented in sect. 15.4 misses a most obvious opportunity to do something for the mental maturity of his students.

15.43. Each point of the region displayed in Fig. 15.1 represents a triangular shape. (For diagrams surveying in an analogous way ellipsoidal and lenticular shapes see G. Pólya and G. Szegö, *Isoperimetric inequalities in mathematical physics,* Princeton University Press, p. 37 and p. 40.) Thus, Fig. 15.1 may prepare the student for a certain way of using diagrams in science, for instance, indicator diagrams in thermodynamics. Moreover Fig. 15.1 offers nonpointless practice in the geometrical interpretation of linear inequalities.

15.44. Here are some of the facts to which experimentation with decimal fractions may lead.

All three types of decimal expansions represent rational numbers and, conversely, the decimal expansion of a rational number must belong to one of the three types. The difference between the types depends on the prime factors of the denominator of the represented rational number; according as all these prime factors divide 10, or none of them divides 10, or there is one prime factor that divides and another that does not divide 10, the decimal fraction is terminating, or purely periodic, or mixed. (In speaking about *the* denominator b of a rational number a/b we have supposed that a/b is in lowest terms, that is, that the integers a and b have no common divisor greater than 1, and $b \geq 1$. We have disregarded two obvious cases: the case $b = 1$ (integers) and the case of the infinite decimal expansion of such rational numbers as can be represented also by a terminating expansion. *Cf.* I. Niven, *Numbers: rational and irrational,* Random House, pp. 23–26, 30–37.)

The length of the period is independent of the numerator.

If the denominator is a prime number p, the length of the period is a divisor of $p - 1$. (More generally, the length of the period divides $\varphi(b)$, the number of positive integers not exceeding the denominator b and relative prime to b. What can you say about mixed expansions?)

If the denominator is a prime and the length of the period an even number, each digit in the second half of the period completes the corresponding digit in the first half to 9. (For instance, in the expansion

$$\tfrac{1}{7} = 0.\overline{142857}$$

$$1 + 8 = 9 \qquad 4 + 5 = 9 \qquad 2 + 7 = 9$$

The knowledge of this fact may save much work in computing decimal fractions.)

If the denominator is not divisible by 3, the sum of the digits in the period is divisible by 9. For instance

$$\tfrac{15}{41} = 0.\overline{36585}$$

$$3 + 6 + 5 + 8 + 5 = 27$$

The reader should check these statements on examples. The proofs are easy, provided that some knowledge of the theory of numbers is available—to rouse the reader's interest in this theory is one of the aims.

15.45. Observation:

$$9 \times 11 = 99, \quad 27 \times 37 = 999, \quad 99 \times 101 = 9999, \quad 271 \times 369 = 99999$$

Explanation: Therefore, for instance,

$$27 \times 0.037037 \cdots = 0.999999 \cdots = 1$$

We should not hesitate to compare little things with great things (*parva componere magnis*): such a comparison may be instructive. The step we have taken from "observation" to "explanation," from noticing a regularity to noticing the underlying connection, is infinitely smaller in scale than, but similar in nature to, the step from Kepler to Newton; *cf.* MPR, vol. 1, p. 87, ex. 15.

15.46. Except for the last statement (concerning the sum of the digits in the period) to each result of ex. 15.44 there corresponds a parallel result in the binary system. For instance, in the *binary* expansion

$$\tfrac{3}{5} = 0.\overline{1001}$$

the length of the period is $5 - 1$ and

$$1 + 0 = 1, \qquad 0 + 1 = 1$$

15.47. Challenging arithmetical work, practice in decimal fractions and factorization. Broad background: concept of real number ("And what about the decimal expansion of $\sqrt{2}$, or of π?"). A prelude to number theory. On the general cultural level: wide opportunity for inductive reasoning, even for the construction of a comprehensive theory starting from an experimental basis.

Take a detail; ex. 15.45 offers an exceptionally neat opportunity to confirm a guess based on observation by a proof, by the understanding of the underlying connection.

15.48. See ex. 15.49.

15.49. In Fig. 15.11, $t(n)$ is the number of dots with abscissa n. We find that $t(n) = 1$ for 1, 2, 4, 8, 16.

When p is an *odd* prime $t(p) = 2$.

Even after these significant hints (and even after a comparison of Fig. 15.11 with Fig. 15.12) it may take longer experimentation and some acumen to discover the rule: $t(n)$ *equals the number of odd divisors of n.* The reader should prove this rule; he may profit from the following remarks.

(1) The trapezoidal representation displayed in ex. 15.49 is equivalent to the equation

$$2n = r(r + 2a - 1)$$

(2) Of the two factors r and $r + 2a - 1$, one is odd, the other is even, and the odd factor must divide n.

(3) The smaller of these two factors is r, the number of rows.

(4) If n and r are given, a is uniquely determined.

15.50. We use the symbol $\tau(n)$, as defined in ex. 9.12. The rule distinguishes five cases:

(1) If n is odd and not a square, $s(n) = \tau(n)/2$.
(2) If n is odd and a square, $s(n) = [\tau(n) + 1]/2$.
(3) If n is even and not divisible by 4, $s(n) = 0$.
(4) If n is divisible by 4 and not a square, $s(n) = \tau(n/4)/2$.
(5) If n is divisible by 4 and a square, $s(n) = [\tau(n/4) + 1]/2$.

To prove this rule, observe that

$$n = (2a + 1) + (2a + 3) + \cdots + (2a + 2r - 1) = r(r + 2a)$$

If n is divisible by 4, observe that

$$\frac{n}{4} = \frac{r}{2}\left(\frac{r}{2} + a\right)$$

15.51. Let us compare the present project with that assessed in ex. 15.47. In the present case the problem is more artificial, the background less rich, the law more difficult to guess, but the proof, although challenging, needs very little preliminary knowledge; I think that the project is very much worthwhile.

Figure 15.11 yields a nontrivial, instructive example of a binary relation (between the two positive integers n and r; n is the sum of r successive positive integers) and of its representation by a diagram. For Fig. 15.12, which represents a better known and more important binary relation, see Leibnitz, *Opuscules*, p. 580. The study of these diagrams could profitably precede the introduction of the concept "binary relation."

15.52. The required quantities can be computed without explicit use of the integral calculus (Cavalieri's principle, Pappus' rules). They are displayed in the table:

	Double Cone	Sphere	Cylinder	
V	$\dfrac{2\pi a^3}{3}$	$\dfrac{4\pi a^3}{3}$	$2\pi a^3$	$1:2:3$
S	$2\sqrt{2}\pi a^2$	$4\pi a^2$	$6\pi a^2$	$\sqrt{2}:2:3$
A	$2a^2$	πa^2	$4a^2$	$2:\pi:4$
L	$4\sqrt{2}a$	$2\pi a$	$8a$	$2\sqrt{2}:\pi:4$
X_A	$\dfrac{a}{3}$	$\dfrac{4}{3\pi}a$	$\dfrac{a}{2}$	$\dfrac{X_A}{X_L} = \dfrac{2}{3}$
X_L	$\dfrac{a}{2}$	$\dfrac{2}{\pi}a$	$\dfrac{3}{4}a$	

$$S = \sqrt{2}\frac{dV}{da} \qquad S = \frac{dV}{da} \qquad S = \frac{dV}{da}$$

$$L = \sqrt{2}\frac{dA}{da} \qquad L = \frac{dA}{da} \qquad L = \frac{dA}{da}$$

For a generalization of the observation about X_A/X_L (which, by the way, turned up in a class discussion) see a paper by C. J. Gerriets and the author, *American Mathematical Monthly*, vol. 66, 1959, pp. 875–879.

15.53. For $n = 1, 2, 3, \ldots$

$$(\sqrt{2} - 1)^n = \sqrt{m + 1} - \sqrt{m}$$

where m is a positive integer which depends on n. Proof by (easy) mathematical induction. See *American Mathematical Monthly*, vol. 58, 1951, p. 566.

No solution: **15.23, 15.24, 15.39, 15.40, 15.54, 15.55.**

APPENDIX

Problems

Problems 1.19.1, 1.19.2, and 1.19.3 appropriately follow (in this order) ex. 1.19, problem 2.27.1 follows ex. 2.27, etc.

1.19.1. Triangle from α, r, R.
[*Could you think of other data more appropriate to determine the unknown? Could you exchange just one datum for a more suitable one?*]

1.19.2. Triangle from a, h_a, r.
[*Could you derive something useful from the data?*]

1.19.3. Triangle from a, r, $a + b + c$.

2.27.1. *How long did Diophantus live?* The problem is proposed in the form of an alleged inscription on Diophantus' tombstone. The original is in verse. (*Cf.* B. L. van der Waerden, *Science Awakening*, p. 278.)
Here is the tomb of Diophantus. This stone tells you his age if you have mastered his art. The gods gave him to live one sixth of his life as a boy. After one twelfth more his beard began to show. Then one seventh of his life passed till his wedding day. After five years of marriage a boy was born. Alas, his beloved son came to an untimely end attaining only one half of the age he himself was permitted to live. After this bereavement he sought consolation in mathematics for four years and then he too ended his earthly career.

2.35.1. From a vertex of a certain triangle draw the altitude, the angle bisector, and the median. Being given that these three lines divide the angle α at the vertex into four equal parts, find α.
[You may wish to know also the shape of the triangle. Pay attention to *each part of the condition.*]

2.40.1. Of a right triangle, given the length of the hypotenuse c and the area A. On each side of the triangle, describe a square exterior to the triangle and consider the least convex figure containing the three squares (formed by a tight rubber band around them): it is a hexagon (it is irregular, has one side

185

in common with each square, and one of its remaining three sides is obviously of length c).

Find the area of the hexagon.

2.40.2. In a right triangle, c is the length of the hypotenuse, a and b are the lengths of the two other sides, and d is the length of the diameter of the inscribed circle. Prove that

$$a + b = c + d$$

[*Restate the problem:* Given a, b, and c, compute d.]

2.50.1. Here is a problem of solid geometry analogous to ex. 2.45.

Start by dissecting the three-dimensional space into equal cubes.

First pattern: Each cube is associated with a concentric sphere that touches its six faces.

Second pattern: "Each second" cube is associated with a concentric sphere that touches its 12 edges. (That is, of two cubes that have a face in common, one does and the other does not contain the center of an associated sphere.)

Compute the percentage of space contained in the spheres for each pattern.

2.52.1. A cake has the shape of a right prism with a square base; it has icing on the top as well as on the sides (that is, on the four lateral faces). The altitude of the prism is $\frac{5}{16}$ of the side of its base. Cut the cake into 9 pieces so that each piece has the same amount of cake and the same amount of icing. One of the 9 pieces should be a right prism with a square base with icing only on the top: Compute the ratio of its altitude to a side of its base and give a clear description of all 9 pieces.

2.55.1. Five edges of a tetrahedron are of the same length a, and the sixth edge is of the length b.

(1) Express the radius of the sphere circumscribed about the tetrahedron in terms of a and b.

(2) How would you use the result (1) to determine practically the radius of a spherical surface (of a lens)?

2.55.2. *The carbon atom* has four valencies which we picture as directed symmetrically in space. From the center of a regular tetrahedron draw four lines to the four vertices and compute the angle α included by any two of them.

2.55.3. *Photometer.* A lamp L has the candle-power I, another lamp L' the candle-power I', and their distance is d. Find the position of a screen that is so placed between these lamps, perpendicularly to the line joining them, that it is equally illuminated from both sides.

(If the candle-power of a point-source L of light is I, the illumination of a surface, which is at the distance x from L and is perpendicular to the distance, is I/x^2. To understand this thoroughly, you should consider two spheres with center L: one with radius 1, the other with radius x.)

3.10.1. *Salvaging.* The ship sank—perhaps there is some treasure in it worth lifting. Your plan has failed—perhaps there is an idea in it worth saving.

In sect. 3.2, our first plan to compute S_2 (notation of sect. 3.3) failed ignominiously: The procedure that worked for S_1 did not work at all as we tried to apply it to S_2. Where does the fault lie? Perhaps we applied the procedure too rigidly. What about applying it less rigidly? Applying it with some modification? Or applying it to some other case?

Such consideration may lead to trials, and it is quite natural to try the procedure on S_k. What is the essential point of the procedure? Combining two terms equally removed from the ends: one term as distant from one end as the other is from the other end. Such are the terms j^k and $(n - j)^k$ of S_k. If addition does not work, subtraction may—and eventually we may adapt the combination to the nature of k:

$$(n - j)^k - (-j)^k = n^k - \binom{k}{1}n^{k-1}j + \binom{k}{2}n^{k-2}j^2 - \cdots + (-1)^{k-1}\binom{k}{k-1}nj^{k-1}$$

Write this successively for $j = 0, 1, 2, \ldots, n - 1, n$:

$$n^k - (-1)^k 0^k = n^k$$

$$(n - 1)^k - (-1)^k 1^k = n^k - \binom{k}{1}n^{k-1}1 + \binom{k}{2}n^{k-2}1^2$$
$$- \cdots + (-1)^{k-1}\binom{k}{k-1}n1^{k-1}$$

$$(n - 2)^k - (-1)^k 2^k = n^k - \binom{k}{1}n^{k-1}2 + \binom{k}{2}n^{k-2}2^2$$
$$- \cdots + (-1)^{k-1}\binom{k}{k-1}n2^{k-1}$$

$$\cdots \cdots \cdots \cdots \cdots \cdots \cdots \cdots$$

$$1^k - (-1)^k(n - 1)^k = n^k - \binom{k}{1}n^{k-1}(n - 1) + \binom{k}{2}n^{k-2}(n - 1)^2$$
$$- \cdots + (-1)^{k-1}\binom{k}{k-1}n(n - 1)^{k-1}$$

$$0^k - (-1)^k n^k = n^k - \binom{k}{1}n^{k-1}n + \binom{k}{2}n^{k-2}n^2$$
$$- \cdots + (-1)^{k-1}\binom{k}{k-1}nn^{k-1}$$

By adding and using the notation of sect. 3.3 (but writing S_0' for $S_0 + 1$) we obtain

$$S_k[1 - (-1)^k] = n^k S_0' - \binom{k}{1}n^{k-1}S_1 + \binom{k}{2}n^{k-2}S_2$$
$$- \cdots + (-1)^{k-1}\binom{k}{k-1}nS_{k-1}$$

Explore the cases $k = 1, 2, 3$ of this result and then try to assess it generally.

3.40.1. In how many ways can the positive integer n be the sum of positive integers? In how many ways can n be the sum of a specified number t of terms (all positive integers)? Let us regard two sums that differ in the order of terms as different.

It is natural to start exploring such a problem by experiments and by trying to arrange neatly the material produced by the experiments. Here are the sums found for $n = 4$ and $n = 5$:

	4	1 + 3	2 + 1 + 1	1 + 1 + 1 + 1
		2 + 2	1 + 2 + 1	
		3 + 1	1 + 1 + 2	

5	1 + 4	3 + 1 + 1	2 + 1 + 1 + 1	1 + 1 + 1 + 1 + 1
	2 + 3	1 + 3 + 1	1 + 2 + 1 + 1	
	3 + 2	1 + 1 + 3	1 + 1 + 2 + 1	
	4 + 1	1 + 2 + 2	1 + 1 + 1 + 2	
		2 + 1 + 2		
		2 + 2 + 1		

Do you notice a pattern?
Prove your guess!
Could a geometric figure help you?

3.40.2. *Fibonacci numbers.* By adding the numbers along the oblique lines of Fig. A3.40.2 we obtain the sequence of the Fibonacci numbers

$$1, 1, 2, 3, 5, 8, 13, 21, \ldots$$

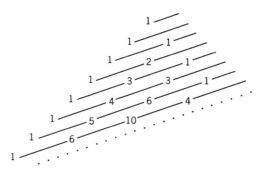

Fig. A3.40.2. An oblique approach to Fibonacci.

We let F_n stand for its nth term, the nth Fibonacci number; thus $F_1 = 1$, $F_2 = 1$, $F_8 = 21$.

(1) Express F_n in terms of binomial coefficients.

(2) Prove that for $n = 3, 4, 5, \ldots$

$$F_n = F_{n-1} + F_{n-2}$$

3.40.3. (*Continued*). The sequence of numbers

$$1, 1, 1, 2, 3, 4, 6, 9, 13, \ldots$$

is analogously generated (see Fig. A3.40.3). Let G_n stand for its nth term; thus $G_9 = 13$.

(1) Express G_n in terms of binomial coefficients.

(2) Prove that for $n = 4, 5, 6, \ldots$

$$G_n = G_{n-1} + G_{n-3}$$

(3) Generalize.

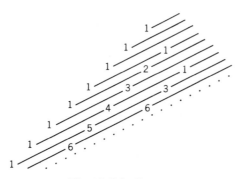

Fig. A3.40.3. Steeper.

†**3.60.1.** Consider the table

$$
\begin{array}{llllll}
 & & 1 \cdot 1 & & & = 1 \\
 & 1 \cdot 3 - & 2 \cdot 2 + & 3 \cdot 1 & & = 2 \\
 1 \cdot 5 - & 2 \cdot 4 + & 3 \cdot 3 - & 4 \cdot 2 + & 5 \cdot 1 & = 3 \\
1 \cdot 7 - \quad 2 \cdot 6 + & 3 \cdot 5 - & 4 \cdot 4 + & 5 \cdot 3 - & 6 \cdot 2 + & 7 \cdot 1 = 4
\end{array}
$$

Guess the general law suggested by these examples, express it in suitable mathematical notation, and prove it.

3.65.1. If x and n stand for positive integers the expression

$$\frac{x^2(x^2 - 1)(x^2 - 4) \cdots [x^2 - (n - 1)^2]}{(2n - 1)!n}$$

represents an integer.

†**3.88.1.** Another solution of ex. 3.86, which is simpler in some respects, uses differential calculus. Find it.

†**3.88.2.** Observe that

$$
\begin{array}{ll}
1 \cdot 1 + 2 \cdot 1 & = \quad 3 \\
1 \cdot 1 + 2 \cdot 2 + 3 \cdot 1 & = \quad 8 \\
1 \cdot 1 + 2 \cdot 3 + 3 \cdot 3 \ + 4 \cdot 1 & = \quad 20 \\
1 \cdot 1 + 2 \cdot 4 + 3 \cdot 6 \ + 4 \cdot 4 \ + 5 \cdot 1 & = \quad 48 \\
1 \cdot 1 + 2 \cdot 5 + 3 \cdot 10 + 4 \cdot 10 + 5 \cdot 5 + 6 \cdot 1 = 112
\end{array}
$$

Guess the general law suggested by these examples, express it in suitable mathematical notation, and prove it.

4.15.1. Define y_k for $k = 2, 3, 4, \ldots$ by the recursion formula

$$y_k = \frac{y_{k-1} + y_{k-2}}{2}$$

and set

$$y_0 = a, \qquad y_1 = b$$

Express y_k in terms of a, b, and k.

5.19.1. *Examining the solution* of ex. 5.19, we may observe that the number found has a simple interpretation: It is the number of ways of choosing v boxes out of $n + v$ boxes. We should be able to see such a simple thing by a simple argument.

Imagine the $n + v$ boxes in a row—each occupying, if you wish, a subinterval of length 1 of the interval $0 \leq x \leq n + v$. What has the proposed problem to do with marking v out of these $n + v$ boxes?

6.13.1. Prove that no number in the sequence

$$11, 111, 1111, 11111, \ldots$$

is the square of an integer.

6.17.1. "How many children have you, and how old are they?" asked the guest, a mathematics teacher.

"I have three boys," said Mr. Smith. "The product of their ages is 72 and the sum of their ages is the street number."

The guest went to look at the entrance, came back and said, "The problem is indeterminate."

"Yes, that is so," said Mr. Smith, "but I still hope that the oldest boy will some day win the Stanford competition."

Tell the ages of the boys, stating your reasons.

7.2.1. There is an interpretation of the diagram in Fig. A7.2.1. which is of historic interest. Can you recognize it?

†9.11.1. *Any solution will do.* Show that there exists a couple of divergent series.

$$a_1 + a_2 + \cdots + a_n + \cdots \qquad b_1 + b_2 + \cdots + b_n + \cdots$$

with positive decreasing terms

$$a_1 > a_2 > a_3 > \cdots \qquad b_1 > b_2 > b_3 > \cdots$$

such that the series

$$\min(a_1,b_1) + \min(a_2,b_2) + \cdots + \min(a_n,b_n) + \cdots$$

is convergent. As usual, $\min(a,b)$ denotes the lesser (not greater) of the two numbers a and b.

[Not all couples of series satisfying the condition stated are required, but just one such couple (any one). Thus Leibnitz's advice quoted in ex. 9.11 is applicable: Narrow down the region of search without increasing your difficulty.]

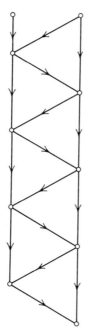

Fig. A7.2.1. Where?

12.2.1. Prove the proposition: If a side of a triangle is less than the average (arithmetic mean) of the two other sides, the opposite angle is less than the average of the two other angles.

(What are the principal parts? *Express them in mathematical language,* by using the usual symbols of trigonometry.)

12.5.1. *Relevant knowledge.* A quadrilateral with sides a,b,c,d and area A is both inscribed and circumscribed (inscribed in a circle and circumscribed about another circle). Then

$$A^2 = abcd$$

[To prove this proposition may be easy or difficult according as you do or do not know a certain related proposition.]

12.9.1. *Do you know a related problem?* Solve the following system of three equations for the unknowns $x, y,$ and z ($a,b,$ and c are given):

$$x^2y^2 + x^2z^2 = axyz$$
$$y^2z^2 + y^2x^2 = bxyz$$
$$z^2x^2 + z^2y^2 = cxyz$$

(We have here a system of 3 equations with 3 unknowns. The best known systems of this kind are linear: Can we "linearize" the proposed system? We may hit upon the following form:

$$1/y^2 + 1/z^2 = \frac{a}{xyz}$$

$$1/x^2 \qquad + 1/z^2 = \frac{b}{xyz}$$

$$1/x^2 + 1/y^2 \qquad = \frac{c}{xyz}$$

which is linear in x^{-2}, y^{-2}, z^{-2} *if* we regard (wishful thinking) xyz as known. The solution will be of the form

$$\frac{1}{x^2} = \frac{A}{xyz} \cdots$$

and opens a new perspective—can you see it?)

12.9.2. *Go back to definitions.* We consider three circles f, f', and v and their centers F, F', and V, respectively. The circles f and f' are fixed, v is variable, f' and v are inside f but outside each other. Prove the proposition: If the variable circle v touches both fixed circles f and f', the locus of its center V is an ellipse. [What is an ellipse?]

12.9.3. *Exploring the neighborhood.* Did you like the foregoing problem (ex. 12.9.2)? Did you like its solution? Then examine its neighborhood—you found a nice ripe apple on this tree, there may be more.

Vary the problem: You may consider a generalization, or particular cases, limiting cases, analogous cases. There is a chance to find something interesting as well as a chance to learn to do research.

The reader should find the various loci for V that result from the following modifications of the assumptions concerning the fixed circles f and f' and the variable circle v.

(1) *Specialization.* The circles f and f' are concentric.

(2) *Limiting case.* Let f be a fixed straight line and f' a fixed point; v touches f and passes through $f' = F'$.

(3) *Analogy.* The two circles f and f' are outside each other and v touches f and f' in the same way, it is either outside both or it contains both.

(4) Limiting case of (3). Let f and f' be two distinct points and let v pass through both.

Consider other particular, limiting, or analogous cases.

14.1.1. *Leap years.* An ordinary year has 365 days, a leap year 366 days. The year n which is not a centenary year is a leap year if, and only if, n is a multiple of 4. A centenary year, that is, a year n where n is a multiple of 100, is a leap year if (and only if) n is a multiple of 400. Thus 1968 and 2000 are leap years, 1967 and 1900 are not. These rules were established by the Pope Gregory XIII.

So far we spoke of "civil" years which must consist of a whole number of days. The astronomical year is the period of time in which the earth completes a full revolution around the sun. If the Gregorian rules would precisely agree with it, what would be the length of the astronomical year?

15.2.1. [Sect. 15.5.] Let a, b, and c denote the lengths of the sides of a triangle and d the length of the bisector of the angle opposite to the side of length c, terminated on the side.

(1) Express d in terms of a, b, and c.

(2) Check the expression obtained in the four cases illustrated by Figs. 15.2 to 15.5.

15.52.1. *One more high-school level research project,* which could also be proposed as part of a term paper in a class for teachers.

A point (x, y, z) in three-dimensional space is characterized in the usual way by its three rectangular coordinates x, y, and z.

We consider four sets of points, C, O, I, and H. Each set is characterized by a system of inequalities (which may consist of just one inequality): Those points (and only those points) belong to the set the coordinates of which satisfy simultaneously all inequalities of the system.

We list the four systems of inequalities defining our sets:

(C) $|x| \leqq 1$ $|y| \leqq 1$ $|z| \leqq 1$

(O) $|x| + |y| + |z| \leqq 2$

(I) all four inequalities listed under (C) and (O)

(H) $|y| + |z| \leqq 2$ $|z| + |x| \leqq 2$ $|x| + |y| \leqq 2$

Describe carefully the geometric nature of the four sets, mention all relevant features (do not forget symmetries) arranging them clearly in suitable tables.

Describe also the relations between the four figures.

Find the volume V and the surface area S of each figure.

Which generalizations does this work suggest to you?

(Cardboard models may be helpful. *Cf.* ex. 2.50.1, HSI p. 235, ex. 8.)

15.53.1. Observe also that

$$2 - \sqrt{3} = \sqrt{4} - \sqrt{3}$$
$$(2 - \sqrt{3})^2 = \sqrt{49} - \sqrt{48}$$
$$(2 - \sqrt{3})^3 = \sqrt{676} - \sqrt{675}$$
$$(2 - \sqrt{3})^4 = \sqrt{9409} - \sqrt{9408}$$

try to generalize and prove your guess.

Solutions

1.19.1. From the center of the circumscribed circle, draw a line to one of the endpoints of the side a and a perpendicular to a. You thus obtain a right triangle with hypotenuse R, angle α, and opposite leg $a/2$. In our case R and α are given and so you can construct a. Construct the desired triangle from the a so found and the α and r originally given by using ex. 1.19.

1.19.2. Divide the desired triangle into three triangles by joining the center of the inscribed circle to the three vertices. Consideration of the areas yields

$$\tfrac{1}{2} r(a + b + c) = \tfrac{1}{2} a h_a$$

Thus you can construct the length of the perimeter $a + b + c$ from the given a, h_a, r and so reduce the proposed problem to ex. 1.19.3.

1.19.3. From the center of the inscribed circle, draw a line to the vertex A and a perpendicular to the side b (or c). You thus obtain a right triangle in which one of the acute angles is $\alpha/2$ and the opposite leg r. Let x denote the other leg. Now

$$a + b + c - 2a = 2x$$

and so you can first construct x from the given $a + b + c$ and a, and then construct α from x and the given r. Construct the desired triangle from the α so found and a and r originally given by using ex. 1.19.

2.27.1. Let x stand for the number of years Diophantus lived. From

$$\frac{x}{6} + \frac{x}{12} + \frac{x}{7} + 5 + \frac{x}{2} + 4 = x$$

we find $x = 84$.

2.35.1. (*Cf. American Mathematical Monthly*, vol. 66, 1959, p. 208.) Let β be the larger and γ the smaller of the two remaining angles. If β is an acute angle, the five lines c, h_a, d_a, m_a, b starting from the vertex A (notation ex. 1.7) follow each other in this order. From the right triangles into which h_a divides the triangle considered

$$\beta = \frac{\pi}{2} - \frac{\alpha}{4}, \qquad \gamma = \frac{\pi}{2} - \frac{3\alpha}{4}$$

From the triangles into which m_a divides the triangle considered

$$\frac{a/2}{m_a} = \frac{\sin \dfrac{\alpha}{4}}{\sin \left(\dfrac{\pi}{2} - \dfrac{3\alpha}{4} \right)} = \frac{\sin \dfrac{3\alpha}{4}}{\sin \left(\dfrac{\pi}{2} - \dfrac{\alpha}{4} \right)}$$

and so

$$\sin \frac{\alpha}{4} \cos \frac{\alpha}{4} = \sin \frac{3\alpha}{4} \cos \frac{3\alpha}{4}$$

$$\sin \frac{\alpha}{2} = \sin \frac{3\alpha}{2}$$

$$\frac{\alpha}{2} = \pi - \frac{3\alpha}{2}$$

$$\alpha = \frac{\pi}{2}$$

2.40.1. (Stanford 1965.) The hexagon consists of three squares and four triangles and these triangles have all the same area A. Therefore the hexagon's area is $2c^2 + 4A$.

2.40.2. (Stanford 1963.) Divide the given right triangle into three triangles with a common vertex at the center of the circle inscribed in the given. From comparing the areas

$$\frac{d}{2} \frac{a+b+c}{2} = \frac{ab}{2}$$

$$d = \frac{2ab}{a+b+c} = \frac{2ab(a+b-c)}{(a+b)^2 - c^2} = a+b-c$$

2.50.1. $100\pi/6$ and $100\pi\sqrt{2}/6$ or approximately 52.36% and 74.07%, respectively. *Cf.* the solution of ex. 15.52.1, part (6).

2.52.1. (Stanford 1964.) Let C stand for the given prism (the cake) and D for the desired prism (with icing only on the top). For a side of the base and the altitude, respectively, let

$$s \text{ and } h \text{ in } C$$
$$x \text{ and } y \text{ in } D$$

stand. The conditions defining D are expressed by the equations

$$x^2 = \frac{s^2 + 4sh}{9}$$

$$x^2 y = \frac{s^2 h}{9}$$

$$h = \frac{5s}{16}$$

which yield

$$x = \frac{s}{2} \qquad y = \frac{5s}{36} \qquad \frac{y}{x} = \frac{5}{18}$$

Carve out D from C so that either the sides or the diagonals of its top are parallel to the sides of the top of C, and in either case so that the prisms C and D have the same 4 planes of symmetry, which cut the remainder of C into 8 congruent pieces each having the same volume and the same amount of "icing" as D.

2.55.1. (Stanford 1962.) Let C denote the center and r the radius of the circumsphere. There are two "key plane figures," two cross sections of the tetrahedron, one through the edge b and the midpoint of the opposite edge, the other through this latter edge and the midpoint of the edge b. These two cross sections are perpendicular to each other; their intersection d joins the two midpoints considered and contains C.

Let x denote the perpendicular distance of C from b (its other end is the midpoint of b) and h the altitude of one of the two faces that are equilateral triangles; therefore

$$h^2 = \frac{3a^2}{4}$$

From right triangles in the two cross sections we find the equations

$$h^2 = d^2 + \left(\frac{b}{2}\right)^2$$

$$r^2 = x^2 + \left(\frac{b}{2}\right)^2$$

$$r^2 = (d - x)^2 + \left(\frac{a}{2}\right)^2$$

and now we have four equations to determine our four unknowns: h is immediately obtained from the first, then d from the second. It is convenient to find x after d (subtract from each other the last two equations). Finally,

$$r^2 = \frac{a^2}{4} \frac{4a^2 - b^2}{3a^2 - b^2}$$

(*Check:* If $b = a\sqrt{3} = 2h$, then $r = \infty$.)

Possible application: Two rigid equilateral triangles with side a and a common side as hinge can be so inclined to each other that all four vertices are on a concave spherical surface; then by measuring b we obtain r. A convex lens demands a somewhat more sophisticated gadget.

2.55.2. If in ex. 2.55.1 we take $a = b$, the tetrahedron becomes regular,

$$r^2 = \frac{3a^2}{8}$$

and

$$\sin \frac{\alpha}{2} = \frac{a/2}{r} = \frac{\sqrt{6}}{3}$$

$$\alpha = 109°28'$$

This may be supposed to be the angle between any two valences of a carbon atom symmetrically bound (as the one in CH_4).

2.55.3. Let x be the perpendicular distance from L to the screen. Then

$$\frac{I}{x^2} = \frac{I'}{(d - x)^2}$$

and so

$$x = \frac{d\sqrt{I}}{\sqrt{I} + \sqrt{I'}}$$

(The practical problem is somewhat different: We are given I, we measure d and x; hence we have to determine I'.)

3.10.1. Here are the first three particular cases:

$$2S_1 = n(n + 1)$$

$$0 = n^2(n + 1) - 2nS_1$$

$$2S_3 = n^3(n + 1) - 3n^2S_1 + 3nS_2$$

The case $k = 1$ yields the evaluation of S_1 by a method which differs but little from the "method of little Gauss" (sect. 3.1).

The case $k = 2$ leads, by a roundabout way, again to S_1.

The case $k = 3$ yields S_3 provided that S_1 and S_2 are already known.

In general, the result allows to compute S_k from the foregoing S'_0, S_1, $S_2, \ldots, S_{k-1}$ *when k is odd,* but not when k is even. This explains to some extent (by the generalization of a modification) why the method that succeeded for S_1 in sect. 3.1 failed for S_2 in sect. 3.2. Morever, by comparing our result with sects. 3.2 to 3.4 and exs. 3.6 to 3.10, we may learn a little—and somebody could use it someday for some purpose.

3.40.1. I hope the reader has explored also the cases $n = 1,2,3$.

Guess: There are $\binom{n-1}{t-1}$ different ways of representing n as the sum of t positive integers.

The cases $t = 1, n$ are trivial, the cases $t = 2, n - 1$ fairly obvious. For a general proof, consider on the number line the interval $0 \leq x \leq n$ and on it the points $x = 1,2,3,\ldots,n-1$. In choosing any $t-1$ out of these $n-1$ points as points of division, we cut the interval into t successive subintervals of integral length and so we make n a sum of t successive terms of the desired kind.

For teachers: The Cuisenaire rods can attractively present this problem and its solution.

3.40.2. Check relations on Fig. A3.40.2 for what values of n you can.

(1) $F_n = \binom{n-1}{0} + \binom{n-2}{1} + \binom{n-3}{2} + \cdots$

(2) From the recursion formula, see sect. 3.6(2). *Cf.* ex. 4.15.

3.40.3.

(1) $G_n = \binom{n-1}{0} + \binom{n-3}{1} + \binom{n-5}{2} + \cdots$

(2) From the recursion formula.

(3) Changing the slope leads to the sequence $y_1, y_2, y_3, \ldots$, dependent on one parameter (the slope, or the positive integer q) satisfying the recursion formula (difference equation, ex. 4.14)

$$y_n = y_{n-1} + y_{n-q}$$

In the case $q = 1$ the slope is 0 and

$$y_n = 2y_{n-1}$$

See ex. 3.32.

3.60.1. Guess:

$$1(2n - 1) - 2(2n - 2) + 3(2n - 3) - \cdots + (2n - 1)1 = n$$

To prove this, consider the coefficient of x^{2n-2} in the product

$$(1 + 2x + 3x^2 + 4x^3 + \cdots)(1 - 2x + 3x^2 - 4x^3 + \cdots)$$
$$= (1 - x)^{-2}(1 + x)^{-2}$$
$$= (1 - x^2)^{-2}$$
$$= 1 + 2x^2 + 3x^4 + 4x^6 + \cdots + nx^{2n-2} + \cdots$$

3.65.1. (*Cf.* Stanford 1963.) Observe that

$$\frac{x}{n} = \frac{(x + n) + (x - n)}{2n}$$

Hence the proposed expression equals

$$\binom{x+n}{2n} + \binom{x+n-1}{2n}$$

and binomial coefficients are integers.

3.88.1. Differentiate both sides of

$$1 + x + x^2 + \cdots + x^n = \frac{1 - x^{n+1}}{1 - x}$$

Ex. 3.87 and 3.88 can be similarly treated.

3.88.2. Guess:

$$1\binom{n}{0} + 2\binom{n}{1} + 3\binom{n}{2} + \cdots + (n+1)\binom{n}{n} = (n+2)2^{n-1}$$

(The difficulty in arriving at this guess may have been recognizing the products $3 \cdot 1, 4 \cdot 2, 5 \cdot 4, 6 \cdot 8, 7 \cdot 16$.)

For a proof, first differentiate both sides of the identity

$$\binom{n}{0}x + \binom{n}{1}x^2 + \binom{n}{2}x^3 + \cdots + \binom{n}{n}x^{n+1} = x(1+x)^n$$

and then set $x = 1$.

4.15.1. The equation in r

$$2r^2 - r - 1 = 0$$

has the roots $r = 1$ and $r = -1/2$. Therefore, by ex. 4.14,

$$y_k = c_1 + \frac{(-1)^k c_2}{2^k}$$

By using the initial conditions (the cases $k = 0$ and $k = 1$), we obtain c_1, c_2, and, finally,

$$y_k = \frac{a + 2b}{3} + (-1)^k \frac{a - b}{3 \cdot 2^{k-1}}$$

5.19.1. We put a mark (a multiplication point if you wish) on v out of those $n + v$ boxes in a row. In each box preceding the first marked box we put a factor x_1, in the unmarked boxes between the marks no. 1 and no. 2 the factor x_2, in the boxes between the marks no. 2 and no. 3 the factor $x_3, \ldots$, in the boxes between the marks no. $(v-1)$ and no. v the factor x_v, in each box following the last marked box the factor 1. Thus to any choice of v boxes out of those $n + v$ corresponds a product of the form $x_1^{m_1} x_2^{m_2} x_3^{m_3} \cdots x_v^{m_v}$ with $m_1 + m_2 + m_3 + \cdots + m_v \leqq n$ and so a term of the polynomial: This is an intuitive argument.

Cf. ex. 3.40.1.

6.13.1. (Stanford 1949.) The problem is: Find a positive integer x such that all digits of x^2 are 1.

(1) *Keep only a part* (a small part) *of the condition:* The last digit of x^2 is 1. As the last digit of x^2 depends only on the last digit of x it is enough to consider

the one-digit numbers; their squares are

$$0, \quad 1, \quad 4, \quad 9, \quad 16, \quad 25, \quad 36, \quad 49, \quad 64, \quad 81$$

respectively. [Observe that x^2 and $(10 - x)^2$ have the same last digit.] Thus only numbers ending in 1 or 9 are eligible.

(2) *Keep a part* (the next larger part) *of the condition:* The last two digits of x^2 are 11. Enough to consider two-digit numbers, of the two numbers x and 100-x just one, and so, by virtue of (1), finally just the ten numbers 01, 11, 21, ... 91. None of them has a square with the ending 11: This proves the assertion.

Moral: It may be advantageous to transform a "problem to prove" into a "problem to find."

6.17.1. [Stanford 1965.] It is understood that for "age" we admit only a whole number of years. Here is a complete list of the decompositions of 72 into three positive integral factors each followed by the sum of the three factors:

$1 \cdot 1 \cdot 72$	74	$2 \cdot 2 \cdot 18$	22
$1 \cdot 2 \cdot 36$	39	$2 \cdot 3 \cdot 12$	17
$1 \cdot 3 \cdot 24$	28	$2 \cdot 4 \cdot 9$	15
$1 \cdot 4 \cdot 18$	23	$2 \cdot 6 \cdot 6$	**14**
$1 \cdot 6 \cdot 12$	19	$3 \cdot 3 \cdot 8$	**14**
$1 \cdot 8 \cdot 9$	18	$3 \cdot 4 \cdot 6$	13

The only sum of three factors that occurs more than once is emphasized by bold print. The remark about *the* oldest boy yields a distinction between two otherwise equally eligible cases: The boys are 8,3, and 3 years old.

7.2.1. See ex. 3.91. The points represent the quantities

$$C_6 \qquad I_6$$
$$C_{12}$$
$$\qquad I_{12}$$
$$C_{24}$$
$$\qquad I_{24}$$
$$C_{48}$$
$$\qquad I_{48}$$
$$C_{96}$$
$$\qquad I_{96}$$

9.11.1. (*Cf. American Mathematical Monthly,* vol. 56, 1949, pp. 423–424.) Choose in advance

$$\min(a_n, b_n) = \frac{1}{n^2}$$

(Any other convergent series with decreasing positive terms would do just as well.) The series

$$(1) + \left(\frac{1}{4}\right) + \left(\frac{1}{4} + \frac{1}{4} + \frac{1}{4} + \frac{1}{4}\right) + \left(\frac{1}{49} + \frac{1}{64} + \cdots + \frac{1}{42^2}\right) + \left(\frac{1}{42^2} + \cdots\right)$$

$$(1) + (1) + \left(\frac{1}{9} + \frac{1}{16} + \frac{1}{25} + \frac{1}{36}\right) + \left(\frac{1}{36} + \frac{1}{36} + \cdots + \frac{1}{36}\right) + \left(\frac{1}{43^2} + \cdots\right)$$

are constructed in corresponding "segments" indicated by parentheses. In one segment each term equals the minimum chosen in advance; in the corresponding segment of the other series all terms are equal to each other and to the last term of the preceding segment, and their sum is 1. These two kinds of segments alternate in both series.

These series do not yet fulfill quite the condition stated: The terms are decreasing in the sense "$\geq$" and not yet in the sense "$>$." Yet there is an easy modification: In segments where the consecutive terms are all equal, subtract from them consecutive terms of an arithmetic progression with sufficiently small positive initial term and difference, especially with sum $< \frac{1}{4}$.

12.2.1. (Stanford 1952.) *What is the hypothesis?*

$$a < \frac{b + c}{2}$$

What is the conclusion?

$$\alpha < \frac{\beta + \gamma}{2}$$

which is equivalent to

$$2\alpha < \pi - \alpha$$
$$\alpha < \pi/3$$

Relevant knowledge:

$$a^2 = b^2 + c^2 - 2bc \cos \alpha$$

$$\cos \alpha = \frac{b^2 + c^2 - a^2}{2bc}$$

$$> \frac{b^2 + c^2 - (b + c)^2/4}{2bc}$$

$$= \frac{3(b^2 + c^2)}{8bc} - \frac{1}{4}$$

$$\geq \frac{6bc}{8bc} - \frac{1}{4} = \frac{1}{2} = \cos \frac{\pi}{3}$$

which proves the assertion.

12.5.1. (Julius G. Baron; see *Mathematics Magazine* **39,** 1966, p. 134 and p. 112.)

The problem is substantially: Find the area A of a quadrilateral which is both inscribed and circumscribed, given its sides $a,b,c,$ and d.

A closely related problem was raised in India twelve centuries ago. You stand a good chance to remember it if you have ever heard of it and ask yourself earnestly: *Do you know a related problem? Do you know a problem with the same kind of unknown?*

In fact, that related problem has the same unknown and the same data as the proposed; it has even one-half of the most striking clause of the proposed condition; it is: *Find the area* A *of an inscribed quadrilateral being given its sides* a,b,c, *and* d. Its solution is (MPR, vol. 1, pp. 251–252, ex. 41)

$$A^2 = (p - a)(p - b)(p - c)(p - d)$$

where $2p = a + b + c + d$.

In possession of this information, it is sufficient to observe: If the quadrilateral is circumscribed and side a is opposite to c, side b to d, then

$$a + c = b + d = p$$

12.9.1. Set

$$b + c - a = 2A \qquad c + a - b = 2B \qquad a + b - c = 2C$$

Then

$$\frac{1}{x^2} = \frac{A}{xyz} \qquad \frac{1}{y^2} = \frac{B}{xyz} \qquad \frac{1}{z^2} = \frac{C}{xyz}$$

By multiplying, we obtain

$$xyz = ABC$$

and so

$$x^2 = BC \qquad y^2 = CA \qquad z^2 = AB$$

For a full treatment there remains the discussion of the case when one or more of the unknowns x,y, and z vanish.

12.9.2. The usual definition of the ellipse mentions the foci. The consideration of this definition may lead to the question "Where are the foci?" and so eventually to guessing a *stronger proposition which is easier to prove:* Under the given assumptions concerning f, f', and v, the locus of V is an ellipse with foci F and F'. In fact, the definition of the ellipse is satisfied. As the figure easily shows

$$FV + F'V = r + r'$$

where r is the radius of f and r' that of f .

12.9.3. (1) Concentric circle with radius $(r + r')/2$. (2) Parabola with directrix f and focus F'. (3) Hyperbola with foci F and F'. (4) Perpendicular bisector of the segment with endpoints $f = F$ and $f' = F'$.

Some other cases:

(5) Limiting case of ex. 12.9.2, or of (3): f' is a point on f, the locus is a straight line.

(6) Particular case of (3): $r = r'$, the locus is a straight line.

There are more questions.

About (3): What is the direction of the asymptotes, where is their point of intersection? Direction and location must be determined by the fixed circles f and f', but how? And why?

About (5): Is the whole straight line a limiting case of an ellipse? Or that of a hyperbola? Or is some part of the line this and the other that?

And so on.

To the teacher: The case in which f' is a point is beautifully presented by four films in the series *Animated Geometry* by J. L. Nicolet.

14.1.1. If 400 consecutive Gregorian years precisely agreed with 400 astronomical years, the length of one astronomical year would be

$$\frac{97 \cdot 366 + 303 \cdot 365}{400} = 365 + \frac{97}{400}$$

days, that is, 365 days, 5 hours, 49 minutes, and 12 seconds, which is only 26 seconds longer than the length derived from astronomical observations. The discrepancy is small, but it would amount to a day in 3323 years.

15.2.1. (*Cf.* Stanford 1964.)   (1) The segments into which d divides c are proportional to the adjacent sides.   Therefore

$$\left(\frac{ac}{a+b}\right)^2 = a^2 + d^2 - 2ad \cos \frac{\gamma}{2}$$

$$\left(\frac{bc}{a+b}\right)^2 = b^2 + d^2 - 2bd \cos \frac{\gamma}{2}$$

Straightforward elimination of γ yields

$$d^2 = \frac{ab[(a+b)^2 - c^2]}{(a+b)^2}$$

(2) If $a = b = c$, then $d^2 = 3a^2/4$.   If $a^2 + b^2 = c^2$, then

$$\frac{d}{a\sqrt{2}} = \frac{b}{a+b}$$

and this proportion is visible from easily constructed similar triangles.

If $a = b$, then $d^2 = a^2 - (c/2)^2$.

If $a + b = c$, then $d = 0$.

15.52.1. The formulation of the problem leaves considerable latitude and it does so intentionally: "real" problems may be rather indeterminate at the start. I detail some points deserving consideration.

(1) The points of each set fill the interior and the surface of a *polyhedron*, see Figs. AS1, AS2, AS3.

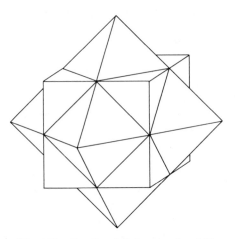

Fig. AS1. See C and O, imagine I and H.

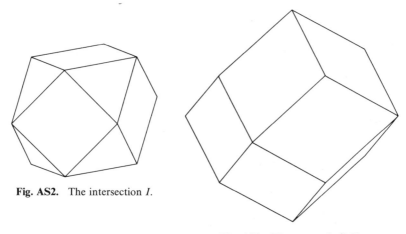

Fig. AS2. The intersection I.

Fig. AS3. The convex hull H.

C is a cube, its faces are squares.

O is a regular octahedron, its faces are equilateral triangles (*Cf.* ex. 5.5).

I is the *intersection* of C and O; it is called the *cuboctahedron*. It has 14 faces: 6 faces are squares each cut out from a face of C, and 8 faces are equilateral triangles each cut out from a face of O.

H contains both C and O; it is, in fact, the smallest convex set containing both, their *convex hull*. Its faces are rhombi, it is called the *rhombic dodecahedron*.

We pass from C to I by slicing off from C 8 congruent tetrahedra.

We pass from C to H by adding to C 6 congruent pyramids.

(2) We list the vertices of our four polyhedra:

C	$(\pm 1, \pm 1, \pm 1)$		
O	$(\pm 2, 0, 0,)$	$(0, \pm 2, 0)$	$(0, 0, \pm 2)$
I	$(0, \pm 1, \pm 1)$	$(\pm 1, 0, \pm 1)$	$(\pm 1, \pm 1, 0)$
H	vertices of both C and O.		

(3) The following table exhibits F, V, and E, the number of faces, vertices, and edges, for each polyhedron.

	F	V	E
C	6	8	12
O	8	6	12
I	$6 + 8$	12	24
H	12	$8 + 6$	24

(4) C, O, and I are inscribed in H. Eight among the 14 vertices of H are the vertices of C, the remaining 6 the vertices of O, the centers of the 12 faces of H are the vertices of I.

Each edge of C is coupled with an edge of O: they bisect each other, they are the diagonals of the same face of H, their point of intersection is a vertex of I.

(5) All four polyhedra have the same kind of symmetry; we describe it in considering the most familiar figure C.

There are *planes of symmetry* of two different kinds in the cube:

Three are parallel, and midway between, a pair of opposite faces of the cube. Six pass through a pair of opposite edges.

All 9 planes of symmetry pass through the center of the cube and cut it into 48 congruent tetrahedra.

There are *axes of symmetry* of three different kinds. (*cf.* HSI, ex. 8, pp. 235, 239, 244–245) in the cube:

Six connect the midpoints of a pair of opposite edges, each one is the intersection of 2 planes of symmetry.

Four connect a pair of opposite vertices, each one is the intersection of 3 planes of symmetry.

Three connect the midpoints of a pair of opposite faces, each one is the intersection of 4 planes of symmetry.

All 13 axes of symmetry pass through the center of the cube. If the axis is an intersection of n planes of symmetry, then the cube rotated about the axis through an angle $360°/n$ coincides with itself.

(6) The two patterns of ex. 2.50.1 involve C and H, respectively. In the first pattern, each sphere is inscribed in a cube and all these cubes fill the whole space without gaps and overlapping. In the second pattern, each sphere is inscribed in a rhombic dodecahedron and all these rhombic dodecahedra fill the whole space without gaps and overlapping.

(7) To compute V [not the same notation as under (3)!] for I and H, we can advantageously start from C. If the polyhedron is circumscribed about a sphere there is a connection between V and S.

C	$S = 24$	$V = \frac{1}{3} 1 \cdot S = 8$
O	$S = 16\sqrt{3}$	$V = \frac{1}{3} \frac{2}{\sqrt{3}} S = \frac{32}{3}$
I	$S = 12 + 4\sqrt{3}$	$V = \frac{20}{3}$
H	$S = 24\sqrt{2}$	$V = \frac{1}{3} \sqrt{2} S = 16$

(8) The example may serve to introduce several general concepts, for instance, systems of linear inequalities, convex hull, symmetry in space, etc.

Some more particular questions: Are there other pairs of polyhedra so coupled as C and O? Other space filling polyhedra? Etc.

15.53.1. Let a, b, and D be positive integers, D not a perfect square, so linked that

$$a^2 - b^2 D = 1$$

$a = 2, b = 1, D = 3$ is an example. Let n be a positive integer; then there are integers A and B such that

$$(a - b\sqrt{D})^n = A - B\sqrt{D}$$

There follows

$(a + b\sqrt{D})^n = A + B\sqrt{D}$

$$\begin{aligned} A^2 - B^2 D &= (a + b\sqrt{D})^n(a - b\sqrt{D})^n \\ &= (a^2 - b^2 D)^n \\ &= 1 \end{aligned}$$

and

$$\begin{aligned} (a - b\sqrt{D})^n &= \sqrt{A^2} - \sqrt{B^2 D} \\ &= \sqrt{A^2} - \sqrt{A^2 - 1} \end{aligned}$$

Just a little change is needed to similarly generalize ex. 15.53, or to merge it with the present problem into a common generalization.

BIBLIOGRAPHY

I. Classics

EUCLID, *Elements*. The inexpensive shortened edition in Everyman's Library is sufficient here. ("Euclid III 20" refers to Proposition 20 of Book III of the Elements.)

PAPPUS ALEXANDRINUS, *Collectio,* edited by F. Hultsch, 1877; see v. 2, pp. 634−637 (the beginning of Book VII).

DESCARTES, *Œuvres,* edited by Charles Adam and Paul Tannery. (For remarks on the "Rules"—the work which is of especial interest for us—and the way of quoting it, see ex. 2.72.)

LEIBNITZ (or LEIBNIZ) (1) *Mathematische Schriften,* edited by C. J. Gerhardt. (2) *Philosophische Schriften,* edited by C. J. Gerhardt. (3) *Opuscules et fragments inédits,* collected by Louis Couturat.

BERNARD BOLZANO, *Wissenschaftslehre,* second edition, 1930; see v. 3, pp. 293−575 (Erfindungskunst).

II. More modern

E. MACH, *Erkenntnis und Irrtum,* fourth edition, Leipzig, 1924; see pp. 251−274 and *passim.*

J. HADAMARD, *Leçons de Géométrie plane,* Paris, 1898; see Note A, *Sur la méthode en géométrie.*

F. KRAUSS, Denkform mathematischer Beweisführung. *Zeitschrift für mathematischen und naturwissenschaftlichen Unterricht,* v. 63, pp. 209−222.

WERNER HARTKOPF, *Die Strukturformen der Probleme;* Dissertation, Berlin, 1958.

I. LAKATOS, *Proofs and Refutations, The Logic of Mathematical Discovery,* Cambridge, 1976.

FRANZ DENK, WERNER HARTKOPF, and GEORGE POLYA, Heuristik, *Der Mathematikunterricht,* v. 10, 1964, part 1.

III. Related work of the author

Books

1. *Problems and Theorems in Analysis,* 2 volumes, Berlin, 1972, 1976; jointly with G. SZEGÖ.
2. *How to Solve It,* second edition, Princeton, 1957. (Quoted as HSI; references are to the pages of the second edition but, for the convenience of the users of former printings, titles are added.)
3. *Mathematics and Plausible Reasoning,* Princeton, 1954. In two volumes, entitled

Induction and Analogy in Mathematics (vol. I) and *Patterns of Plausible Inference* (vol. II). (Quoted as MPR.)

4. *Mathematical Methods in Science,* lectures edited by Leon Bowden, New Mathematical Library, vol. 26, Washington, The Mathematical Association of America, 1977.

5. *The Stanford Mathematics Problem Book,* New York, Teacher's College Press, 1974; jointly with J. KILPATRICK.

Papers

1. Geometrische Darstellung einer Gedankenkette. *Schweizerische Pädagogische Zeitschrift,* 1919, 11 pp.

2. Wie sucht man die Lösung mathematischer Aufgaben? *Zeitschrift für mathematischen und naturwissenschaftlichen Unterricht,* **63,** 1932, pp. 159−169.

3. Wie sucht man die Lösung mathematischer Aufgaben? *Acta Psychologica,* **4,** 1938, pp. 113−170.

11. Die Mathematik als Schule des plausiblen Schliessens. *Gymnasium Helveticum,* **10,** 1956, pp. 4−8; reprinted *Archimedes,* **8,** 1956, pp. 111−114. Mathematics as a subject for learning plausible reasoning; translation by C. M. Larsen. *The Mathematics Teacher,* **52,** 1959, pp. 7−9.

12. On picture-writing. *American Mathematical Monthly,* **63,** 1956, pp. 689−697.

13. L'Heuristique est-elle un sujet d'étude raisonnable? *La Méthode dans les Sciences Modernes* ("Travail et Méthode," numéro hors série) 1958, pp. 279−285.

14. On the curriculum for prospective high school teachers. *American Mathematical Monthly,* **65,** 1958, pp. 101−104.

15. Ten Commandments for Teachers. *Journal of Education of the Faculty and College of Education, Vancouver and Victoria,* nr. 3, 1959, pp. 61−69.

16. Heuristic reasoning in the theory of numbers. *American Mathematical Monthly,* **66,** 1959, pp. 375−384.

17. Teaching of Mathematics in Switzerland. *American Mathematical Monthly,* **67,** 1960, pp. 907−914; *The Mathematics Teacher,* **53,** 1960, pp. 552−558.

18. The minimum fraction of the popular vote that can elect the President of the United States. *The Mathematics Teacher,* **54,** 1961, pp. 130−133.

19. The Teaching of Mathematics and the Biogenetic Law. *The Scientist Speculates,* edited by *I. J. Good,* 1962, pp. 352−356.

20. On Learning, Teaching, and Learning Teaching. *American Mathematical Monthly,* **70,** 1963, pp. 605−619.

21. On teaching problem solving. *The Role of Axiomatics and Problem Solving in Mathematics,* Boston, Ginn, 1966, pp. 123−9.

22. Methodology or heuristics, strategy or tactics. *Arch. Philos.* **34,** 1971, pp. 623−9.

23. As I read them. *Developments in Mathematical Education, Proceedings of the 2nd International Congress in Mathematical Education,* ed. A. G. Howson, Cambridge, 1973, pp. 77−8.

24. Guessing and proving. *Two-Year College Mathematics Journal* **9**(1), 1978, pp. 21−7.

25. More on guessing and proving. *Two-Year College Mathematics Journal* **10,** 1979, pp. 255−8.

(The numbers omitted are quoted in MPR, v. 1, pp. 279−280, and v. 2, pp. 189−190.)

Films and Videotapes

1. *How to Teach Guessing,* Mathematical Association of America (film).
2. *Guessing and Proving,* Open University Educational Media (film).
3. *How to Teach Mathematics,* Mathematical Association of America (videotape); jointly with P. HILTON.
4. *Guessing is Good; Proving is Better,* Mathematical Association of America (videotape).

IV. Problems

Among the examples proposed for solution, some are taken from the *Stanford University (Stanford-Sylvania) Competitive Examination in Mathematics.* This fact is indicated at the beginning of the solution with the year in which the problem was given as ''Stanford 1957.'' These problems, with hints and solutions, appear in *The Stanford Mathematics Problem Book* listed above.

The USSR Olympiad Problem Book, by *Shklarsky, Chentzov,* and *Yaglom,* contains many unusual and difficult elementary problems proposed in Russian competitive examinations. English translation, edited by *I. Sussman,* San Francisco, 1962.

Hungarian Problem Book, New Mathematical Library, vols. 11-12, contains interesting elementary problems, careful solutions, instructive notes on the subject matters involved, and remarks on a competitive examination which essentially contributed to the development of mathematics in Hungary.

Challenging Mathematical Problems with Elementary Solutions, by A. M. Yaglom and I. M. Yaglom, 2 vols., San Francisco, Holden-Day, 1964, 1967.

International Mathematical Olympiads 1959-1977, compiled and with solutions by Samuel L. Greitzer, New Mathematical Library, vol. 27.

V. On Teaching

On the Mathematics Curriculum of the High School (memorandum with 65 signatures). *American Mathematical Monthly,* **69,** 1962, pp. 189−193; *The Mathematics Teacher,* **55,** 1962, pp. 191−195.

MARTIN WAGENSCHEIN, Exemplarisches Lehren im Mathematikunterricht. *Der Mathematikunterricht,* **8,** 1962, part 4.

A. I. WITTENBERG, *Bildung und Mathematik,* Stuttgart, 1963.

A. I. WITTENBERG, SOEUR SAINTE-JEANNE-DE-FRANCE, and F. LEMAY, *Redécouvrir les mathématiques,* Neuchâtel, 1963.

ROY DUBISCH, *The Teaching of Mathematics,* New York, 1963. Contains also extensive annotated references to recent literature in English.

INDEX

This index contains references to both volumes of the present work and, moreover, references to selected parallel passages of my related works. **Heavy print** is used for distinguishing from each other the five volumes in question:

1 for volume 1 of *Mathematical Discovery*
2 for volume 2 of *Mathematical Discovery*
H for *How to Solve It,* second edition
I for volume 1 of *Mathematics and Plausible Reasoning (Induction and Analogy in Mathematics)*
II for volume 2 of *Mathematics and Plausible Reasoning (Patterns of Plausible Inference).*

The numbers in ordinary type, following a sign in heavy type, refer to pages.

Italics are used either to emphasize certain sentences whose role is discussed in Chapter 12 or to indicate that the entry is the title of a publication.

The index does not (and could not reasonably) give exhaustive references to topics treated virtually throughout all five volumes such as "Analogy" "Generalization" "Guess" "Induction" "Unknown" etc.

The index has been revised for this printing so that it includes a more comprehensive covering of the topics from *Mathematical Discovery, Volumes* **1** and **2**. It retains all of the index entries from the original publication, including those references to the other three books mentioned above.